2013-2014

Common Read Project

Compliments of Office of Undergraduate Education

THE UNWANTED SOUND OF EVERYTHING WE WANT

ALSO BY GARRET KEIZER

Help

The Enigma of Anger

God of Beer

A Dresser of Sycamore Trees

No Place But Here

THE UNWANTED SOUND OF EVERYTHING WE WANT

A Book About Noise

GARRET KEIZER

PUBLICAFFAIRS
New York

Published in the United States by PublicAffairs™, a member of the Perseus Books Group.

Printed in the United States of America.

PublicAffairs books are available at special discounts for bulk purchases in the U.S. by corporations, institutions, and other organizations. For more information, please contact the Special Markets Department at the Perseus Books Group, 2300 Chestnut Street, Suite 200, Philadelphia, PA 19103, call (800) 810-4145, ext. 5000, or e-mail special.markets@perseusbooks.com.

Text set in 11.5 point Arno Pro by the Perseus Books Group

Library of Congress Cataloging-in-Publication Data
Keizer, Garret.
The unwanted sound of everything we want : a book about noise / Garret Keizer.—1st ed.
p. cm.
Includes bibliographical references and index.
ISBN 978-1-58648-552-8 (alk. paper)
1. Noise—Psychological aspects. 2. Sound—Psychological aspects. I. Title.
BF353.5.N65K45 2010
363.74—dc22
2010005391

First Edition
10 9 8 7 6 5 4 3 2 1

For Kathy and Sarah

CONTENTS

Sitting Quietly at the Back: A Set of Resources

PART I

What We Talk About When We Talk About Noise: A Basic Introduction

CHAPTER 1

Noise Is Interested in You

You may not be interested in war,
but war is interested in you.

—ATTRIBUTED TO LEON TROTSKY

Noise is not the most important problem in the world. Compared to the disasters of famine, war, and global climate change, the existence of "unwanted sound" hardly counts as a problem at all. It rarely emerges as a public issue in countries struggling with the worst forms of poverty and violence. So far as I am aware, there is no Society for the Suppression of Unnecessary Noise in the cities and villages of Afghanistan and the Congo.

Even among societies with levels of political stability and industrial commotion sufficient to raise an organized cry for quiet, that cry can quickly be silenced by a crisis. The so-called Roaring Twenties included a number of initiatives in both the United States and Europe to address the Jazz Age roar of motorcars and radios. But with the Great Depression, followed by the Second World War, the issue of noise all but vanished from the public agenda. Established in 1929, the New York Noise Abatement Commission was dissolved in 1932, by some accounts one of the worst years of the Depression. Noise did not again become a prominent issue, even in New York, until the 1960s.

Noise might be called a small or a "weak" issue, in some cases a fussy issue, which may be what certain individuals had in mind when they asked me why on earth I wanted to write a book about it. My father asked me when he first heard of my plans, as did several of my friends and acquaintances. "Why noise?"

It was a good question, and this was my best answer: I chose to write a book about noise because it is so easily dismissed as a small issue.

And because in that dismissal I believe we can find a key for understanding many of the big issues.

Noise reminds me of a Norse myth in which the god Thor is invited to wrestle with a giant king's decrepit old foster mother. Though Thor is one of the mightiest of the Norse gods, he is unable to gain any advantage over the crone. He cannot lift her, throw her, best her in any way. Only later is he told that he was wrestling with Old Age itself. Noise is a lot like Thor's mysterious opponent. It appears lightweight and even frail at first glance, but once you try to pick it up, you discover that you are trying to heft the whole world.

A "WEAK" ISSUE BECAUSE IT AFFECTS "THE WEAK"

To say that noise is a relatively weak issue because it is less momentous than world hunger or global climate change is to make an incomplete statement. Noise is a weak issue also because most of those it affects are perceived, and very often dismissed, as weak. The ones who dismiss them, in addition to being powerful, are often the ones making the noise.

In using the word *weak* I am not referring to personal capabilities, to someone's IQ score or muscle mass, though these factors may come into play. I am thinking rather of a person's social standing and political power. Make a list of the people most likely to be affected by loud noises (though not all noise is loud), either because of their greater vulnerability to the effects of loud sound or because of their greater likelihood of being exposed to it, and you come up with a set of members whose only common features are their humanity and their lack of clout. Your list will include children (some of whom, according to the World Health Organization, "receive more noise at school than workers from an 8-hour work day at a factory"), the elderly (whose ability to discriminate spoken speech from background noise is generally less than that of younger contemporaries), the physically ill (cancer patients undergoing chemotherapy, for example, are often more sensitive to noise), racial minorities (blacks in the United States are twice as likely, and Hispanics 1.5 times as likely, as whites to live in homes with noise problems), neurological minorities (certain types of sound are especially oppressive to people with autism), the poor (more likely than their affluent fellow citizens to live next to train tracks, highways, airports), laborers (whose political weakness has recently been manifested in weakened occupational safety standards), prisoners (noise, like rape, being one of the unofficial punishments of incarceration), members of the Armed Forces (roughly one in four soldiers returning from Iraq has a service-related hearing loss)—or simply a human being of any description who happens to have less sound-emitting equipment than the person living next to her (who might for his part have car speakers literally able to kill fish) and no feasible way to move.

Consider a toddler holding a toy capable of emitting 117 decibels* (on a par with the sound pressure of a rock concert or a sandblaster) at the length of her stubby arms and a combat-fit Marine exposed to weapons fire and explosive devices that may produce sound levels as high as 185 dB and you seem to be looking at two very different categories of human strength and weakness. Take a closer look and you see two human beings who have less say than many of us do about what goes into their ears.† Consider an elderly person living in a noisy tenement, a patient in the notoriously noisy wards of certain hospitals, a studious undergraduate living in a typical college dorm; then consider the likelihood that any one of them could improve his or her situation

---

* A decibel (dB) is a unit for measuring sound pressure. It has been in common use since the 1920s. Strictly speaking, it is not a measure of loudness, of what you "subjectively" hear with your ears. The universal rule of thumb is that an increase of 10 decibels will be heard as a doubling of loudness. In other words, a sound that increases from 50 to 60 decibels will be *heard* as twice as loud. A person's *actual exposure* to noise doubles with every 3-decibel increase. At levels where prolonged exposure can damage human hearing, generally in excess of 80 to 90 decibels, a person's risk of hearing loss approximately doubles for each 3-decibel increase in exposure. For further details see the appendices "Common Terms Used in Discussions of Noise" and "Decibels in Everyday Life and Extraordinary Situations" at the back of this book.

† In regard to the child, the safety standards of the American Society for Testing and Materials allow toys to emit sounds with a *peak* level of 138 dB, while OSHA safety rules require employees to wear ear protection for exposure to noise levels *averaging* above 85 dB. Your ears may be safer in a factory than in a daycare center.

by complaining. What they rightly perceive as helplessness, some others around them will readily perceive as entitlement. A person who says "My noise is my right" basically means "Your ear is my hole."

For about the past year and a half I've been corresponding online with men and women who've gotten that message—as the saying goes—loud and clear. With the help of a former student of mine, I set up noisestories.com, which invites its visitors to submit a firsthand experience of noise. I have received stories from people as close as fifty miles away and as far as New Zealand, stories from farmers, bartenders, physicists, zookeepers, bus drivers, performance artists (in fact, one from a woman who is both a bus driver and a performance artist), and several from people who are possibly out of their minds. Some of my correspondents are dealing with industrial noise at work or in their neighborhoods, others with noisy entertainment venues, and many with a noisy individual or family living next door. Most seem glad that someone is interested in their predicament. A few have mistaken my credentials and asked if I could make the noises stop.

One of the recent stories I've received came from a stay-at-home mother in Texas whose name I've chosen to change out of regard for her safety. Kaiying Keller was born in China and speaks a slightly accented English. She is married to an American she met in China but is not a U.S. citizen herself. Two years ago her neighbor, then on friendly terms with her family, built a swimming pool and located the pump and filtration system near the border of Keller's property. The pump is audible inside Keller's house. A year ago Keller gave birth to a daughter. She is afraid to take her baby into the backyard because of the noise and because her neighbor's two large, aggressive dogs are often butting their heads against the dividing fence, a noise that "scares my baby to cry." Keller has spoken to her neighbor about the pump noise and gotten nowhere. She has spoken to the police and to her town government and gotten nowhere. Like many of my other correspondents, she has sent me a great sheaf of letters and documents. She has been told that the city noise ordinance lists 85 dB as the maximum allowable threshold and that since the motor noise measures "only" 76.1 dB there is nothing that can be done—even though the city ordinance also prohibits "any noise of such character, intensity and continued duration, which substantially interferes with comfortable enjoyment of private homes by persons of ordinary sensibilities."

Keller admits that she may be more sensitive to noise than other people, but she does not see herself as having abnormal sensibilities. Not long ago, when Keller tried to repair the loosened dividing fence, her neighbor ordered her not to touch it. When she refused to obey, he washed her down with a garden hose. This time the police told her she could press charges for third-degree assault, but her husband has pleaded with her not to make trouble. The neighbor is a volatile, Hummer-driving bully who claims to be a Dallas police officer (though he is not identified as such on the Dallas PD website) who "owns tons of guns." Keller's husband worries what will happen when he's away at work.

I don't think of Keller and her husband as weak people, but the experience seems to have made her feel that way. She is seeing a doctor about the stress caused by the constant noise. She says she is depressed. Her story is of a type I have heard many times now. Some correspondents tell me they have moved; very few tell me they have won. Keeping in mind the infamous telephone survey that predicted Dewey would beat Truman in the election of 1948 (it turned out that people well-off enough to have telephones were more likely to vote Republican), I should add that the women and men who've written to me all have access to e-mail and to the English language. I suspect that some of the worst noise sufferers lack those advantages, that they are hunkered down, locked in (Keller writes of her house having become "a jail"), debilitated by what psychologists call "learned helplessness."

In addition to affecting "weak" people, noise tends to interfere with weak pursuits, by which I mean any activity not likely to generate news or money. One of the first casualties of noise is conversation. The human voice seems to have evolved to be audible in the natural environment. Most sources put the volume of normal human speech between 55 and 60 decibels.* That's a

* The World Forum for Acoustic Ecology reported in 1999 that the average decibel level for an "animated conversation between typical Americans" was about 65 dB, up ten decibels from the previous decade.

good fit with a primeval forest, even a chattering rain forest, but a poor match for a gun or a jet ski. If the decibel readings police took at Kai-ying Keller's house are accurate, it's even a poor match for some pool filters. But talk is cheap, after all; jet skis and swimming pools are not. Sleep is sort of cheap too—just about any "loser" can do it, even homeless people manage to do it—though it is estimated that the yearly U.S. cost in lost worker productivity due to sleeplessness is around $18 billion. It has also been estimated that of the 3 billion prescriptions dispensed annually in the United States, something like 3.9 million are mistakes, about a third of these are injurious, and many result from a combination of like-sounding names for different medicines and the noisy environments in which many doctors, nurses, and pharmacists have to work.

Statistics like these get our attention; nobody wants to lose all those billions in productivity, and nobody wants to get a dose of Maalox when the doctor was calling for morphine. But let the issue be the construction of a new box store or a new electrical generation facility, and let someone start talking about bird songs and the simple pleasure of taking a quiet walk, and he or she is instantly met with condescension and scorn. Up against such big issues as shopping opportunities and cheap kilowatts, can you imagine some joker talking about wanting to hear the birds?

For that matter, can you imagine some joker giving a damn about the birds themselves? That's where the logic of the loud always leads. The dismissal of the small pleasure in the small sound ends inevitably in the dismissal of the life, or the way of life, producing that sound. It turns out that the birds are also affected by noise, which reduces the audibility of their mating calls and thus their ability to reproduce. One 2007 study shows the deleterious effects of industrial noise on the pairing success of ovenbirds. Another from 2006 shows significant modifications in the "anti-predator behavior" of California ground squirrels living near large wind turbines. Though research has shown the adverse effects of noise on a variety of animals, including large sea mammals, dogs, and horses, smaller creatures seem to be especially vulnerable. For one thing, smaller animals generally require more sleep than larger animals. They are also more likely to be eaten by larger animals. The reflexive defenses of the desert kangaroo rat, for example, which can normally hear the sound of a sidewinder rattlesnake at the critical distance of thirty inches, is temporarily disabled by the noise of a dune buggy.

Perhaps the most famous of all zoological wimps is the Tennessee snail darter, a fish whose threatened extinction in the 1970s nearly halted the building of the Tellico hydroelectric dam. This was not a noise issue per se, at least insofar as the fish were concerned, but it sparked a noisy debate that went all the way to the Supreme Court (which ruled for the fish, though the dam was built with congressional approval anyway). It also served to illustrate prevailing notions of what constitutes "the small stuff" and the big, as well as the prevailing tendency to subordinate "quieter" concerns like the preservation of obscure species to the booming imperatives of boundless development. "The awful beast is back," bellowed Senator Howard Baker of Tennessee. "The Tennessee snail darter, the bane of my existence, the nemesis of my golden years, the bold perverter of the Endangered Species Act is back." Another Tennessee senator who joined with Baker in working to see that the fate of the "awful beast" did not prevent the completion of the dam would later go on to achieve great and well-deserved renown for his work on behalf of melting ice caps and drowning polar bears. Like most of us—including those who say to hell with ice caps and polar bears—he had bigger fish to fry.

THE CRUX OF THE MATTER

Why in fact are we faced with those issues we call "big"? Why do we have genocidal wars and melting ice caps? There are many answers one might give, ranging from the slowness of evolutionary change to the fickleness of fate. I would suggest that we face many of these problems because of our contempt—and you can say our "sinful" contempt, if you like, or our naturally selected, genetically preconditioned contempt, though I would be happy simply to say our *noisy* contempt—for anything we regard as weaker or smaller than ourselves: for the "nonproductive" segments of the population (often the very old, the very young, or the very disabled), the "weaker" sex, the "inferior" races, the "backward" cultures, the "useless" parts of the environment, the "impractical" solutions that do not enlist the entrepreneurial arrogance and technocratic grandiosity that caused us to need solutions in the first place.

Not least of all we are undone by our noisy contempt for "harmless" (i.e., "unimportant") pleasures that do not involve consumption and speed, a regulation outfit and a blood-curdling whoop. In short, that do not produce much in the way of "volume."

Perhaps nothing points so tellingly to the crux of the matter as the fact that in English we use the same word to denote the space occupied by an object or body and the loudness of a sound. The major political and environmental issues of our generation all come down to the basic question of how much sustainable "volume" a single human being can and should occupy in his or her society and on his or her planet.* The volume of the noise we produce is both a part of that question and a signifier of that question.

To recap: Noise is a weak issue compared to problems of greater consequence. That said, we also think of noise as a weak issue because it disproportionately affects "the weak." In our contempt for what we perceive as weak—disadvantaged people, but also small creatures and harmless pleasures—we have created many of the economic and environmental problems we acknowledge as "big." As a civilization we will either deal with noise, the underlying causes of noise, and the bigger problems of human "volume" that noise signifies, or we will ultimately arrive at a place where people scarcely make a peep.

SOME BACKGROUND AND BIASES

A person who talks like this is obviously not approaching his subject with a great deal of objectivity. No surprise there, given the fact that noise itself is not an entirely objective phenomenon. Nor is it, as noisemakers famously like to claim, "all subjective." Too many hard data say otherwise. But whatever I might know of those hard data and of the history of unwanted sound, my point of view is still influenced by my own needs, biases, and history.

---

* Even the number of hours we work corresponds to a measure of environmental "volume." It takes around twenty-three acres of land to support an American working twenty to forty hours a week, twenty-eight acres to support someone working more than forty. Why? The second person has more money to spend and less time to engage in more sustainable practices like buying locally grown food and preparing it at home.

I was raised with a keen awareness of noise. As a child I was told "Keep your voice down" whenever my voice was likely to disturb "the neighbors." When my family came home late from a night at the movies or a day at the shore, my father would insist that we latch the car doors as quietly as possible and then press them fully closed. This taboo against slamming a car door at night was part of a code whereby holding down a job and getting oneself to work on time were sacred. A man who couldn't get himself moving at the crack of dawn wasn't even a man. Interfering with a neighbor's sleep was something akin to horse thievery on the old frontier, an assault on another person's livelihood, a hanging offense.

Given my upbringing, I can't help but be amused by those postmodern cultural historians who frame the impetus for noise suppression as a case of bourgeois rectitude "privileging" itself over the raucously "subversive" propensities of the working classes. Admittedly, there are legitimate reasons for such a construction: Many of the anti-noise activists of the nineteenth and early twentieth centuries were middle-class reformers self-righteously horrified by the "loudness" of immigrant organ grinders and proletarian rowdies—not unlike our own upper-class "environmentalists" who go wild at the sight of a beer can tossed by the side of the road but think nothing of spewing a thousand miles worth of carbon emissions to sample an artisan beer. Nevertheless, I would suggest that one deeply ingrained characteristic of working people is that they need their sleep. They need their sleep for the simple reason that they need to work. Along the same lines, I have a strong hunch that aside from the sound of a whip cutting into human flesh, the noisiest, most frequent, and most bitterly hated sound on an antebellum plantation was not the unauthorized "sounding off" of African slaves but the party-hearty clamor of those big-house denizens who had the privilege of sleeping late the next day—and of making some particularly nasty mischief whenever their "transgressive" exuberance took them to the slave quarters at the edge of the woods.

I have never heard a noise that menacing or been that helpless to stop it. All I can claim is some limited experience of being awakened by noise and of feeling, initially at least, the limits of my power. The first home my wife and I shared together was in an apartment complex designated as "married student housing" by the university we both attended as graduate students. It was located directly across the street from the off-campus freshman dormitories of another college. Like many freshman dormitories then and now, these amounted to a de facto ghetto within the larger campus.

These dorms were famously—and proudly—loud. One locally popular bumper sticker proclaimed "I survived a [college name]'s keg party!" At least some of the noise seemed intentionally directed at the married students across the way. I can remember one night being awakened by someone shouting "We're getting the same thing over here that you guys are getting in there." Presumably he did not mean a good night's sleep.

We had a neighbor who did not care for this noise and he happened not to be weak. He had previously attended college on a football scholarship (as a lineman) and was now studying for a master's degree while also working to support a wife and two small children. He was a large man in every sense, including large-hearted. One of my fondest grad school memories is that of him strolling into the community gardens near the apartments like Eden's God "in the cool of the day," beer in one hand and baby in the other, calling mightily through the vegetation: "Heeey, Gar."

One night after he and his entire family were awakened by a stereo blasting from one of the dorms across the road, this neighbor pulled on his clothes and walked over to deal with the disturbance. He tramped up the stairs and ducked through the doorway of a room containing five or six young men and a stereo going full tilt. "Look, I've got two little kids across the way who can't sleep because of your music, and I need you to turn it down." Turn it down they did. No sooner was he out of the building and onto the street than the volume of the music rose even louder than before. Back up the stairs he went, this time with an appeal that, as it was later told to me, included an offer to "kick the shit out of the bunch of you." This time the volume went low and stayed there. Nothing woke his children until the morning.

I have sometimes wondered what would have happened if this neighbor—who for all his size and strength was one of the least overbearing men I have ever known—had made good on his threat. Would he have unintentionally killed or crippled some 19-year-old? Would the dorm community have mustered enough able-bodied and sufficiently sober young men to throw him down the stairs or over the third-floor balcony and onto the street? Noise has long been associated with violence—with the sounds of war and destruction—but in more pacific times and places it has been a cause of violence, sometimes bringing about the death of the noisemaker, sometimes making a fatality of the person who complains. Twenty-year-old Kathy Jackson of Omaha, Nebraska, was the latter. At 3:00 in the morning in May 2008 she went upstairs to ask two women to turn down their music, a confrontation that ended in an argument. About an hour later there was a knock at her door; when she opened it, she was shot in the head. She died two days later.

Earlier that same year, in a confrontation in Fort Worth, Texas, it was the noisemaker who died. Police had been called thirteen times in three months to the house of 22-year-old James Eckenrode in response to the loud parties he liked to hold in his garage. Joseph Rosier, a 41-year-old father of two children who lived nearby, finally ended the noise—and his neighbor's life—with a blast from a shotgun. The hundred-plus e-mailed comments that appeared in response to the online version of the *Fort Worth Star-Telegram* news story amount to a noise treatise all by themselves. Reading between the lines you can see the potential for another dozen or so incidents just like the one being commented on. "I say Rosier deserves a standing ovation," says one person, to which another replies, "[I]t takes a real piece of crap to say anyone deserves to die of something so stupid like music."

Certainly no one deserved to die in the dormitories across from our first apartment; as it turned out, no one had to. Less courageous than my neighbor but no less desperate, I took an approach to the noise that ultimately proved more effective than his. One sleepless early morning I reached across my wife to the nightstand on her side of the bed, fumbled for the phone book, and looked up the home number of the president of the college theoretically responsible for the dorms. After identifying myself by name to the groggy voice who demanded "Who is this?" I posed a single question and made a solemn vow. The question was how the president liked being awakened at two o'clock in the morning. I will leave you to guess the vow.

I never had to keep it. The next day we had a visit from a college PR official who promised to address the situation—and urged me to remember "what you did when you were young." Apparently he knew all about that. But he was true to his word. The college's solution consisted of requiring the dorm students to hold their parties on the back side of the dormitories instead of the front. However lame this might seem, the sound-mitigating properties of a three-story brick building, added to those of the flat expanse of undeveloped terrain behind the dorms (little reverb, in other words), proved to be surprisingly effective. The results weren't perfect but they succeeded in moving "unlivable" a few notches toward "live and let live."

Sometimes a noise issue is so intractable and complex that anything like a solution seems beyond reach. Class antagonisms, cultural contradictions, economic imperatives, and acoustical conundrums combine in such a way as to make a satisfying outcome seem next to impossible. We build quieter jets, but then we put more of them in the sky. In her book *Mechanical Sound,* Dutch historian Karin Bijsterveld writes that "the history of noise problems" has largely been "a tragic story. Despite the many attempts to control noise, noise still features prominently on the Western world's public agenda." By temperament I am biased toward tragic interpretations, of this problem and many others.

And yet, noise stories like the one I've just told remind us of certain historical instances, admittedly few, when something we regarded as part of our tragic destiny was altered for the better without anybody getting shot or beaten up. The wall that divided a city comes down with a cheer, the detention camp is closed by the stroke of a pen, sound barriers are erected along an interstate repaved with a quieter type of asphalt, and once again people are out in the park feeding the pigeons, pushing their kids in the swings, and wondering to themselves and to one another how they ever put up with all that commotion.

THE FALLACY OF "MAKING NOISE"

Looking back, I'll give this much to the students in those dormitories: At least they were making noise. True, some of the noise came from their stereos, but the shouting and the general carrying on—they *made* that. In the larger context of our loud society, that should count for something.

In general, the phrase "making noise" is a quaint anachronism, a verbal antique akin to "half-cocked" or "hold your horses." What horses? Most of the noise we hear does not come from people shouting, stamping their feet, or beating pots and pans in celebration or protest. Most of the noise we hear—from mechanical engines to electronic amplification—is automatic. People flip a switch, hand over their tickets, and let the mechanism do the rest. In fact, most people said to be "making noise" could drop dead and the noise would continue after their bodies had turned cold. Even the more "organic" noise of an endlessly barking dog functions this way. People buy a dog as they would buy a TV, because, after all, a family should have a "family dog," even if they are hardly ever at home or hardly recognizable as a family, tie the animal to a tree in the yard, and drive away. Woof, woof, woof. Woof. Woof, Woof, WOOOOF. The only difference between the dog and a TV is that people usually turn a TV off before they leave the house. That, and the rarity with which an abandoned TV becomes frantic or depressed.

This separation between the human person and the object that makes noise on his or her behalf is crucial. We often like to think of our noise as visceral, raunchy. We're the earthy sort, you see, so we make all kinds of noise. But much of the noise in our culture has a weirdly disembodied, spiritualized quality. It is the noise of ghosts. Three-quarters of our *boom* is *boo*.

Make a list of the loudest or most prevalent noise sources in our world: airplanes, automobile traffic, weaponry, power tools, "thrillcraft." Almost every one of them has to do with reducing (or "killing") time, space, or labor. Even our recording and amplifying devices exist to minimize the importance of one's physical or temporal proximity to a musician. You don't have to attend the performance. You don't have to *make* any noise by clapping or shouting for an encore. All you need to do is shell out the money and crank up the sound. Even the verb *crank* grossly exaggerates the physical motion required.

What are time, space, and labor from a human perspective if not the conditions of having a body? Time and space are just abstract words for skin and bones. It should come as no surprise that noise carries a number of adverse and well-documented physiological consequences: deafness, tinnitus, high blood pressure, heart disease, low birth-weight, even statistically significant reductions in life span. Why shouldn't the effort to break out of our physical bodies do some damage in the process?

This brings me to another of my biases. I have a bias for the body. I have a bias for any activity that gives sensuous pleasure in the absence of noisy equipment. Forced to adjudicate between the rights of a man with a vibrating pool pump and a woman with a sleeping baby, I will always rule for the woman and the baby. To be biased in this way is to be at odds with a good deal of what we in the West call civilization. We probably don't even need to say "West." Much of the "heroic" enterprise, West and East, has been about trying to transcend the body, whether the means be aristocratic (having other people do your work), ascetical (reducing the body to a shell), or technological (subordinating matter to whim). The puerile pop history that says once upon a time people went around wearing hair shirts in the hope that their souls would go to heaven but then guys like Newton and Darwin came along and we started to wear iPods instead is rather missing the point. The monastery bells ringing out over the medieval village and the jet plane flying over the twenty-first-century metropolis are communicating much the same message vis-à-vis the mortal limitations of the flesh: namely, that you can do a whole lot better. You can grow wings. I say this as someone who occasionally flies in the planes and frequently heeds the bells, but with doubts in both cases. Of this I am convinced beyond any doubt: To say that you want to live in a less noisy world, and to say it with any depth of conviction, is in essence to say that you'd like to have your body back. But would you?

HOLD YOUR HORSES

We should always be wary of drawing pat moral analogies between noise and evil, quiet and good. By all accounts Adolf Eichmann was "a quiet person" who "rarely raised his voice" during his interrogation for the crime of murdering millions of people. Serial killer Ted Bundy's landlords remembered him as "a quiet and helpful tenant." No loud parties from the Tedster. David Berkowitz, "The Son of Sam" killer, was tormented by the howls of neighborhood dogs, believing them to be "messages from demons," while the arguably saner but no less deadly Unabomber Ted Kaczynski is a famous hater of noises and of the technologies that produce them. "There's a little bit of the

Unabomber in most of us," Robert Wright wrote in *Time* magazine; it would be useful for some of us to remember that.

The ridiculousness of moralizing about noise and noisemakers does not mean, however, that noise is free of moral implications. It does not mean that there is no food for thought in Hitler's memorable statement that without the loudspeaker the Nazis never would have conquered Germany. Indeed, what device could serve better than a loudspeaker to symbolize what a Nazi is? When Adolf talks, people listen; they listen because they have no choice—because like every other dictator he has jacked up the volume so high that they can't hear anything else.

Put that down as one last bias of the author: I am biased against whoever is barking an order, even if the command amounts to something as innocuous as a woofer-enhanced rendition of "Look at me!" Even if the command is as supportive of my preferences as "Quiet down!"—what I heard the other day on the designated "silent" floor of a university library when two students, audibly engaged in solving a mathematics problem, were gruffly scolded by another library patron—I still don't like it.

WHERE WE'RE HEADED

I also happen not to like books in which the author keeps telling me where I've been and where I'm going, as if following his road map is of greater interest to me than what I can see for myself out the window. To avoid doing too much of that here, I've tried to see that my chapters are always clear in their intent, and I've placed a number of subheadings within each chapter to serve as signposts. I've also tried to keep my main ideas—most of which I've already stated—to the fore.

Still, I caution the reader to remember that a subject like noise is different from a subject like the Battle of the Bulge or the workings of the sensory nervous system. There is no built-in narrative, no given organic structure. There is only noise, which almost by definition "makes no sense" in and of itself. That means the connections are ours to make as we explore the subject. We will need a plan certainly, but we will also need a plan that leaves us room to breathe, that includes the silence necessary to hear ourselves think.

Part I of this plan, "What We Talk About When We Talk About Noise," attempts to define a phenomenon that has fascinated men and women for a long time and from a broad array of backgrounds. Examining noise, putting it on trial if you will, requires calling expert witnesses from many fields: physicians, physicists, musicologists, engineers, philosophers, psychologists, artists, historians. To anyone who needed to ask me a second time why I was writing this book—"Yeah, but why noise?"—I said that whenever you have ten noise experts in a room you have something like a renaissance. And a renaissance tends to make any room more interesting.

Also in Part I, we will look at some of the political implications of noise, which emerge at around the same time as politics do. Another way to say this is that certain sounds became too harsh for human ears around the same time as certain human beings became too big for their britches. The political implications of noise inform much of this book and determine a good deal of its purpose.

Part II, which I've named "Laetoli Footprints" after one of the earliest human fossil remains, is a brief history of the phenomenon defined in Part I. We will begin on the plains of Tanzania and end somewhere on the outskirts of Paterson, New Jersey, both having a claim as my birthplace, depending on whether you take a long or a short view.

"Lighter Footsteps," Part III, is an attempt to give a broader perspective to the information presented in the first two sections. To do so I've chosen two broad areas to explore, the first being "Loud America." I concentrate on America not to minimize the problem of noise in other parts of the world, or the impressive progress other countries have made in dealing with it (both discussed throughout the book and especially in Chapter 6), but rather to acknowledge how many sources of noise have originated in America and seem so innately "American" in their symbolic clamor. A culture attempting to imitate America rarely grows quieter. If we equate noise with power and clout—or, if you prefer, with the ability to generate "shock and awe"—then America is the loudest country in the world today, probably the loudest that has ever existed. And yes, I love my country, even as I also love midtown Manhattan, my chain saw, and the Rolling Stones.

The second area of emphasis in Part III is the relationship of noise to issues of sustainable living, which, in keeping with other emphases in this book, also includes the possibility of more convivial living. My overall contention is that the more sustainable, equitable, and convivial a society becomes, the less noisy it will be.

Finally, since the third part is also the last part, the book contains a conclusion, which I hope will accomplish two equally desirable objectives: summarizing the book's main arguments, and adjusting the volume on any argument that might have gotten too loud.

After that, there's a considerable amount of back matter. Some of it (a time line, for instance) is included for the sake of general interest, but most of it (such as a list of anti-noise organizations) is there for the reader's practical use. Like any number of helpful people, the most helpful parts of books are sometimes found sitting quietly at the back.

YOU MAY NOT BE INTERESTED IN NOISE . . .

But noise is interested in you. Or it will be. Noise becomes "interested in you" when it goes from being an occasional annoyance to being an overpowering fact of your life. Suddenly—to use an example recounted to me by acoustical consultant Karl Searson in the Temple Bar district of Dublin—there are low-frequency vibrations coming from the pub next door that are literally rippling the surface of the water in the glass on your kitchen table. And the reason you have a glass of water on your kitchen table is that you are preparing to swallow yet another aspirin to get rid of your headache. This happens more often to the "weak" than to the "strong," but it can happen to anyone.

Noise can also become "interested in you" when you're suddenly made aware that you've been making a lot of it. A neighbor bangs on the wall between your adjoining apartments with his fist, or, more serendipitously, the overnight mail carrier knocks on your door just after you've spoken with a person like Chris Peeler of Greensboro, North Carolina, who fought, unsuccessfully, to prevent the airport expansion of a major FedEx hub near her home.

In all these examples noise brings a heightened awareness of your connection to other people. Your happiness and well-being are seemingly at odds with their happiness and well-being, but only because, on the deepest social level, your happiness and well-being are connected to theirs. You may not be interested in neighborhood, but neighborhood is interested in you. Exploring that connection is much of what this book is about.

In recent years a new field of inquiry has arisen called "the science of happiness." Researchers have gathered some fascinating data on what makes people happy, on the relative levels of happiness in various countries and cultures, and on what individuals can and can't do to be happier than they are.* Much as I respect these efforts—and much as I want to be happy—I would prefer a science that explores how we can stop making one another miserable. I suspect that in the long run such a science would actually prove more conducive to happiness. I also suspect that any scientist who took up the study would soon be interested in noise.

---

* Not surprisingly (to noise researchers anyway), Daniel Nettle reports in *Happiness: The Science Behind Your Smile* that one of the things you *can't* do to increase your happiness is "get used" to a distressing noise. If anything, the noise will make you unhappier as time goes on.

CHAPTER 2

The Unwanted Sound of Everything We Want

It was a sound you were obliged to take personally.
—IAN MCEWAN, *ATONEMENT*

At this moment you are engaged in what is probably one of the quietest activities of your life. Aside from sleeping, it may be *the* quietest activity of your life. Possibly you are doing it in a noisy place. You could be reading by an airport gate or in a doctor's office, straining to parse the words on the page as voices blare from a television overhead. You could be seated in a library next to someone with his "personal" listening device turned up to a distracting (for you) and deafening (for him) volume. But except for the sound you make turning a page, except for the sigh, snort, or chuckle you might make in response to something I've written, you are scarcely making a sound.

Two thousand years ago you would have been more audible. (Your book would have looked different, too.) Though there is evidence that people in the ancient world were capable of reading silently, the practice does not seem to have become normative in the West until about the ninth century of the Common Era. Most ancient Greeks and Romans—those who were literate, that is—did their important reading aloud. The same norm seems to have held for the ancient civilizations of the East. Written words were meant to evoke heard speech and were considered inadequate until they did so, like tea leaves before the addition of hot water.

As recounted in his *Confessions*, the fifth-century Christian theologian Augustine of Hippo was amazed to observe his mentor, Ambrose of Milan, reading silently.

> [W]hen he was reading, his eyes ran over the page and his heart searched out the meaning, but his voice and his tongue were at rest. . . . We hazarded conjectures as to his reasons for reading thus; and some thought that he wished to avoid the necessity of explaining obscurities of his text to a chance listener. . . . But the preservation of his voice, which easily became hoarse, may well have been the true reason of his silent reading.*

Reading would become quieter, and thus faster, more private, and potentially more subversive, as the centuries went on.

Now here are you, almost definitely reading this page in silence. People would find you peculiar—and noisy—if you did it any other way. But along with those sounds that lie historically behind your silent reading, you might consider the enormous amount of sound energy that lies beneath it. I mean the noise that goes into making a modern book.

When I first set about writing this one, I went to watch my neighbor Chris Devereaux, a self-employed logger who was harvesting a stand of soft-wood trees a few miles from my home. Instead of using a chain saw, Devereaux was seated in the cab of a steam-shovel-sized "hydro-buncher," a machine with a pair of enormous jointed pincers above a circular saw blade turning parallel to the ground. Making a sound loud enough to require him to wear ear protection (though some decibels lower than a crew of men with chain saws would have made), the machine would grip a clump of cedars, cut them off clean at the base, hoist them into the air with their tops waving, and stack them in a pile. The motion suggested a dinosaur-sized beetle learning to write with a feather pen—and managing it rather neatly. Nearby an even louder log crane drew the trees through a "slasher" that removed their limbs. Eventually the sectioned trunks would be trucked to a chipping plant, the next noisy step

* It has been suggested that as a provincial from North Africa, Augustine probably was amazed by all sorts of "oddities" that a more sophisticated Roman would have taken for granted, but most historians agree that Ambrose was at least somewhat the exception Augustine thought he was.

on their way to becoming paper—soft paper in this particular case, though book pages and toilet tissue both begin the same way.

Shortly after visiting the woodlot, I drove to a large paper mill in central Maine. It did not produce the paper for this book, but if you are indeed seated in an airport or a doctor's office, it may have made the paper in some of the glossy magazines lying at your elbow. After we had each worked a pair of foam plugs into our ears, my guide and I walked alongside a mechanical behemoth that turns the wet mash of saturated wood pulp into eight-foot-long rolls of paper at the rate of 4,100 feet per minute. One of the largest of its kind in the world, this machine was about as big as one side of the main street of an average New England town. You could easily have fit a pharmacy, bank, and barber shop plus several other stores and a diner into its frame. We walked its length on thick rubber pathways that absorbed the vibration, shouting at each other head-to-head, reading lips as much as we heard each other's words. We took off our winter jackets in the humid heat. The men operating the machine were safely encased in a rectangular box about the size of a small house trailer. I did not ask to take a sound reading, but I knew the sound must have been prodigiously loud—first, because I could feel it in my body and, second, because my subsequent requests for a decibel level were never answered. (In an industrial setting, asking about decibels is a bit like asking a dinner guest how much money he makes.) I did learn that the workers at the plant are given yearly hearing tests. My guide said he had worked there for twenty-seven years without any measurable loss.

My old friend Freeman Keith could not have made the same claim. After years of printing high-quality, limited-edition books, many of them destined for repositories of tomb-like silence, he was one of the most hard-of-hearing people I have ever known. Some would have called Freeman "old school," less attuned to health hazards than people are now, yet even today noise-induced hearing impairment remains "the most prevalent irreversible occupational hazard" in the world.

So, in spite of immediate appearances, your quiet reading is based on a great deal of noise. The same can be said for my quiet writing. I wrote most of this book in an old farmhouse, tucked among the hills of northeastern Vermont. If you could see the place where I was sitting when I wrote this sentence, you would in all likelihood think it enviably quiet, the ideal place for a writer to commune with his thoughts. Yet, this quiet occupation of mine depends on a great deal of noise: from the vehicles that bring my ink and paper up the interstates, the power plants that generate electricity for my lamp and laptop, the substations that see my Internet signals go through to New York—to say nothing of the travels I made to gather my material and the people I disturbed in the process.* In a book of many ironies, none has impressed me more than the noise I made to write it.

Noise forces us to ask knotty questions about what we want, what we don't want, and how we negotiate between the two. It forces us to consider how those trade-offs work for us and for other people, not least of all for the people whose labors make "what we want" possible. Noise is the fine print in our contract with the world. Small wonder if our attempts to understand it should yield some conflicting conclusions, even in regard to so basic a question as the one we turn to now: What is noise?

NOISE IS EASY TO DEFINE

Like Justice Potter Stewart, who famously said that although he could not define obscenity, he knew it when he saw it, most of us feel confident in our ability to identify noise. We know it when we hear it. Something about the way it intrudes on our awareness, not unlike a pornographic image does—except that our ears have no counterpart to eyelids, no comparable "off" switch when we turn our heads—puts the word to the sensation with no trouble at all. When one of my correspondents on noisestories.com writes to tell me her version of the "neighbor from hell" story, about living under an alcoholic kindergarten teacher who expresses her animosity by drilling holes in her floor in the middle of the night, I am certain she is talking about noise.

One of my favorite depictions of noise comes from Kiran Desai's 2006 novel, *The Inheritance of Loss*. Desai is a master of description who can create an unforgettable image with just a sentence or two; in her hands the graffiti inside a gum-studded Manhattan phone booth becomes "the sick sweet rot-

* According to the World Health Organization, "Transportation noise is the main source of environmental noise pollution, including road traffic, rail traffic and air traffic."

ting mulch of the human heart." When her character Biju, a young Indian immigrant, encounters New York taxicabs, she writes,

> They harassed Biju with such blows from their horns as could split the world into whey and solids: paaaaaaWWW!

Obviously Desai does not require typographical gimmicks to create vivid impressions. Having those taxi horns "split the world into whey and solids" is impressive enough. Nevertheless, she chooses to break up the uniformity of her typeface and to have one nonword stand out amid scores of carefully wrought sentences, outrageously demanding our attention, because *that is exactly what noise does.*

Of course, noise does not have to be loud to have that effect. Harold Pinter's darkly comic play *The Homecoming* contains a passage about the ticking of a clock during a sleepless night. Says his character Lenny: "All sorts of objects, which, in the day, you wouldn't call anything but commonplace. They give you no trouble. But in the night any given one of a number of them is liable to start letting out a bit of a tick." It's possible Lenny suffers from *hyperacusis,* a condition in which certain sounds are perceived as painfully loud, though he has other troubles to keep him on edge. It's also possible that someone else would be reassured by the ticking. In a song by rock group Death Cab for Cutie, comfort comes from the sound of a leaky faucet.

Noise does not even have to originate from an acoustical source. If Lenny was one of the millions of people who suffer from tinnitus (50 million in the United States alone, of whom at least 12 million have symptoms serious enough to require medical intervention), he might hear a ringing or buzzing in his ears, or a sound like crickets, a constant hiss, or an unceasing roar. He might hear it even if he were deaf.

Though tinnitus was believed to indicate mystical awareness in some ancient cultures, I have not heard of anyone who regards the condition as other than a curse. Like many tinnitus sufferers, a friend of mine has experienced bouts of depression along with his aural symptoms, which he likens to "a radio test of the Emergency Broadcast System." In a recent *New Yorker* article Dr.

Jerome Groopman tells of a 64-year-old retired machine repairman whose sensation of "a high-pitched squeal most of the time" became so oppressive that he went outside during a thunderstorm and stood next to a metal flagpole in the hopes that he would be struck dead.

Tinnitus has as many causes as it has classifications, ranging from vascular disease to the side effects of certain drugs, from traumatic neck and shoulder injuries to brain tumors. Not surprisingly, one of the major causes is exposure to loud noise. The Department of Veterans Affairs reports that almost 70,000 of the 1.3 million U.S. soldiers who have been deployed in Iraq and Afghanistan are receiving disability for tinnitus. "They answered the call," as we in safer circumstances like to say, and heard the ringing ever after.

Combat-induced tinnitus is perhaps the best example of how one person's peace, quiet, or security can be purchased at the cost of another person's noise.* It also provides us with a good metaphor for the culture we inhabit. The poet W. H. Auden spoke of his era as "The Age of Anxiety"; we might call ours "The Age of Tinnitus." Few of us ever go for long without something ringing, beeping, buzzing, throbbing, chattering, or screaming into our ears.

When those sounds annoy us, when they prevent or distract us from hearing another sound we'd prefer to hear instead, we call them noise. Noise is usually defined as "unwanted sound," what Biju, Lenny, and millions of tinnitus sufferers would call it, too. The word *noise* possibly derives from the same Latin root that gave us the word *nausea*.† Over the centuries *noise* has lost some of the meanings it once had in English, including those of "quarreling," "slander," "reputation," and "a band of musicians"—lost the meanings but not the ties. A reputation, for example, can sometimes operate like a noise, drowning out the day-to-day actions of a "reputable" or "notorious" person. The reputation is so loud, we hear nothing else. What we call *hype*, English-speakers in the fifteenth century would have called *noise*.

Except for deafening blasts and the ringing they can make in our ears, what counts as "unwanted sound" will vary from person to person and place to

---

* A less extreme example is provided by the fact that tinnitus rates are rising among preschool teachers.

† The *Oxford English Dictionary* raises doubts about the accuracy of that derivation but still gives it.

place. What's noise in a nursery might not count for noise in a dance club. (An obvious example, yes, but neonatal hospital units turn out to be quite noisy environments and "preemies" are especially susceptible to the negative effects of noisy monitoring devices and gabby personnel. This sonic overload during a key developmental period may explain why children born before the completion of the third trimester are often so easily distracted in later life.) Adapting the anthropologist Mary Douglas's definition of dirt as "matter out of place," the British physicist G. W. C. Kaye suggested in 1931 that noise might be defined as "sound out of place."

No doubt, there can be such a thing as too clean. Most of us would not care to live in a totally sterile environment, with kitchens like operating rooms; most of us would not want to live in total silence, either. In fact, tinnitus can be triggered in some people with normal hearing by placing them in a silent environment. Even rural quietness can be unsettling to people who are not used to it. One acoustical consultant I talked to told me that his mother-in-law "actually finds noise quite comforting. It gives her the sense that there is safety and security around her by hearing other human activity. When we have gone out into the woods camping in the past, she kind of freaks out when there's no sound at all to assure her that there's another human being nearby." In Isaac Bashevis Singer's short story "The Letter Writer," a suddenly jobless bachelor returns to his apartment and is comforted by the human commotion around him.

> Never before had Herman Gombiner enjoyed his apartment as he did on that winter day when he returned home after the closing of the publishing house. . . . From the neighboring apartments he could hear the laughter of children, women talking, and the loud voices of men. Radios were turned on full blast. In the street, boys and girls were playing noisily.

In a survey that asked people to describe the ideal soundscape, Canadian researcher Catherine Guastavino found that most respondents gave a positive rating to "the sounds of other people." In other words, like Herman Gombiner, they did not think of those sounds as noise.

Within such a broad category, however, are many gradations. On the one hand, few of us would like to work in a place where no one ever spoke. (After employees complained about the sound-suppressing acoustics of its new offices, the BBC agreed to install a noise machine that would play "mutter" to reassure them.) On the other hand, most of us have met or worked with someone whose constant verbal output seemed more about filling up pauses than making conversation. It has been suggested that technically advanced, media-saturated societies condition their members to find any kind of quietness unnatural and every kind of activity deficient that lacks a musical soundtrack.

This "colonization of silence," to use the phrase of contemporary American composer Andrew Waggoner, like other forms of colonialism, is often driven by commercial aims. Recently while shopping at a J. C. Penney's, my wife found herself standing in a place where she was aware of three different music tracks, each coming from a different department in the store, like overlapping stations on a radio dial. She asked one of the cashiers if this effect ever bothered her. "Try working here eight hours," the cashier said, adding that it was "company policy" to play different soundtracks in different parts of the store—presumably because that was what the customers wanted.

NOISE IS NOT SO EASY TO DEFINE

Some people are not satisfied with calling noise "unwanted sound."

One of them is Les Blomberg, founder and director of the Noise Pollution Clearinghouse in Montpelier, Vermont. Out of his small two-person office, to which he travels each day by bike, Blomberg maintains what is probably the largest accessible noise-related database in the world. For Blomberg noise is best defined by the name of his organization: It's a pollutant. "Do we define air pollution as 'unwanted particulates'?" he once asked me. On another occasion, he said that if he could go back and name his organization all over again, he'd get rid of the word *noise*.

With degrees in both physics and philosophy, Les Blomberg is the first person who helped me to understand noise as more than an annoyance. Though Blomberg's interest in noise began when he was awakened by garbage trucks emptying dumpsters in his neighborhood at 4:00 in the morning, he claims not to be among the 12 to 15 percent of the general population who are acutely noise sensitive.* For him noise is not "personal" the way it is for many anti-noise activists, but it is serious—too serious to be defined as "unwanted sound."

Defining noise in this way is relatively new, Blomberg told me. It dates from the early decades of the twentieth century, when scientists and engineers were developing the electronic communication devices that would determine so much of our modern acoustic environment. (For a history of this period he referred me to Emily Thompson's fascinating *The Soundscape of Modernity 1900–1933*.) To these experts, noise was primarily interference, static. It was a technical problem rather than a health issue or a social injustice. Ironically, this highly technical agenda gave us what Blomberg regards as an overly subjective definition. "Do we really want *desire* in science?"

To make his point, Blomberg gave the illustration of a kid who loses some of his hearing at a rock concert, something people have been doing in spite of repeated warnings for well over a generation. Rock concerts can reach sound levels in excess of 120 decibels,† the equivalent of a jet at takeoff. (By way of comparison, the Occupational Safety and Health Administration requires that hearing protection be worn by workers with prolonged exposure to sounds exceeding 85 dB.) Most of us would say that the kid in Blomberg's example was partially deafened by noise. But can we say that he was deafened by "unwanted sound" when he wanted to go to the concert, paid a lot of money to go, and may also have wanted it to be loud?

Probably he wants his MP3 player to be loud as well, a preference that has been blamed for contributing to the hearing losses of some 5 million children in the United States. As of now, one American child in eight has noise-induced hearing loss. Effects like these are trivialized, in Blomberg's view, when we define noise in terms of desire.

---

* Some sources put the figure as high as 25 percent. The difference may depend on how we define sensitivity, as self-reported noise annoyance or as noise annoyance sufficient to make a person seek a remedy.

† In 1976 The Who set a concert-loudness record of 126 dB at 32 meters. The record has since been broken by other groups (e.g., Manowar's 129.5 dB in 1994). Perhaps in the interest of public health, the *Guinness Book of World Records* no longer keeps a record for loudest concert.

Blomberg might like the suggestion of Australian musicologist Jamie Kassler, who defines noise as "sonic abuse." She notes that "[a]lthough the ear itself is structured to minimize damage from loud sounds, modern electronics introduces a new factor in the history of humankind. It makes readily available the technology for reproducing steady-state and high-intensity impulse stimuli, thus increasing the risks to hearing not only of individuals but of large groups of people." In fact, the average "normal" hearing threshold for a 60-year-old man in an industrialized society is 19 decibels above that of his counterpart in a nonindustrialized society. In other words, you would have to raise your voice another 19 decibels for him to hear you.

While respecting both Blomberg's point and Kassler's suggestion, I'm not sure we need to jettison "unwanted sound." I think that it can be a workable definition for noise if we consider "unwanted" in relation to the entire human being, including the human body. Noise exposes the delusion of regarding ourselves as disembodied consciousness, a will that points and clicks. Instead, we might think of our ears as "deciding" how much they will hear and for how long before they "refuse" to hear any more. When they have had enough, they start to shut down, temporarily when they have been fatigued, permanently when they have been damaged. According to noise researcher Karl Kryter, "It has been conjectured that perhaps [an] organism somehow senses when it is being stimulated by a sound such that the sensorineural receptors in the cochlea are being, or will be, unduly fatigued and that loss in hearing sensitivity may ensue. It is perhaps this sensing that gives rise to the *unlearned sensation of unwantedness, noisiness, or annoyance,* independent of and in addition to, any negative, or positive learned meanings that the sound in question may convey to the listener" (emphasis added).

With that in mind, we could modify the definition of noise from "unwanted sound" to "repulsive sound," the repulsion taking place either on the conscious level of what we reject for reasons peculiar to our experience and temperament or on the unconscious level of what our bodies can sustain. It so happens that our auditory system has neural processing centers in the upper brain, midbrain, and brain stem, thus allowing us, in Kryter's words, "to respond to sound at a more primitive level." On one level or another, noise repels us. Noise, we could say, is the sound of the unsustainable.

However we define noise, the place we give to our bodies in the definition has practical consequences for the environment. It may even predict our aptitude for preserving the environment. In his *Theory of Moral Sentiments* (1759) the Scottish economist Adam Smith raised the interesting hypothesis of "a man of humanity in Europe" who learns that an earthquake in China has just killed a hundred million of his fellow human beings. He regards the prospect with appropriate sorrow, but it does not keep him from his dinner. Let the same humane European learn that he will have his little finger amputated tomorrow, however, and he will not be able to sleep the night. Noise gives us Smith's hypothesis with a twist. What if the humane European doesn't even care about his finger? More to the point, what if he doesn't care about the loss of his hearing? How much will he care about the loss of tree frogs in Brazil?

NOISE IS OBJECTIVE

First with our bodies, and foremost with our ears, we have a reasonable basis for calling noise an "objective" problem. Of course, much of what we call noise has a subjective component, too. We have already considered examples of noise that clearly depend on a listener's subjective evaluation: the dripping faucet that comforts one person in the night but would drive another batty, the noisy neighbors who give comfort to a lonely old man and would make another feel helpless and alone. But on the most practical level, on the level of the street, it is important to be able to say that "noise is objective," if only to counter the claim that "noise is subjective," "all in your head." Subjectivity does not make you deaf.

Noise causes hearing loss in one of two ways. The first is by a loud impulsive sound like that of an explosion or a gunshot. The second way is through prolonged exposure to loud sounds of lesser volume but probably above 85 decibels, give or take. Even lesser impulsive sounds can be damaging given a significant rate of exposure. A 2009 study published in the *British Medical Journal* attributes hearing loss in one male subject to his prolonged exposure (three times a week for eighteen months) to the high-decibel sound (128 dB) of a golf ball being struck by one of the new-style titanium clubs. (The older, and cheaper, stainless-steel models are apparently quieter.) In both types of hearing loss, the noise bends or breaks off some of the roughly 16,000 fine hairs of the inner ear. These hairs do not grow back.*

Of the estimated 120 million people in the world with disabling hearing disabilities, some have *presbycusis* (age-induced hearing loss), others *nosocusis* (hearing loss caused by diseases of the ear), and many others *sociocusis* (noise-induced hearing loss, or NIHL). Cross-cultural studies of elderly populations indicate that much of what we used to take for *presbycusis* is actually *sociocusis*. It has been suggested that eliminating *sociocusis* from the industrialized world would require an estimated drop of 10 dB in the overall average sound level, an increment equal to the difference in sound pressure between a face-to-face conversation (about 60 dB) and a vacuum cleaner (70 dB). How or if such a reduction could occur is interesting to ponder. The Canadian composer R. Murray Schafer, who coined the word *soundscape* in the mid-1970s, said that it could happen in only one of two ways, either through a cultural shift in appreciation for the sonic environment, a process he called "ear cleaning," or through a global energy crisis. Needless to say, the jury is still out on Schafer's prediction.

Aside from its adverse effects on hearing, noise has other negative effects on human health, most owing to the fact that noise is a stressor, even at levels well beneath those that can damage the ear. This is why pharmaceutical companies sometimes use noise in testing stress-reducing medications. It is why an ancient Chinese text from the third century B.C.E. recommends the use of noise as a punishment for capital crimes ("Ring, ring the bells without interruption until the criminals first turn insane then die") and why it is used as a form of torture today. It is also why you and I are here. Our survival as a species owes in part to the naturally selected ability to respond energetically to sounds. Noise is known to elevate so-called stress hormones like cortisol and adrenaline, useful chemicals in a "fight-or-flight" situation, but needless and costly physiological expenditures in those instances where no threat exists but the noise itself. Scientist Bart Kosko expresses it very well when he writes:

* In 1988 scientists discovered hair cell regeneration in birds; more recent studies provide evidence of the same phenomenon in the auditory and vestibular portions of the inner ears of mammals. Scientists hope to find ways to stimulate significant hair cell regeneration in the human ear.

"A hunter-gatherer's sensitivity to high-decibel noise does not promote Zen calm or good digestion on a Monday morning while walking against the sidewalk crowd in New York."

What it does seem to promote is elevated blood pressure. A recently concluded study that followed 10,872 sawmill workers in British Columbia for eight years found statistically significant increases in hypertension for workers exposed to noise over 85 decibels. The results confirm those of other recent studies, including one of Austrian schoolchildren showing that those living in noisier areas exhibited elevated resting systolic blood pressure, and another of people living near four major European airports showing that noise events raised subjects' blood pressure *even when it did not wake them*. Noise from military air bases has been linked to lower birth weights in Japan; and highway traffic noise, to increased risk for heart attack in Sweden. Within ten years noise may become the number-one "burden of disease" in the Netherlands.

One obvious but not easily quantifiable effect of noise is sleep disturbance. According to Professor Michael Chee at Duke-NUS Medical School, "Sleep deprivation could be *the* silent killer of the 21st century, but in a different way from how hypertension was similarly labeled in the 20th." He might have said "different *ways*." Lack of sleep not only taxes the immune system, impedes the growth of new neurons, and possibly makes our brains more susceptible to "oxidative stress" (from free radicals that form whenever the body metabolizes oxygen), it also reduces higher visual cortex activity (our ability to make sense of what we see), thus raising the risk of accidents. The National Highway Traffic Safety Administration estimates that fatigued or drowsy drivers cause more than 100,000 motor vehicle accidents each year. Apparently even the loss of an hour's sleep can be significant: On the Monday after the weekend when we move our clocks forward by an hour, there will be a 17 percent increase in traffic accidents over the Monday before. While it's hard to determine how much sleep disturbance results from noise as opposed to other factors, it's equally hard to ignore the double-jeopardy effect of noise pollution. People living next to a highway and sleeping fitfully as a consequence are apparently more likely to get killed on one, too.

The same double jeopardy affects the child who goes from a noisy home or neighborhood to a noisy school. As long ago as 1975, New York researcher

Arline Bronzaft discovered that the reading scores of students on the train-track side of a public school were as much as a year behind those taught in classes on the quieter side of the building. Subsequent studies have confirmed her findings, along with showing that the students most likely to be adversely affected by noise are those whose academic abilities already lag behind those of their peers. One of the physical characteristics of sound, reverberation, has a parallel social effect. What wakes people from their sleep and halts their conversations reverberates in their health, their education, and, ultimately, their prospects for ever living farther from the tracks.

Determining what, if any, cause-effect relation exists between noise and mental illness remains problematic, but a correlation between the two is fairly well established. Prescriptions for psychotropic drugs and admissions to psychiatric hospitals are more numerous in populations exposed to major noise sources and among individuals who self-report noise annoyance. It is also known that noise is more likely to annoy people with psychological disorders. One observed effect of prolonged solitary confinement, for instance, is a tendency to become enraged even at slight noises.

The consensus in the literature is that while noise does not lead directly to any mental illness, it can aggravate the symptoms. On the positive side, quiet places of natural beauty have been demonstrated as mentally restorative, confirming the wisdom—or should we say, the common sense—of those urban planners who saw the need for parks.

Attempts to prove objective connections between noise and health effects beyond hearing loss continue to generate controversy. Popular news reports tend to proclaim certainties that a closer examination of the research throws into some doubt. If people living next to a noisy highway sleep less but also smoke more, if they have fewer years of education and more unpaid bills than their counterparts over the hill, then what is the cause or chief cause of their elevated blood pressure? In some cases researchers are able to make statistical adjustments for non-noise factors, but no study is as fine-tuned as its human subjects. I have talked to researchers and activists who chafe against proving that noise is quantifiably bad for you. "Why do we have to keep producing these dead bodies?" I heard one Norwegian scientist exclaim during an international session on noise and health. John Stewart, founder of the UK Noise Association, told me: "There clearly are health implications, but actually the bottom line should be, are people so annoyed and upset that it affects their quality of life?"

In other words, what proves so vexing for scientific research need not frustrate a political examination—especially if we focus on what separates people who can expect a good night's sleep from people who can't. I am not so much interested in what noise *causes* as in what noise *announces*. Where there's smoke there's fire, and where there's noise there's often a complex of social, economic, and environmental disadvantages, the eradication of any one of which would likely reduce the effects of the others.

A second reason for calling noise "objective" has to do with the physical aspects it shares with other sounds. Sound is a vibration that we sense with our ears. "One could imagine an alien species that does not have ears, or that doesn't have the same internal experience of hearing that we do," writes Daniel Levitin in *This Is Your Brain on Music*. "But it would be difficult to imagine an advanced species that had no ability whatsoever to sense vibrating objects."

Our species is also able, with the aid of instruments, to measure sound vibrations for their pressure (what we register with our eardrums), their rate of vibration (or *frequency*), and the energy that propels these waves for a certain time and through a certain space of air. The behavior of sound in a world of air, rocks, structures, and bodies of water can consequently be described and estimated. Sound bounces off reflective surfaces, which means that the concrete corridors of urban environments can channel noise so that it is not reduced as much with increasing distance as it might be in open spaces. Sound can be affected by variations of terrain and temperature, which means that on-site testing can be of considerable help in predicting the real impact of a proposed noise source.

Noise can also be *attenuated* by passing through or over certain substances, natural ones like grass and snow, human-made ones like medieval tapestries and the sound-absorbing attire of nineteenth-century ladies in hoop skirts and multiple petticoats. Of course, we understand this better than our ancestors did, though how much better remains a source of speculation. Why is it that one is able to hear a pin drop in the orchestra of an ancient Greek amphitheater even from the outermost seats, or that medieval cathedrals seem so well suited to the singing of medieval church music? One of the more fascinating conversations I had in preparing this book was with acoustical archeologist David Lubman, who cites the construction of Mayan ball courts and temples in support of his controversial hypothesis that ancient peoples understood acoustics far better than we believe and constructed their sacred sites to create awe-inspiring special effects.

If sound can be measured and described, it means that what we call noise can often be physically remedied, a third justification for insisting on the word *objective*. Thanks to improved technology, today's commercial jets are much quieter than the first generation of jet passenger planes that took flight in the late 1950s—though the gain in quietness has been offset to a large degree by the phenomenal growth in air travel (a 438 percent increase since 1960). The paper plant I visited in Maine makes use of acoustical solutions first developed for Hollywood movie sets in order to reduce sound levels outside the plant. Since the early 1980s European countries have been paving their highways with porous road surfaces that significantly reduce the generation and propagation of traffic noise. (They also have fewer SUVs, whose larger weights and wheels generate more noise than standard-sized cars.) Equipment is now available that allows bar owners to monitor noise coming from their establishments as it's heard in the surrounding neighborhood—but without stepping from behind the bar.

Though my emphasis in this book is mostly on the social aspects (sometimes called "the soft side") of noise, writing it has given me a great respect for those who work on the technical side. As expert witnesses in a state or municipal permitting process, they are probably no better or worse than other experts; that is to say, they know their facts and one of the facts they know is who pays them. But as servants of the community at large, they could prove to be, if not "the unacknowledged legislators of the world," as Shelley called the poets, then the unacknowledged diplomats. Especially in the design of multiple housing units and public spaces, acoustical science can cover a multitude of sins.

Obviously the goal of building for quiet is often at odds with that of building for profit, probably one reason the militantly "pro-business" Reagan administration cut funding for the U.S. Office of Noise Abatement and Control in 1982. Noise research and policy in the United States have not been the same since.

NOISE IS SUBJECTIVE

As much as it depends on any acoustical feature, noise depends upon how we interpret it. Especially in the case of sounds that do not threaten to damage our hearing, interpretation can make all the difference between a noticeable sound and a distressing one.

Fred Woudenberg, head of Environmental Health in the Netherlands, gave me a good example when I visited with him in Amsterdam. He had done some mediation work with an industry in Rotterdam that periodically expelled pressure from one of its operating systems by releasing large, noisy flares. During a designated "neighborhood talk" between industry officials and the community, neighbors complained about these periodic bursts of fire and their noise. As the discussion continued, it became evident that a primary concern was the fear that the bursts might indicate potentially dangerous malfunctions in the plant.* People also assumed that the bursts occurred only at night (probably because the sound was masked by other sounds in the noisier daytime), leading them to wonder if the company might be trying to hide its mechanical defects. Representatives of the plant were able to ease these concerns, and effectively reduce the level of noise annoyance, by explaining the purpose of the flares, inviting community members to tour the plant, providing a schedule for the flares (so there would be no surprises), and adjusting the timing of the bursts so as not to coincide with the hours when most people were getting to sleep. The community was still not in control of the situation, but they had exercised an influence over the people who were.

* Similarly, fear of crashes has been cited as a factor in the perception of aviation noise.

The research is at least thirty-five years old that shows lack of control to be a major factor in noise annoyance. This explains why you're less likely to be distressed by the sound of your food blender than by that of the table saw across the street, even though the blender is much louder to your ears. You can turn off the blender. You always know when the blender will start up again. And of course it's *your* blender, preparing *your* food.

Woudenberg cited other examples of Dutch industries reducing noise complaints by addressing subjective factors in a respectful way, but he added two words of caution. First, he insisted on the real effects of subjectivity. "I spend a lot of time explaining to people that when psychological factors determine if an effect is going to occur, it doesn't mean that the effect itself is subjective. What if a subjective factor is shown to increase your likelihood of getting cancer? The cancer is still cancer. It's the same with noise. High blood pressure is high blood pressure. I think one of the basic reasons noise doesn't get the attention it deserves is that there is no compassion for victims because [people think] 'they do it to themselves. They're weenies.'"

Second, Woudenberg cautioned against assuming from his examples that noise annoyance can be managed entirely by a tactful manipulation of people's subjective concerns. "I have found that the successful application of these non-acoustical factors is almost only successful in combination with an acoustical measure"—in other words, when an industry is willing to put its money where its mouth is. Acoustical measures can also pay subjective dividends. An industry that takes concrete steps to remedy a noise problem is contributing to what sociologists call the "fair process effect."

This means that the way we're treated in a process influences how we react to its outcomes. The pertinence of the fair process effect to noise annoyance has been supported both anecdotally, as in the case of the Rotterdam factory, and with human subjects in the lab. Even animal responses to noise are determined in part by the way the animal is handled during an experiment. The same applies when a community is being "handled" by a government agency or a corporate developer. If an entity has secured its right to make noise through a process that people know or feel was rigged from the start, or if that entity makes predictions of noise levels that prove false or disingenuous, then people will have a more negative experience of the noise—with potentially *objective* effects on their well-being. In contrast, Swedish and Dutch scientists have recently completed a study suggesting that "people who benefit economically from wind turbines have a significantly decreased risk of annoyance, despite exposure to similar sound levels."

Sometimes subjective reactions to noise are rooted in cultural differences. Anyone who's seen the Japanese film *Tampopo* will recall those scenes where diners slurp their long noodles with great gusto, especially the males. This is an instance not of bad manners but of good: Japanese boys are taught to "eat more noisily, like a man." Anyone who's been to Japan also knows that the intrusive public cell phone use Americans increasingly take for granted is regarded as a breech of etiquette there.

In his book *Different People,* American expatriate Donald Richie, who's been writing about his adopted Japanese homeland for more than half a century, tells of an ongoing noise dispute with an obstreperously noise-sensitive neighbor in his apartment building. The police officer who arrives in response to the neighbor's complaints as much as admits that she's being unreasonable (among other demands, she wants Richie to flush his toilet less), but adds, smiling: "Still, Japan's a small country. We all have to get on peacefully together somehow." I heard the same geographical explanation for noise awareness in the Netherlands. In a tight space, people are perhaps more conscious of their acoustic footprint.

Climate may also play a part in what people regard as noise. Some have suggested that the outdoor cultures of warmer countries place less emphasis on the propriety of an "indoor voice" than northern cultures do. One public health officer in London told me that a frequent subject of noise complaints there had to do with "Australians and their barbecues," with West Indians running neck-and-neck with the Aussies. Still, a "street culture" with a different sense of noise is not the same thing as a culture with no sense of noise. John Stewart of the UK Noise Association told me that though he found the streets of Istanbul noticeably noisier than those of northern European cities, he had the feeling that "a single dominant noisemaker"—someone blasting a stereo, for instance—would not be tolerated there. "In Serbia it was quite noisy, too," he added, "but as in Turkey you felt *the street belonged to the people.*" Contrast his impression with the statistic that one person driving a motor scooter through the nighttime streets of Paris potentially wakes as many as 200,000 sleepers. To whom does *that* street belong?

Where different cultures meet, there is the potential for ethnic or racial prejudice to masquerade as noise sensitivity—or, depending on your point of view, for legitimate noise complaints to be dismissed as ethnic or racial prejudice. Also in Great Britain I heard the infuriating story of a hardworking Greek-Cypriot immigrant whose repeated insistence on the quiet enjoyment of his home was written off as a "cultural difference" with his noisy neighbors. A dour lot, those Greek Cypriots. On the other side of the coin, an announcement of plans for a new mosque in County Armagh, Northern Ireland, raised objections from Ulster Unionists about "the wailing noise calling these people to worship," though church bells have been splitting the air of Irish cities since the days of the Vikings.

Not infrequently, the complaint about the noisy "others" amounts to a dismissal of their music as noise. That's what American nativists called the traditional music of immigrant Chinese laborers, what critics called the emergence of that quintessentially American music known as jazz.* It's what some French people have called the popular North African music known as *Rai*.

I took up the question of music, prejudice, subjectivity, and noise with a man selling hip-hop CDs out of a booth on lower Broadway in New York. I was drawn by the music coming from under his tent, a respite from the din of traffic, like free lemon ices on a sweltering summer's day. Earlier that morning, on a sidewalk outside the "José Tailor Shop and Variedads" in Queens, literally in the shadow of the elevated train tracks, I'd felt the same attraction: small audio speakers opposing the balm of Latin rhythms to the infernal screeching overhead.

As tentatively as I had introduced myself at the Maine paper mill, I stepped up to the table full of CDs, struck yet again by how dicey it feels to approach a logger, biker, or rapper and open with "Hi there, I'm writing a book about noise." Recently, though, I'd picked up a copy of Tricia Rose's provocative

* Historian Emily Thompson notes "a curious conjunction of racism and antimechanism" in this reaction. In other words, jazz was simultaneously dismissed as "uncivilized" and as a noisy echo of modern civilization.

study of hip-hop music and culture, *Black Noise,* and let her title serve for a segue.

The man behind the table, who identified himself as Anderson, works nights as a bouncer (successfully, from the look of him) and was manning the booth that afternoon on behalf of his relative Lenny M, a local DJ connected somehow to a more famous practitioner known as Funkmaster Flex. Anderson saw hip-hop music as a mixed bag of "good and bad, like all music," but he also saw an unfairness in the media's exclusive focus on "the bad" in rap. The music he was selling "gives people a lift with what they're going through," he told me, and was "not just about the sound; it's poetry." After honoring my request for a beginner's course by recommending DJ Alemo's anthology album *Back in the Days* (perhaps thinking the title suitable to my years), Anderson mentioned a kind of music that was new to him, a promiscuous sampling of materials known as *mashup.* "It sounds like a lot of noise to me," he confessed, "but to somebody else. . . ."

Whether or not this was meant to serve as the moral of our discussion, I took it as such. But was Anderson saying that noise is all subjective, that such a thing as noise doesn't exist? Not by his actions. After I'd said hello, he immediately turned down the volume on his box so we could talk without raising our voices. No less than in the music, there was poetry in the gesture.

NOISE IS NEW

Like mashup in the ears of a hip-hop aficionado, or the call of a muezzin in an Irish town, noise sometimes takes the form of a sound we've never heard before. The newness is partly what defines the sound as a noise.

In his groundbreaking 1977 book *Noise: The Political Economy of Music,* French cultural historian Jacques Attali argued that the sounds of a new historical development tend to be heard as noise by the old order, only to become sanctioned by the order that replaces it. As his subtitle indicates, Attali's book is focused almost exclusively on music; "our music foretells our future," he says. But Attali's thesis can be applied to some nonmusical sounds as well.

Trains, for example. When they first appeared on the scene, some people were horrified, or at least sorely annoyed, by their commotion and its impact on the rural landscape. Nathaniel Hawthorne wrote about having his reflections interrupted by the startling passage of a train. Thoreau, whose famous cabin on Walden Pond was within earshot of the train tracks, described the railroad as a "devilish Iron Horse, whose ear-rending neigh is heard throughout the town." Several generations and hundreds of blues songs later, many of us hear the long lonesome note of a train whistle with affection—especially if it's in the distance.* One acoustical engineer told me of being approached by an old man wondering if a highway noise barrier being erected near his home would also block out the sound of trains running parallel to the interstate. It probably would, said the engineer, believing he was about to kill two birds with one stone. "That's too bad," said the man. "I love that sound."

Still, it's hard to believe that anyone will ever come to be fond of the current generation of railroad "air horns," or those loopy flying-saucer car alarms, or that a cell phone going off during a Bar Mitzvah will ever bring tears of nostalgia to anyone's eyes. Hear, O Israel—the theme from *Cats*. Noise may be the ever-changing "sound of the new," but as Jamie Kassler pointed out when she defined noise as "sonic abuse," our ability to generate perpetual and deafening noise from a multitude of sources takes "new" to a whole *new* level.

On a more positive note we could say that the definition of noise as "unwanted sound" is also new, not just in the technologically based history we discussed before, but in its progressive implications. The word *unwanted* applied to a phenomenon as basic as sound suggests a worldview in which human beings have other options besides bowing to their fate. It implies a world that can be changed.

NOISE IS OLD

A noise may be new, but the power of sound to repel us is primeval. It has always been with us, like a love of sweets and the fear of death.

It is sometimes assumed that noise became an issue with the Industrial Revolution—the result of loud machines combined with the bourgeois fas-

* Of course, many Americans welcomed the railroad; even Hawthorne and Thoreau were not altogether opposed to it.

tidiousness of the people who owned them—and it is certainly during this period that the issue took on importance. But there are older substrata. Over the past several years I've heard a number of noise stories that imply a seemingly modern conflict between one person's quality of life and another person's way of earning a living, complaints about dry cleaners, airports, frost fans, and Jake brakes. These are all recent inventions, yet I've also read a Sudanese folktale that tells of a holy man whose prayers to Allah are repeatedly disturbed by the ringing of a woodcutter's ax.

As for the ringing we call tinnitus, it is described in ancient Egyptian papyri dating from the sixteenth century B.C.E. The first record of a noise complaint is 1,000 years older than that. The knowledge that noise can damage hearing is no more recent than the Renaissance.

Even the subjective side of noise has historic antecedents. Queen Elizabeth I of England is reported to have liked the sound of her subjects ringing church bells, which they were wont to do whenever they could steal into a church and grab hold of a rope. For her it was a sign of the kingdom's vitality. But for the women of Aurillac, France, in 1896, the sound of their church's tenor bell was annoying enough to merit a petition. According to the petitioners, the bell was "battering us about the head and instilling sadness and grief in our hearts, banishing the sweet thoughts and tender feelings that we harbor toward sex." I'm guessing the town fathers made fast work of that bell.

NOISE IS MEANINGLESS

Near the end of Shakespeare's play *Macbeth*, the villainous king makes a famous speech about the meaninglessness of an existence in which "Tomorrow, and tomorrow, and tomorrow creeps in its petty pace from day to day." Of such a life he says:

It is a tale
Told by an idiot, full of sound and fury,
Signifying nothing.

He means that life is a bunch of noise.

According to some scientists the universe is, too. It was born as waves of energy, a few of which we can manipulate to send signals, but most of which "signify nothing." For these scientists noise is defined as anything that interrupts a signal, such that noise exists relative to whatever signal we're trying to hear. Bart Kosko, who subscribes to that definition and therefore does not restrict noise to audible sounds (he also counts computer spam, for example), summarizes the relativity of noise this way: "To God all is signal."

I like that statement—that generous leap of the imagination by someone who seems to dislike noise every bit as much as I do. I also like what it implies about human limits and prerogatives. *Only* to God is every sound a signal. To human beings, some sounds are just noise. Some sounds interrupt their sleep, damage their hearing, raise their blood pressure, slow their children's progress at school, and banish the sweet thoughts and tender feelings they harbor toward sex. Those sounds are unwanted. And part of what makes them unwanted, aside from the harm they do, is that they strike us as oppressively meaningless, an extra helping of absurdity in a universe that can seem plenty absurd without them.

And yet. . . .

NOISE IS SELDOM WITHOUT MEANING

The most primal human noise is also the most basic signal: the one we make when we come wailing into the world. Here am I. Don't neglect me. Don't leave me shivering in the cold. The philosopher Hegel (who spoke of "the noisy din of the World's History") believed that the essential human need was for acknowledgment. In Hegel's view we strive, individually and as a species, for mutual recognition. If he was right, then some of the noise we smugly dismiss as "attention-getting behavior" may be exactly what we say it is, and it may merit more appreciation than we suppose. Why does a particular person crave attention, and why does noise seem like the best or only way to get it? Would we be paying any attention to him or her otherwise? And what of our own noisemaking—the writing of books, for example—and the unmet needs it implies? These questions point to one possible meaning of noise.

There are others. Since the 1880s engineers and mechanics have recognized noise as the sound of mechanical inefficiency. Most of us take this insight for granted. We call our mechanic and attempt to describe the noise our car is making the way we try to describe a symptom to our doctor. My do-it-yourself father was fond of saying that ignoring a noise in your car is like ignoring a pain in your body. No good can come of it. Something "isn't right," and probably not just for the car. Wasted energy can result in needless pollution. Noise is often the sound of excess: excess energy, excess wealth, excess time, excess frustration. Waste.

But it can also be the sound of something missing. After years of advising on noise disputes through the Noise Pollution Clearinghouse, Les Blomberg feels that most noise disputes indicate some kind of breakdown of community. The sounds of people we know and like seldom strike our ears as "unwanted." I find that I like the sound of my neighbors' sons riding their four-wheelers or rumbling by in their throaty pickup trucks. I have watched them grow from babies, and it pleases me to think of them having fun. At the same time, they obviously take some care not to disturb the old curmudgeon when they drive past his house: They usually open their throttles farther down the road. There is an unspoken covenant between us, sealed by little more than a wave. But populate the same landscape with young men I do not know, and I'll be hearing more noise—psychologically to be sure, and perhaps measurably as well.

The composer John Cage said in his book *Silence* that when we listen to noise "we find it fascinating." We can also find it revealing, including politically revealing. I was reminded of this when along with a record number of other Americans I watched the 2008 Democratic presidential convention on TV. Like the majority of those who voted, I was excited by the candidacy of Barack Obama, and not merely because his 1995 autobiography *Dreams of My Father* contains a nuanced passage about noise and community:

> That night, well past midnight, a car pulls up in front of my apartment building carrying a troop of teenage boys and a set of stereo speakers so loud that the floor of my apartment begins to shake. I've learned to ignore such

> disturbances—where else do they have to go? I say to myself. But on this particular evening I have someone staying over; I know that my neighbors next door have just brought home their newborn child, and so I pull on some shorts and head downstairs for a chat with our nighttime visitors. As I approach the car, the voices stop, the heads within all turn my way. "Listen, people are trying to sleep around here. Why don't y'all take it somewhere else."

I was also excited to see a speaking roster that gave acknowledgment to "ordinary Americans," including personal caregivers, union reps, teachers—the "weak," and those who serve the weak—the sorts of people whose voices are easily drowned out in a noisy, celebrity-obsessed culture and especially in a campaign that features such a charismatic candidate.

What interested me almost as much as what these people had to say, however, was the volume of convention-floor noise that increased noticeably whenever one of them spoke: the noise of delegates conversing, stretching their legs, getting to the restrooms. This was in striking contrast to the respectful hush that fell over the floor whenever the lights dimmed for a video or when the applause subsided after an "important" speaker came to the rostrum.

At least three messages were coming through to the TV audience, one implicit, one explicit, and one in the form of "meaningless" noise. Only the first two were intended. The implicit message said, "This political party cares about these sorts of people." The explicit message said, "People like us feel supported by this political party and that's why we endorse this candidate." The noise said, "When's Bill coming on?"

From an engineering perspective, the inattentive commotion of the delegates coming through so loud and clear to the TV audience was undoubtedly noise in the classic sense of unwanted sound. From a purely political perspective, though, it carried a clear message, honest, thought-provoking, and more than a little sad.

Here, then, is another reason why noise is unwanted sound. Silence, even the innocent silence of an hour's silent reading, can lie. It can tell us that we're quieter than we really are. It can tell us that our seemingly "quiet lifestyle" disturbs nobody. Noise, on the other hand, has an uncanny way of telling the truth. Much of the truth it tells is political.

CHAPTER 3

The Noise of Political Animals

Wherever Noise is granted immunity from human intervention, there will be found the seat of power.

—R. MURRAY SCHAFER, *THE SOUNDSCAPE*

Noise, politics, and cities came of age together. They grew up on the same streets. When Aristotle famously said that "man is a political animal," he was thinking of "man" as a creature who lives in a *polis*, or "city state." That is where human beings make "civilization" (from the Latin *civitas*, for "city"); it is also where they begin to make lots of noise. Metalworking, professional musicianship, and wheeled transportation all appear with the first urban settlements.

So do first-class seats and second-class citizens. When Aristotle's teacher Plato said that the ideal *polis* would consist of about 5,000 people, he was not counting women, children, or slaves. The pornographic acrobats and stuttering orators who occasionally provided the entertainment at all-male Athenian *symposia* were probably not counted either. Aristotle and Plato had very different ideas than we do about what makes a democracy, or a "civil" society, or a noise. The ability to amplify music, or to travel faster than the speed of sound, or to detonate city-leveling explosions from a thousand miles away—these did not belong to their universe. Neither did cultural pluralism, international law, or animal rights.

Still, when we talk about noise today we are never far from issues that were already at the center of politics in Aristotle's time: issues such as the rights of citizens, the distribution of wealth, and the proper exercise of power. These remain useful avenues for understanding noise. No less important, noise can prove a useful avenue in understanding our political selves.

NOISE IS POWERFUL

The most powerful forces in nature are loud. At least *what we perceive* as the most powerful forces in nature are loud. Gravity and electromagnetic attraction are potent and silent, but thunder and the crashing of the sea, cyclones and earthquakes, all make an awesome sound. So it was probably inevitable that awesome sounds would become a favorite insignia of the powers that be. When a royal personage steps onto the stage in a Shakespeare play, the directions might call for *hautboys* (from the French for "haughty wood"), a flourish of horns. In a Chinese court the job would have been done by a gong. The clatter of horses' hooves on the pavement, the bass beat throbbing at the mouth of the street—hark, our betters are coming.

For centuries the loudest human-made sound in the Christian world was the church bell, just as the most powerful institution was the Church. In time (and in step with the development of cannon), some church bells would become enormous. The second tenor bell of Notre Dame in Paris, smashed during the French Revolution, weighed 25,000 pounds. The audible range of the bells literally "sounded" one's home and identity. (A "Cockney," for example, is still defined as anyone born within earshot of the bells of St. Mary-le-Bow in Cheapside.) Symbols of power, the bells were also believed to possess power. It was widely held that they could banish demons and stop thunderstorms. As Alain Corbin describes in his wonderful book *Village Bells*, the French political history of the nineteenth century, from revolution to reaction, is inextricably woven with the bells. Disputes over who owned them, church or state, and what information they ought to peal, and when and how often they ought to be rung, were occasions of high passion and public uproar.

Church bells also had a role in proclaiming social status. No bells were rung for those who died before being baptized. Death knells for children who died before their first communion varied from place to place, but the custom seems to have been that bigger bells or more peals were sounded for boys than for girls. In our present world, too, the sounds of mechanical and electronic devices seem to give status to their operators, a noisier signature more often than not announcing a claim to masculine power.

To make a big noise is to overpower other sounds. It is to make them inaudible, of no account. Historically that ability has been linked to an ability to do the same to other people. In his *Soundscape: The Tuning of the World,* R. Murray Schafer speaks of "sound imperialism." Europe and North America, according to Schafer, have "masterminded various schemes designed to dominate other peoples and value systems, and subjugation by Noise has played no small part in these schemes." One recalls all those old jungle movies in which the bone-nosed natives are dispelled by the discharge of the white *bawana*'s omnipotent gun.

For the person on the receiving end of an oppressive noise, the experience is often one of powerlessness. In the 2007 film *Noise,* an attorney named David Owen becomes so incensed at the intrusion of car alarms into his life that he becomes a vigilante—a transformation that fails to silence all the car alarms but does work wonders with his sexual impotence. The symbolism is obvious but not ill-chosen. A sound that interrupts our sleep, or our children's sleep, or our lovemaking (Owen suffers all three) can feel like an attack on our status as adults. What, after all, marks our first callow steps toward adulthood if not the privilege of deciding for ourselves when we will go to bed? The noisemaker rescinds the privilege.

Of course, adults are not the only ones to feel overwhelmed by noise. Among the documented effects of prolonged exposure to residential and classroom noise on children is a syndrome called "learned helplessness," wherein a child begins "giving up." Possibly something like learned helplessness figures in another empirically demonstrated effect of noise: its ability to curb altruistic behavior. In a well-known experiment done over thirty years ago, a person was placed on an urban street in an awkward predicament (dropping a parcel), first in a quiet setting and next near a loud noise source (a jackhammer). Passersby were less likely to help in the noisy scenario. Up until recently I'd assumed that the noise simply made it unpleasant to come near the person in difficulty. After reading the research pertaining to children, though, I've wondered if the noise created a sense that there was nothing much one could do. If I can't even stop the racket in my own ears, why should I think I can make things easier for anybody else?

If a tyrannical regime wished to extend its power and to crush all resistance, what higher priorities could it have than the suppression of its subjects' altruism and sense of self-determination? None. In her 1985 study *The Body in Pain,* Elaine Scarry says that the goal of torture is to "unmake the world" of the victim. Gaining confessions and information is secondary. Torture attempts to do to the prisoner what war attempts to do to a civilization—and what noise seems to do to some children. While it would be an obscene exaggeration to equate the anxieties of someone living in an apartment next door to a compulsive door-slammer with the misery of someone imprisoned in the Russian Gulag, it would be silly not to see the same effects—and perhaps the same intentions—operating on a much smaller scale. In fact, Scarry cites Aleksandr Solzhenitsyn's description of "how in Russia guards were trained to slam the door in as jarring a way as possible or to close it in equally unnerving silence."

We don't need horrendous examples to illustrate the relationship between noise and power. We all know that wedding custom in which guests at the reception bang their glasses with spoons until the bride and groom kiss. When they comply, the banging stops. It's a sweet custom—for about the first ten minutes. It turns obnoxious when some boorish uncle or one of the drunker groomsmen tries to turn the couple into a pair of smooching marionettes. Whatever its original purpose, the custom has the effect of teaching one of the cardinal rules of married life: knowing when to resist the overbearing noises of tribe and kin, the telephoned or televised voice that says, "This is how a good wife acts, this is how a real man behaves, this is how we've always observed Thanksgiving in *our* family." I have been to weddings where the bride and groom's eventual refusal to obey the noise, sealed with a glance as tender as a kiss, was every bit as moving as their vows.*

---

* Noise has also been used to interrupt kisses (note the custom of the wedding-night *charivari* below) or even to stop them, as Queens zookeeper Charlotte Wyatt informs me. One winter when the zoo's two bears (then "in rut") had been mating for most of the day, and after the keepers had waited for an hour past closing time for them to return to their holding areas, the keepers tried noise—"They yelled. They called the bears with a megaphone. They sang and chanted, smashed metal bucket lids together, honked the horn of the utility cart"—all to no avail. Wyatt muses

NOISE IS REVOLTING

There are, of course, noisier things to bang than champagne glasses and reasons more crucial than getting a pair of newlyweds to kiss. The girls and women of Catholic Belfast banged trash-can lids on the pavement to warn of British patrols, to mark the August 9th anniversary of the first invasion of their neighborhoods in 1971, and to mourn the deaths of imprisoned hunger strikers. Who can hear of a prison uprising without imagining (with the help of a few old movies) the ominous clamor of hundreds of tin cups banging on the bars? Noise is the sound of revolt, the refusal to be ignored or silenced. "If I had a hammer," Pete Seeger and Lee Hays wrote in 1949, long before their song would or could become a hit, that hammer would bang morning and night. Because it hammers for justice. "The Lord *roars* from Zion," proclaimed the ancient Hebrew prophet Amos, and why? "Because they sell the righteous for silver, and the needy for a pair of sandals—they who trample the head of the poor into the dust." "Say it *loud*!" James Brown told his people, because in the midst of a white racist society "I'm black and I'm proud" should be said no other way.

Even if there were no gross injustices to shout or hammer down, there would still be a need for an interruption of order, a challenge to reverential silence. On some level we need the sneeze in the sanctuary, the baby who cries "*Human! Human!*" during the sermon about things divine, the dog that barks and piddles during the raising of the flag. The secular American liturgy known as Major League Baseball contains both the national anthem and the Bronx cheer. It wouldn't seem like baseball without them.

But are these noises "interruptions of order" or part of the order? Emphasizing the scientific aspects of noise, Bart Kosko explains how a phenomenon called "stochastic resonance" serves to restore order to an operating system. "Small amounts of noise" can actually improve the clarity of certain signals and even promote healthy physiological responses. His defining example is

---

whether "this behavior reflects a certain determination to breed, or if [the bears] have become so accustomed to unwelcome noise, being in New York, that such agitation is not effective." I'll blame New York for the sake of our subject, but I'm rooting for the bears either way.

the way that raindrops falling at random have been known to calm the waves of the sea. Though most of the literature on stochastic resonance is focused on technology and neurology, the principle has social and political application as well. A stable society will allow a certain amount of noise for its own sanity and preservation. This is a good thing to keep in mind, not only for the crank who thinks of every noise as a needless disruption but also for the cuckoo who thinks of every noise as a revolutionary act. Noise may be revolting, but it also has a conserving effect.

Medieval Europe, not on anyone's top-ten list for progressive societies, seems to have understood that effect quite well. The authorities countenanced such noisy outlets as the Feast of Fools and the Mass of the Ass, the latter a commemoration of the Holy Family's flight to Egypt (mother and child mounted on a donkey) that included "asinine" braying between priest and people. The peasants of Europe also had the custom of "rough music," in French the *charivari,* a procession of noisemakers who would visit selected members of the community, sometimes as a merry form of hazing (as with newlyweds), often as a way of heaping public ridicule on transgressors of various kinds. Readers of Thomas Hardy will remember the "skimmington" in *The Mayor of Casterbridge,* the "din of cleavers, tongs, tambourines, kits, crouds, hum-strums, serpents, rams-horns, and other historical kinds of music" that throws Lucetta Farfrae into an anguished panic over her "sullied" past—and is meant to. Same tradition.

The "rough music" of peasant societies excites the imaginations of scholars hungry for any hints of "the subversive" in the history of popular culture. Historian Bruce R. Smith's comment on "rough music" is a typical if slightly more balanced version of a sentiment to be found throughout academic treatments of noise: "Charivari can be interpreted . . . as a conservative gesture. But rhythmic percussive sounds have their own subversive power." Jacques Attali's influential book *Noise,* which opposes the more anarchic sounds of Carnival with the repressed quietness of Lent, is also in this strain. The book you're reading is not. I'm attracted to peasant culture as much as any other person with two cars to clean and a lawn to keep up, but whenever I hear people holding forth on the subversive exuberance of the irrepressible "folk," I want to ask if they've ever heard of a pogrom.

In her book *Dancing in the Streets: A History of Collective Joy*, Barbara Ehrenreich makes this comment on a 1618 English skimmington that featured "three or four hundred men" and the noise of guns, horns, pipes, and bells: "To the noble listening from his manor house or the cleric hidden away in his rectory, the sound of armed revelry must have been profoundly unnerving." With all respect to Ehrenreich, whom I happen to respect immensely, I doubt it. First, because no "rough music" could have raised a concern for the noble or the cleric that a regiment of horse and a few summary hangings could not have put to rest in half an hour. Second, because here is a fuller account of that same 1618 skimmington:

> [A]t Quemerford, Wiltshire, in 1618 . . . a suspected cuckold and his wife were visited by a party of men estimated by the victims at three or four hundred, some of them armed like soldiers. A drummer led the way. An impersonator of the cuckold, crowned with horns and wearing a smock, rode along. When they reached the victims' house, "the gunners shot off their pieces, pipes and horns were sounded, together with lowbells and other smaller bells, which the company had amongst them, and rams' horns and bucks' horns, carried upon forks, were then and there lifted up and shown. . . . " The crowd threw stones at the windows, forced their way in, fetched the offending wife from the house, and threw her into a wet hole, where she was trampled, beaten, and covered with mud.

The idea that noise is politically subversive, that the established order quakes in its boots every time it hears what might be construed as "rough music," including and especially loud rock music,* is one of the most cherished pieces of cant of my aging generation and its meanest legacy to those who came after. The only people who quaked in their boots in the incident

* It would make an interesting study to compare some of the pop songs of the past fifty years in terms of raucous sound and revolutionary content. The Beatles' "Revolution" (the fast version) and The Who's "Won't Get Fooled Again," for example, contain burn-down-the-house instrumentals and vocals—and decidedly counterrevolutionary themes. Contrast these with the gentle determination of pieces like Thunderclap Newman's "Something in the Air," Bob Marley's "Redemption Songs," Tracy Chapman's "Talkin' 'Bout a Revolution," or even Sam Cooke's "A Change Is Gonna Come."

above were an abused woman (her name was Agnes Mills) and her terrorized family. That revolutions are sometimes noisy—and necessary—I have no doubt. That noise in itself is revolutionary I don't buy.* At its usual worst, "rough music" accompanies some form of rough treatment or social ostracism. At best it is sanctioned "stochastic resonance," a Mass for the Asses that quiets the waves of social discontent but leaves the sea of injustice at full ebb.

For something purportedly so "subversive," noise can inspire the most slavish kinds of acquiescence (learned helplessness perhaps?). "THE MODERN EAR HAS ALREADY GROWN ACCUSTOMED TO THIS AUDIO TAPESTRY, and must learn to deal with more & more sound in the future," writes one of the contributors to an anthology provocatively titled *Sounding Off! Music as Resistance/Subversion/Revolution*. "Our species will not only adapt to but eventually feed off of the AURAL GLUT." I had rather been hoping for a future in which the species—every member of it—would eventually feed off food. "Let them eat noise" is no more subversive than "Let them eat cake."

NOISE IS ABOUT DISTRIBUTION

Noise also highlights one of the main reasons for revolt: injustice. In his monumental *Theory of Justice*, American philosopher John Rawls said that the "pervasive inequalities" that are "presumably inevitable in the basic structure of any society," and which "favor certain starting places over others," are the first things that any enlightened political system needs to address. Justice he called "the first virtue of social institutions." In the social distribution of quietness and noise, it often figures last.

Quietness, the ability to hear the sounds of nature, to converse without raising one's voice, and to replenish one's body and mind with sleep, is a form of wealth. We don't need a philosopher to tell us so. A real estate agent will do. Simply ask her how much money you are likely to get for your house if a

* Apparently, neither did Karl Marx. Here is his one-sentence summary of the destruction of the revolutionary Paris Commune of 1848: "The *serene* working men's Paris of the Commune is suddenly changed into a *pandemonium* by the bloodhounds of 'order'" (emphasis added).

rock quarry opens next door. Even seemingly nonacoustic aspects of property value—frontage, acreage, shrubbery, central air conditioning, construction above and beyond code—often translate into an owner's enhanced opportunities for quiet.

As for those who don't own, their opportunities are usually fewer. In a recent British survey, 30 percent of respondents who lived in detached houses reported being annoyed by loud music from neighbors; that was true for 53 percent of those in flats. The discrepancy was even greater for "shouting and arguments," which affected 20 percent of those in houses, 47 percent of those in flats. Not surprisingly, the same study identified income as a "significant factor" in noise exposure. One in five people with incomes of less than £17,499 regularly heard noise from neighbors; one in eight for those making over £30,000. Those in "social housing" (government supported) fared the worst, with 93 percent reporting that they "could hear their neighbors and were annoyed by their noise."

With an effort, those of us who live on spacious housing lots or in well-constructed apartments might be able to imagine what it's like to hear all of a neighbor's bathroom sounds and to have a neighbor hear our own.* More difficult to imagine is what it's like living in circumstances where *the desire to be considerate* is thwarted as often as the desire for privacy, where your softest footsteps will still be heard by the person living below. This is an aspect of poverty too seldom discussed: how disadvantage mocks those impulses toward generosity that privileged people can fulfill at whim. One of the saddest noise stories I ever heard involved an elderly woman who was repeatedly awakened by the thunderous sounds of running footsteps on the floor over her head at one o'clock in the morning. The runners were children greeting their father when he came home from work. The family was Hispanic; the woman below was African American. The flooring I don't know about, but I imagine it was crap. I'm told the problem eventually came before a judge, who ruled for the woman, but I'm certain that none of the parties ultimately responsible for this needless grief were in court.

---

* From a British Housing Association tenant: "A bath or shower is one of the most tense times in our house, it's one of the most personal things, that you hope to enjoy in private, but our neighbours can hear us, as we can hear them!"

Since it was first studied in 1967, students of public policy have used the term *environmental justice* to refer to the way that environmental "disamenities" are distributed by race and class. From the beginning, noise has been part of that discussion. The literature remains contentious, but the writing is on the paper-thin wall. In a 2004 study of 216 mostly white low- and middle-income third- to fifth-graders in upstate New York, Cornell researchers Gary Evans and Lyscha Marcynyszyn examined such factors as crowding, noise, and housing quality (e.g., cleanliness, hazards, indoor climate). They also took overnight urine samples from the children to determine neuroendocrine indices for chronic stress. After determining that five times as many low-income children as middle-class children were exposed to two or more environmental risk factors, they found that this "cumulative environmental risk exposure was significantly related to overnight urinary neuroendocrine levels in low- but not middle-income children." In other words, the low-income children were not only more likely to experience the risk factors but more likely to be adversely affected by them than their similarly exposed middle-class peers.

Environmental justice has also been examined in regard to the location of industrial and transportation noise sources. In a 2007 study of aviation noise around the airport in Phoenix, Arizona, Robin Sobotta, Heather Campbell, and Beverly Owens found that Hispanic ethnicity was the most important predictor for a household falling within the noise contours that the airport had determined—out of thirty-two possible options—in 1992. Some have argued that noise effects of this kind are situated not so much with the aim of overt race or class discrimination as with an eye toward avoiding the likeliest sources of political resistance. Communities with high reserves of political capital, including social connections to influential persons, access to information, the leisure to organize, and the means to pay attorneys (to say nothing of a few attorneys on the town list), are never the cheapest places to make noise. But even when Sobotta and her colleagues factored in such politically handicapping conditions as lack of a bachelor's degree, single-parenthood (single parents presumably having less time for political activity), low household income, and inability to speak English, their hypothesis of a "barrio barrier" held.

Within this issue lurks a cruel irony: By far the loudest and most pervasive forms of environmental noise come from movement—that is, from transportation—and they have their harshest effects on people with the fewest options to move.

One question that often arises in discussions of environmental justice is whether people travel to the environmental disamenity or the disamenity travels to them. Are noisy airport runways deliberately sited close to disadvantaged populations, or do disadvantaged populations inevitably gravitate to the cheaper properties beside the runways? In the case of the Phoenix study, as in the routing of certain interstate highways, the population was clearly there first; in other cases, the answer is less clear. While one can understand why this sort of precision is necessary to scientific inquiry, the question itself strikes me as an example of how scientific inquiry can be turned into political "noise"—with a masking effect over the still small voices of common sense and social conscience. At what level do the answers even matter? It is like asking for a precise causal analysis of why over 800,000 African Americans happen to be behind bars. Is it because of inferior legal council, a prejudiced judicial system, bad schools, the prevalence of drugs in urban communities—or a lingering, cankerous racism that has always been more convenient to put into a graph than into its grave?*

Acoustical engineer David Towers told me a story that has stuck with him over the years and is perhaps our best last word on noise and environmental justice. Towers was taking sound measurements for a rail project in an urban area of Miami when a man approached and asked him what he was doing. Towers told him he was measuring for noise.

"I didn't know they worried about noise in the ghetto," the man said.

Not sure how to respond, Towers said, "Is that what you call it?"

"You know what they call it."

---

* A modest attempt to address environmental injustice, including the unfair distribution of noise, was made in 1992 by the Clinton administration's Executive Order No. 12898, which requires that federal agencies identify and address the environmental impacts of their "programs, policies, and activities" on minority and low-income populations of the United States and its possessions. The last sentence of the document makes it clear, however, that no judicial review "involving the compliance or non-compliance of the United States, its agencies, its officers, or any other person" is sanctioned by the order. It was a step toward making environmental justice a part of the federal government's vocabulary; how much it changed what disadvantaged people *actually hear* remains doubtful.

"Well, I don't know," Tower said uneasily. "This is the first time I've been here."

"It'll be your last," the man said and walked away.

NOISE IS ABOUT COMMON PROPERTY

Questions of fair distribution include resources that people share in common. Noise Pollution Clearinghouse director Les Blomberg thinks that noise is an issue of "the commons." He refers to the ancient custom of commonly owned agricultural lands on which community members were free to graze their livestock and grow their crops. In his view, the de facto seizure of the soundscape by any imperious noise source, be it a jet ski or an airline industry, has its precedent in the devastating "enclosures" of the sixteenth century, when English aristocrats seized what had for centuries been common lands for their own private use. Two centuries later, when the Industrial Revolution created a large market for wool and a corresponding need for cheap labor, some of the larger landlords evicted their tenant farmers in order to expand their pasturage. The social costs were enormous. Our word *proletarian,* which came to mean an industrial laborer, originally meant a displaced one.

Thinking of the soundscape as common property, like the air and the oceans, is a dicey proposition for a society that sometimes seems incapable of valuing any resource not covered by a bill of sale. Still, the nearly universal capacity to be afflicted by noise keeps alive the idea that some things belong to all of us, an idea with enormous political and environmental implications. Noise may be the poor relation among pollutants—its effects are less devastating than those of water pollution or climate change—yet noise *audibly* reminds us that there are dire consequences whenever the commons are usurped for private gain.

NOISE IS ABOUT HUMAN RIGHTS

At the same time, noise underscores our sense of private rights. It does so on the most intimate level: Our ears are in our heads. If the concept of a "right" has any objective basis whatsoever, it is in the human body, in the universal desire, transcending all creedal or ideological differences, that the skin of one's body not be punctured and its orifices not be penetrated without one's say-so.

It follows that the places where we answer the needs of our bodies, where we eat, sleep, bathe, and make love, ought to share some of the body's inviolability—whether we happen to *own* those places or not. These are surely the most basic rights that exist, rights that even an animal, with no notion of "rights," will fight ferociously to defend. To prattle about your "right" to make noise while you force sound pressure into another person's ears or living space is the ultimate self-contradiction, an appeal to "rights" that undermines the very basis of rights.

In this light it is no wonder that disputes involving noise can become so bitter and even violent. We are most defensive at the thresholds of our flesh and our homes. When the British were determining their bombing strategy for defeating Germany in the Second World War, they rejected the American demand for "precision" bombing of industrial targets in favor of what was called "dehousing." After analyzing the effects of the blitz on Britain, Churchill's science advisor, Lord Cherwell, determined that what most bothered victims was the loss of their homes; "people seem to mind it more than having their friends and relatives killed."

In her study of torture, Elaine Scarry notes that "the unmaking of civilization inevitably requires a return to and mutilation of the domestic, the ground of all making," one reason, she thinks, that common domestic objects—doorknobs, refrigerators, and bathtubs—are often made part of torture's ghastly rituals. Only when noise has penetrated their domestic spaces have my correspondents tended to speak of it as torture.

NOISE IS ABOUT REDRESS

One of those correspondents, a woman in Los Angeles named Wendy Schaal, wrote to tell me of her ongoing battle with a next-door "chain restaurant/bakery" that "illegally receives HUGE truck deliveries between 3:00 and 5:15 A.M." across a narrow alley from her home. "We can't get the police to come and the two times they did, they only warned them; they didn't cite them. We've written letters, made phone calls, put up fliers asking patrons to boycott and one time I went down in my nightgown and raincoat and sat in front of their door so they couldn't make the delivery. Yet they continue. Now we're going to try to get them to mediation. The only thing we have left is the 'very expensive' court system and that's what it's going to take, because the city is pro business and they [the business owners] know it." Schaal's narrative is typical of many others I've received. In a nutshell, she was looking for redress. Less typical was her ability to find it: Eventually she wrote to tell me that the business had agreed to allow mediation through the city attorney's office and to curtail deliveries prior to 6:00 A.M.

Most people, at least most people in democratic societies, see redress as an essential part of their political systems. The First Amendment to the U.S. Constitution, which guarantees freedom of religion, speech, and press, also upholds "the right of the people peaceably to assemble, and to petition the government for a redress of grievances." Though the framers must have been hearing some kind of noise in the back of their minds when they felt the need to qualify *assemble* with *peaceably*, we can't be certain if any of them would have thought of noise itself as a legitimate grievance. Many of their descendents have, not always with success.

A number of factors have stood in the way, not the least of which are economic. Others have to do with the physical properties of sound (it's invisible, varying, often temporary) and the subjective aspects of noise. Where noise ordinances exist, they are often ambiguous. Police are not always trained or equipped to enforce them—or convinced that they merit police attention. The social status of those most likely to be affected is doubtless another factor; I have a hunch that gender may be a factor too. I have no basis for saying so beyond my own anecdotal evidence and the words of one prominent Dutch researcher who told me that the "typical" noise complainer in her country is thought to be "a middle-aged woman," in other words, someone "typically" lacking two prerequisites for being taken seriously: a deep voice and a high-status job.

In the case of entrenched noise sources, redress becomes even more elusive. Speaking mainly of aircraft noise, Mary Ellen Eagan, president of the acoustic-consulting firm HMMH, told me: "Noise is the only pollutant I know of that is regulated by the people who make it." In fact, the DNL (day-night average sound level) of 65 dB that the Federal Aviation Administration uses as its benchmark for acceptable overflight noise—providing noise-mitigation funds only in those cases that exceed or are likely to exceed it—significantly exceeds the level recommended by the World Health Organization and a number of other federal regulatory agencies. The Environmental Protection Agency recommended changing the DNL standard for aircraft noise to 55 decibels as far back as 1974.

One principle often heard in discussions of noise mitigation—and a policy recommended by the World Health Organization—is that "the polluter pays." In practice, that seldom happens. John Moyers, who has worked on public policy issues for many years, says that noise impacts are very often "socialized" while the profits of the noisemaking enterprise—and the savings reaped from cutting acoustical corners—remain in private hands. As an example he cites the HVAC (heating, ventilating, and air conditioning) systems used in many of the upscale housing units that have been built by Donald Trump and others in New York City. The cheaper (to install but not to operate) under-window, through-the-wall units pass costs to the tenants and noise—"this dull roar"—to the street. Of course, one common way that urban dwellers attempt to mask street noise is to turn on *their* air conditioners.

The same "socialization" of noise effects occurs when federal tax money is used to reduce the environmental impacts of airlines that resist upgrading their fleets. I spent some time with Armando Tovar, noise officer for the Raleigh-Durham Airport Authority, who describes himself as "pro-aviation" and admits to having little sympathy for noise complaints from people (mostly upper-income in this context) who choose to move within the noise footprint of an airport when there has been full disclosure of the noise risks involved. But he speaks with some frustration about 30- and 40-year-old DC-9s, which in spite of FAA-approved "hush kits," produce more noise (and are less fuel efficient) than later models. "My fundamental position has always been that airplanes need to be quieter," he told me as we were driving to take noise measurements near a "pool and racquet club" located under one of the flight paths. "When the airplanes get quieter, the public gets quieter. When the public gets quieter, less federal money needs to be spent."

Along with the question of who should bear the costs of noise abatement, Tovar's statement raises the related question of where the noise is more properly abated, at the source or at the receiver. The question remains even in cases where "the polluter pays." Millions of dollars have been spent in the United States on soundproofing houses and other buildings near airports and highways. But a soundproofed house is not a soundproofed home, if yards, sidewalks, and neighborhood streets are considered part of the definition. In some cases, the only feasible redress is home buyout, and airports have done that as well.

It goes without saying that not every noise complaint lends itself to redress. Some are simply out of this world. Helen Matthews, an environmental health officer with the Department of Environment, Farm, and Rural Affairs in London, told me of getting a call from a woman who demanded that the agency reimburse her hairdressing bills. The reason was that a mysterious noise, which she claimed had produced incurable diseases in many of her neighbors and killed all the plant and animal life around her apartment, required her to a wear a defensive tin hat that was wreaking havoc with her hair.*

NOISE IS ABOUT DIVERSITY AND COMMON GROUND

Just as noise can highlight the issues surrounding common and private property, it can also remind us of the complementary values of diversity and common ground. A healthy politics requires both. Without diversity there is no satisfaction in finding common ground—if we're all the same anyway, what difference does it make?—and without common ground there is no basis on which to honor diversity. Without common ground, "who's to say" racism is bad?

Noise compels us to locate common ground in the most basic of our biological needs. We find common ground every time our heads hit the pillow.

* As hilarious and pathetic as this complaint sounded to me at first, I would eventually learn that in the mid-1960s, researchers at London University, investigating an inexplicable acoustical phenomenon known as The Bristol Hum, instructed those troubled by it "on how to construct an aluminum helmet to be placed over the head when the Hum was in operation." It's possible that the preposterous complaint that found its way to one official agency had its origins in the advice of another.

Go to a biker festival some night to learn that different people have different standards for what constitutes "too loud," but go to a biker encampment on the morning after and bang some pots together to test the notion that "too loud" exists only for the straitlaced bourgeoisie. And wear your running shoes.

Noise also speaks to our differences. We do not all have the same sensitivity to noise. No more than 25 percent of us seem to be especially noise-sensitive. Deaf people are unaware of all but the loudest noises, whose vibrations they can feel, though the hard-of-hearing who rely on hearing aids are often frustrated by noise. And not all hearing people are sensitive to the same sounds. One of my correspondents speaks of a syndrome called "soft-sound sensitivity," which in her case means that the sound of someone chewing gum or even that of her husband breathing in his sleep are intolerable. On average, children and teenagers hear higher frequencies than adults do, a difference some of them apply in setting their cell phone ringers for undetected classroom use, and that some retailers have exploited with high-frequency anti-loitering devices.

Many people with autism are especially vulnerable to noise. Scientist and animal rights advocate Temple Grandin says that when she was a child loud noises often felt "like a dentist's drill hitting a nerve. . . . Birthday parties were torture when all the noisemakers went off." Grandin's governess soon discovered how to use this autistic sensitivity as a punishment and "kept a supply of paper sacks on top of the refrigerator so that she could burst them in my face if I misbehaved or drifted away from the world of people." Later, in college, the sound of her roommate's hair dryer "sounded like a jet plane taking off."

How we react to this information will say a lot about our politics or at least about our construction of diversity. "Yes, noise bothers some people, but that's because they're *not normal*!" It would seem that a truly civil society might come to a different conclusion: Justice for all does not mean one size fits all.

That still leaves the question of how many sizes we need. For example, one issue raised by the development of quieter electric cars is their potential danger to visually impaired people, who depend greatly on auditory cues for their safety. "Sound is our sight." Ideally the electric car is one way of quieting our shared life on the planet and of inviting future generations to partake of it. The challenge is how to invite them without rescinding someone else's invitation.*

And do we extend our invitation beyond the human circle, as Temple Grandin would undoubtedly insist that we do? In fact, we have already begun to address the effects of noise on other creatures. The U.S. Fish and Wildlife Service, for example, has required highway construction projects to measure and reduce their noise effects on nesting songbirds like the least Bell's vireo in California. Researchers have determined that most kennels are too loud (and also that dogs are happier when housed in groups as opposed to being caged alone). On the negative side, the U.S. Supreme Court ruled in 2008 that Navy sonar exercises can continue in spite of evidence suggesting that they disorient whales and other marine animals and have caused mass deaths of these creatures in the Bahamas and Canary Islands.†

Acoustic biologist Bernie Krause has gone so far as to suggest a new term, *biophany*, to denote the different aural niches that different species have assumed in order to be heard in the overall soundscape of any given ecosystem. Considering human noise in terms of its disruptive effects on the *biophany* is tantamount to thinking of animals as our neighbors. Since most of us have at one time or another had occasion to think of our neighbors as animals, reversing the comparison seems like a reasonable—and certainly more positive—step.

NOISE IS ABOUT WHAT WE WANT

Simply by calling noise "unwanted sound" we raise a basic political question: What exactly *do* we want? If noise is unwanted sound, then what is *wanted* sound? Not unlike religion, art, and commerce, politics is at bottom about human desire. Look closely at any dispute or piece of research centered on noise, and you will see a larger question about what makes for a more desirable society.

* One training school for seeing-eye dogs has purchased a Prius in order to acclimate the animals to the sound.

† "Even if the plaintiffs have shown irreparable injury from the Navy's training exercises," the Court said, "any such injury is outweighed by the public interest and the Navy's interest in effective, realistic training of its sailors."

Perhaps the most curious and politically suggestive noise study I know has to do with a 1927 experiment by one A. W. Kornhauser, who tried to see if typists working in a quiet office could do more work than typists working in a noisier one. Kornhauser was able to determine that typists working in noise were able to produce 1.5 percent more lines than those in quiet.* After more than fifty years of variations on the Kornhauser experiment, noise has been demonstrated to have a positive effect, a negative effect, and no effect on work performance.†

What makes Kornhauser's experiment so fascinating to me is not his dubious result but his loaded question. He didn't ask which kind of environment typists prefer to work in, or which seems to conduce to the overall well-being of typists, or the larger if admittedly unscientific question of why we work in the first place and what work might possibly mean. Reflecting the dominant concerns of his society (and of the people willing to fund this kind of research), he asked instead: How do we get the most out of the "girls" in secretarial?

Noise continues to reflect our desires and priorities. Take, for example, so mundane a controversy as the one over leaf-blowers. They're noisy things; older models can exceed 70 decibels. Attempts to ban or restrict their use in California have been condemned as anti-Hispanic by landscapers there.‡ No doubt there is much more to the controversy than meets the ear, ethnic prejudice notwithstanding. The enhanced status that loud machinery gives to its operator may be one factor. Working against it is the enhanced pleasure certain property owners would take in the quiet rhythms of their underlings raking leaves "the old fashioned way" on the old hacienda. Another factor may

* Various explanations have been offered for this outcome, including the masking effect of noise on distracting in-office conversations, an effect sometimes called "acoustic perfume."

† Noise does seem to give workers the psychological impression that time is passing more quickly, which is presumably what you want time to do when you're working in a noisy office.

‡ This is not to suggest that the leaf-blower issue is restricted to California and Hispanic landscapers. One of my noisestories.com correspondents tells of a retired couple on her street, with "his and hers" leaf-blowers and an apparent mission to save the neighborhood from every leaf that dares touch the ground, the result being a block of preternaturally tidy lawns and an almost ceaseless drone of noise. Like the Kornhauser experiment, the pathos of this situation implies larger questions about what we desire and how ill-equipped we are—apart from our lawn equipment—to realize it. One wonders when the two retirees get a chance to talk to each other and what they talk about when they do. Leaves, I guess.

be an envious resentment of neighbors with large enough lawns and salaries to afford a "noisy" (i.e., conspicuous) lawn service. Not least of all are the economic factors of the low wages landscapers make per lawn and the larger number of lawns that can be serviced with the aid of machines.

But beyond all this lies a question so much more compelling than whether or not to ban leaf-blowers, namely: What do we want our world to look and sound like? At present we have a society of people too rich and too busy to appreciate the physical pleasure that is a lawn's best benefit, served by people who cannot aspire even to a small lawn of their own without magnifying their meager earning power with machinery, to say nothing of the dispiriting spectacle of human beings got up like extraterrestrials as they move in robotic half-steps through a cloud of dust and deafening noise. In contrast, imagine a society in which everyone has a modest lawn and the leisure* to tend it if they so choose, where no worker lacks a livable wage, and where an unapologetically *elitist* aesthetic awards status to any activity done on foot or by hand.† The soundscape of such a society is not only quieter (though hardly silent) but much more diverse. It includes the voices of workers, both hired and home-based—their cars (or bicycles) indistinguishable at the curb—conversing in their own languages and playing their own music. That's salsa coming from the yard over there, though I'm not sure if it belongs to the landscaper or her obstetrician. Someone is probably going to complain about it sooner or later, but at least the complainer won't have to shout.

If noise raises the basic political question of *what we want*, does that mean silence holds the answer? I don't think so, if only because that would imply that death holds the answer too. Nevertheless, thinking about noise and si-

* Twenty-first-century Americans work an average of 350 hours more per year than their European counterparts. Over 25 million Americans put in workweeks longer than fifty hours; another million exceed sixty hours.

† Some will argue that technological "progress" is not reversible, that there's no such thing as "turning back," a dogma history does not support. The Japanese Samurai class forsook the gun for more than 200 years, even though gunsmiths in that country had achieved the highest levels of technical proficiency. Aside from seeing the weapons as a devaluation of his courage, the typical Samurai seems to have been put off by the ungainly stance required to fire them. In short, he worried that a gun made him look stupid. Imagine applying a Samurai aesthetic to such diverse activities as talking on a cell phone while driving in a full-sized SUV or "raking" leaves with a device out of a Star Wars movie.

lence as extreme zones of human habitation can help us to imagine the kind of society we might most want to live in. Toward that end, I offer the following considerations.

First of all, silence and noise are not zones most people prefer to inhabit for any length of time. Most of us would not be willing to enter the remote Carthusian monastery depicted in the 2007 German documentary *Into Great Silence*. Likewise, most of us are not ready to take up residence in a video arcade (decibel levels as high as 110), though we might have thought it would be fun when we were 14. Most of us want to dwell in a place somewhere in between these extremes, a place where we're free to talk and able to hear other sounds and voices.

That said, many people will feel a need to visit these extreme zones from time to time. People seek the silence of temples, caves, and libraries; they also seek the commotion of casinos, carnivals, and car races. As one young man at a NASCAR event said when I asked him how he'd feel if race cars could be made quieter, "Noise is good!" A healthy society will provide individuals with the opportunity to satisfy needs on both ends of the spectrum.

A healthy society will also recognize those few exceptional people who choose to inhabit either zone for much of their lives, hermits and astronauts in the case of silence, war correspondents and certain kinds of artists in the case of noise. Not infrequently a benefit accrues to the rest of us from their willingness to do so. Without them we would not have the poetry of St. John of the Cross or Keith Richards' incomparable smile.

But any appreciation we have for the zones of silence and noise turns to justified abhorrence as soon as either becomes compulsory. While some of us might admire a Trappist monk who took a vow of silence (along with his other vows of poverty, chastity, and obedience), most of us would or should be outraged at the thought of someone *forced* to be silent (as well as poor, celibate, and submissive). The same holds true for noise, including those forms of noise sanctified by the name of "music." "If music be the food of love, play on," Shakespeare says, but compulsory music (either played or heard) is not music any more than force-feeding is eating or sexual slavery is love.

The Nazis drove home that point when they forced inmates at the Polish death camp of Sobilor to sing a song about how happy they were, whipping those who failed to sing with sufficient "enthusiasm." Kurt Gerron, a renowned German Jewish opera star, was forced to sing the song that had made him famous as he was marched to the gas chambers, while prisoners at Auschwitz were forced to sing as they swung heavy wooden mallets to pulverize the human bones pulled from the ovens. More recently, Amnesty International has documented cases of torture in which the victim will incur further pain if he or she cries out (often as a result of various objects placed in the mouth), and a corresponding technique whereby prisoners are forced to communicate exclusively by shouting.

We want to walk away from such horrific details, and when we do walk away, we head for those in-between zones that might be called "quietness" and, at a slightly higher volume, "conviviality." These are the zones where we conduct most of the business we would call human. They are the zones in which we experience pleasure and exercise preference, with introverts statistically more likely to gravitate toward quietness and their extrovert counterparts drawn to a more robust acoustical mix.

These are also the zones in which we nurture children and within which children are able to thrive. I mean that in the fullest political sense, the sense employed by James Baldwin when he wrote in his book *No Name in the Street*: "It has been vivid to me for many years that what we call a race problem here is not a race problem at all. . . . The problem is rooted in the question of how one treats one's flesh and blood, especially one's children. The blacks are the despised and slaughtered children of the great Western house."

In other words, "the problem" is rooted in—as the zones of quietness and conviviality are also suited to—*adulthood*. What does adulthood mean if not making a society that is good for children, *all* children? This has nothing whatsoever to do with one's individual reproductive status or sexual orientation. It has nothing to do with whether one personally "likes kids" or finds them a pain in the neck. Rather, adulthood has to do with knowing one must die and wanting to be outlived in spite of the fact. It has to do with wanting a world that doesn't rob children of their health with fast-food, their hearing with electronic devices, and their lives with guns.

Seen in this light, a society that moved its cultural center of gravity in the direction of noise would also be a society in danger of losing its adulthood (and its children). Such a society would almost certainly privilege adolescence.* Childhood would consist of obsessively preparing for it, and adulthood of desperately trying to recapture it. Such a society would be unable to manage either its financial or its natural resources—its "allowance," if you will—for the common good. Its popular entertainments would resemble nothing so much as schoolyard hazing rituals (e.g., who are we voting off the show this week?) or masturbatory fantasies. Its public places would have the visual and aural features of video arcades. Its foreign policy would consist of swaggering about the globe picking fights with nations weaker than itself. It would wake up in the harsh morning's light of a new decade or a new political administration cranky and insistent on the need for "moving ahead," never mind the damage of last night. If the nagging for transparency became too intense, it would turn up its "stereo" and lose itself in a deafening dose of media noise.

NOISE IS ABOUT THE LIMITS OF POLITICS

Noise cannot be addressed exclusively through politics. If only in that sense, noise *is* subversive. It subverts the totalitarian notion that with the right form of government every major problem can be solved. Noise reminds us of the different tools we have to engage human problems and demands that we use them all, the political and the apolitical alike.

Noise can and has been addressed legally, through legislation like the U.S. Noise Control Act of 1972, by which Congress declared it "the policy of the United States to promote an environment for all Americans free from noise that jeopardizes their health or welfare."

Noise can also be addressed technically. Any time we squirt oil into a squeaky door hinge or have a new muffler put into a car we acknowledge that physics has as much to say as politics does about the amount of noise we hear.

Not least of all, noise can be addressed culturally. During my lifetime I have seen a change in cultural mores around the acceptability of smoking indoors (not okay) and the use of four letter words in "mixed company" (okay up to

* By *adolescence* I do not mean a stage of life inaugurated by puberty but a cultural construct—"the teen"—virtually unknown to people of other historical periods and societies.

a point) that would have been virtually unthinkable in my childhood. Cultures make collective decisions on what they regard as "polite."

Cultures also develop different strategies to resolve conflicts, including noise conflicts. In the conclusion to her history of twentieth-century noise, *Mechanical Sound*, Karin Bijsterveld recommends cultivating the art of "complaining in style."

> Complaining elegantly was the strategy of [these] columnists and journalists. . . . Many of them complained of noise in a manner full of humor and rhythm, thus making sure that they could raise the issue that bothered them while keeping two back doors open for escape in case the reader would identify them as bores. The first back door was that of irony, the second that of admiration for the virtuosity of their style. The first strategy is a tricky one for those serious about noise, but the second one is highly advisable.

The likelihood that every noise sufferer will have the talent or the disposition to take this approach is another matter, but Bijsterveld never says it is the only approach.

Another cultural way of addressing noise is through the medium of art—that is, by taking those sounds that many people would regard as "nothing but noise" and using them for creative expression.* While so-called sound art can be heard either as subversive (of traditional ideas of music) or defeatist (if you can't beat noise, you might as well join it), the sound artist performs a useful office for the rest of us: first, by holding out the hopeful if tenuous possibility that we might find something salvageable in sonic refuse; and second, by distinguishing art from life. "Sound art" is rarely forced on us, and we know its parameters, its beginning and its end—precisely what we don't know with the noise that comes through our front doors. Perhaps unintentionally, the sound artist is saying, "This dissonant composition that I've just played for you isn't noise. What you live with everyday—that's noise."

---

* Probably the most definitive source for readers interested in the artistic uses of noise is Douglas Kahn's *Noise, Water, Meat: A History of Sound in the Arts* (1999).

There is another side to all this. Noise not only speaks to the limitations of the political; it also works to exaggerate those limitations by driving people away from political engagement. When the public square becomes too loud, we seek the isolation of headphones. When the debate becomes too shrill, only the shrill debate. Extreme noisemaking and extreme noise abatement point to the same extreme position: the republic of one.

FINALLY, NOISE IS ABOUT PAIN

A standard feature of many discussions of noise is a chart listing decibel levels for various sound events, from the least audible (say, the rustling of leaves) to the loudest (say, the launching of a rocket).* Often such charts list a decibel level for the "threshold of pain"; most sources put it around 140 decibels. This is not the lowest level at which noise can damage hearing. This is the level at which noise becomes physically painful.

It is also at this level that we have yet another reason to think about noise from a political point of view. One way we evaluate political systems is in relation to pain: the pain they abolish, the pain they cause, the pain they use to maintain control.† The "threshold of pain" is seldom unrelated to the exercise of power. Given the human body's vulnerability to noise, and the ways that noise affects vulnerable human populations, our best last word on the politics of noise is likely to be *pain*.

One of the most succinct statements I've ever read about pain comes from the composer John Cage, a major influence on our contemporary thinking about silence, music, and noise. As far back as 1937, for instance, Cage predicted that "the use of noise to make music will continue and increase until we reach a music produced through the aid of electrical instruments." That this seems so obvious now testifies to how prophetic it was then.

The prediction is found in Cage's highly original book *Silence*, which contains an even more striking passage about pain:

---

* A chart of this kind appears at the back of this book.

† For a sonic example, we might cite the blasting of the recorded shrieks of dying rabbits during the 1993 siege of the Branch Davidian compound in Waco, Texas. My sources do not say if the volume of this noise reached the "threshold of pain," though its use for the purpose of liberating allegedly abused children does seem to approach the threshold of madness.

> I went to a concert upstairs in Town Hall. The composer whose works were being performed had provided program notes. One of these notes was to the effect that there is too much pain in the world. After the concert I was walking along with the composer and he was telling me how the performances had not been quite up to snuff. So I said, "Well, I enjoyed the music, but I didn't agree with that program note about there being too much pain in the world." He said, "What? Don't you think there's enough?" I said, "I think there's just the right amount."

Cage's remark goes to the heart of our subject as almost no other statement could. A world in which there is "just the right amount" of pain cannot have too much noise, can it? If the world has "just the right amount" of pain, then even if noise can reach the "threshold of pain," that world still has just the right amount of noise. And just the right amount of political progress. You will have to decide your own take on Cage's statement. I expect that by now you can guess mine.

PART II

Laetoli Footprints: A Brief History

CHAPTER 4

What the Python Said

Prehistory to the Eve of the Industrial Age

> *"My dear friend," said the python to the squirrel,*
> *"could you please make less noise."*
>
> —AFRICAN FOLKTALE

It is tempting to begin a history of noise with the Big Bang, the cosmic explosion that brought our universe into being roughly 14 billion years ago. Tempting, but not much to the point. For one thing, what we call a bang may actually have been more of a hum (a Hindu might call it an *Om*) or a hiss. For another, even if it was a bang, it could not have been a noise.

Though scientists speak of the Original Event in terms of sound, and have gone so far as to speculate that the distance between those first sound waves ultimately determined the distance between the galaxies, the low frequencies of the Big Bang—fifty octaves below middle A on the tonal scale, according to one estimate—would not have been audible to human ears. Of course, no human ears were around to hear it, much less to find it pleasant or annoying. We can debate forever that old question about whether a tree that falls in a forest with no one to hear makes a sound, but of one thing we can be sure: It does not make a *noise.*

Perhaps the only interesting thing about the Big Bang from the standpoint of a noise historian is not what took place in a void billions of years ago but, rather, what was taking place in the minds of the people who first named the event. Apparently, nothing less stupendous than a "bang"—and a "big" one at that—would do to describe something so important. Like a "blow-out" car sale or a "blow-away" guitar solo, "Big Bang" must have sounded right to the big guns who gave us the theory.

VOLCANOES AND THOSE WHO FIRST HEARD THEM

Probably the loudest sound on record took place in 1883 with the eruption of Krakatoa in the Indian Ocean. It was heard for a distance of some 3,000 miles—roughly the width of the continental United States—and colored European sunsets for months with its debris. But Krakatoa was certainly not the first and possibly not the loudest volcanic eruption to be heard by human ears.

As if to underscore that point, one of the oldest known fossil traces of human existence is in the form of two and possibly three sets of hominid footprints (the third set is smaller and less distinct) left in volcanic ash in what is now Tanzania some 3.5 million years ago. The so-called Laetoli Footprints are about as moving a human artifact as one could find. Perhaps only the sight of Pompeii's doomed lovers, caught in their embraces by the petrifying ash of a later eruption, comes close.

Where were these proto-humans going, we wonder, when they ambled over that ash, and what sounds did they hear as they went? Had they heard the volcano erupt before the ash was cool enough to walk on? Had they snatched up the one with the little feet and run for their lives? Was a crater still smoking visibly in the distance? Did they whistle walking past it?

I doubt that what they heard from a volcano was "noise" in our sense of the word. It would not have been "unwanted sound." The eruption itself they would not have wanted, but the warning rumble would have been a useful thing for them to hear. You and I live in a world full of bogus noises; our first ancestors did not. Our tendency to experience loud noise as a stressor almost certainly evolved from our ancestors' experiences of actual dangers. Our experience of noise as annoyance is the luxury—and the curse—of having sounds we can *safely* do without.

Even some of our most hated and "unnecessary" noises may have had a redeeming function in the prehistoric past. It turns out that the sound of fingernails pulled over a blackboard is repellant across cultures. Speculating as to why this noise should be so universally hated, some scientists have wondered if the answer lies in its similarity to a predator's shriek. Is some primal reflex triggered by the sound? If so, triggered along with it is the residue of a time when no awful sound would have been a noise, in the strict sense, but a note of warning—perhaps one that made all the difference between getting away or being eaten alive.

THE FIRST UNWANTED SOUNDS

It would be a mistake to say that early humans lived without noise or to assume that noise begins with civilization. Noise becomes a social problem at about the same time as people begin writing historical records. But several kinds of noise precede the urbanization and technical developments that made the human world so loud.

The first unwanted sound was any noise that masked the sounds of approaching danger. One of the casualties of noise is communication, or "speech intelligibility." This was also true for our ancestors, except that they were communicating with the natural environment as much as with each other. Native Americans, for example, were known never to make camp near fast-running water, which would have masked the sound of approaching enemies.

The second kind of noise our ancestors would have heard is also familiar to us: whatever wakes us up "for no good reason." Some African folktales of indeterminate antiquity have to do with talking animals that disturb each other's sleep. In one Yoruba tale a tortoise named Àjàpá is repeatedly awakened by the sound of the Dawn Bird, who for some reason does not confine his announcements to the dawn. With the self-importance that typifies an ostentatious noisemaker, the bird proclaims:

I am the bird,
I am the bird,
I am the bird that calls to the night.
I am the bird!

He's a bird all right, and we still meet with plenty of his feather. As told by folklorist Oyekan Owomoyela, "Àjàpá was angry that his sleep had been disturbed by such a senseless announcement. He could not imagine that the information was of any use to anyone, especially in the middle of the night."* After several futile attempts to quiet the bird, first with entreaties and then with a retaliatory song of his own (including the catchy line "I am Àjàpá that eats the bird"), Àjàpá eventually persuades the Dawn Bird to settle their differences in a swamp, where the tortoise manages to silence his harasser once and for all.

In another African folktale, a sleep-deprived python tries in vain to quiet a chattering squirrel. Adopting a familiar tactic, the squirrel dismisses the python as a sort of elitist: *He* can afford to be sleepy because he's just swallowed some prey, whereas the poor squirrel still needs to gather his nuts. "Nobody denies you the right to eat," the python replies. "But that doesn't mean you have to disturb everyone else while eating." The squirrel finds all this laughable. Drawn by the noise of its incessant chatter, a passing hunter kills both animals and wraps their bodies in a cocoa-yam leaf, which figures in the tale as a third casualty of the dispute. "So it was that the noise made by the squirrel caused the death of all three."

These possibly ancient tales suggest that the features of many present-day community noise disputes—including the desire for attention, the potential for violence, the unwitting effect on third parties, and the tendency to take both a noise and a noise complaint as personal affronts—have a long pedigree.

The stories also suggest that the origins of human noise annoyance might be dated imaginatively to when those foragers who left the Laetoli Footprints bedded down together for the night—perhaps when one of them awakened the other by snoring, what was once referred to in certain regions of England as "driving pigs."

---

* Àjàpá's predicament lives on in the life of Maryland "noisestories" correspondent Kristi Grant, whose sleep is continually disturbed by a "mechanical crow squawker" installed on top of the public housing unit where she lives. After repeated complaints, she has succeeded in getting the volume turned down but not off. Though the device is ostensibly used to scare away flocks of migrating crows, Grant believes it is actually intended to deter the homeless from congregating in the area and to keep tenants off the streets.

EERIE NOISES

The third imaginable prehistoric noise was any natural sound of wind, water, or seismic disturbance that aroused anxiety or dread. The native peoples around Haddam, Connecticut, called the place *machemodus*, "the place of noises," a reference to frequent subterranean tremors that the early English settlers found "fearful and dreadful." Those mythic Sirens mentioned in *The Odyssey*, whose singing lures passing sailors to their doom, may have their basis in the experience of uncanny natural sounds.

My friend and neighbor Howard Frank Mosher, a novelist who's been prospecting for stories in the hills and hollows of northeastern Vermont for the past forty years, told me a local legend about a woman who was supposed to have been driven mad by the weird gurgling of a brook. That she was also reputed to have murdered both her laudanum-addicted husband and a foster child, the latter impregnated by her loutish son, raises the question of what caused what. Mosher takes such stories with a grain of salt; he's still not sure that such a woman ever existed and he has found no written record of her crimes. Nevertheless, just prior to hearing the legend, he had fished the stream near the place where the woman's house was supposed to have stood, noticed a strange sound coming from the water, and found himself "overwhelmed by a physical sickness, similar to the sickening horror that has come over me on the few occasions when I've seen old footage of concentration camps." Some months later he was talking to his grown son and sometimes fishing partner, Jake, and without mentioning either the chilling story or his own past sensations, he asked if Jake recalled a catch of trout they'd taken near that spot years ago. Jake would have been around 9 years old at the time, and he remembered.

"I remember something else about that brook, too," he added. "From the time you and I first fished there, something about it revolted me." Mosher favors an acoustic rather than supernatural explanation—an unusually disturbing collusion of water and hollowed-out stone—though one has to wonder how the sound might have affected an already deranged murderer (shades of Poe's "Telltale Heart" come to mind) or if it could in any way have been a contributing cause of her derangement. One also wonders how the brook would have sounded to a prescientific human being who believed, or was beginning to believe, that the natural world was inhabited by spirits—especially if people with no such beliefs can be so powerfully affected.

Writing of his travels in Switzerland, D. H. Lawrence also gave testimony to the unnerving effect of certain notes of running water: "There was the loud noise of water, as ever, something eternal and maddening in its sound, like the sound of Time itself, rustling and rushing and wavering, but never for a second ceasing . . . something that mocks and destroys our warm being."

Uncanny noises must have played a part in developing the religious imaginations of early humans. Interestingly enough, when these sensations begin to take on the attributes of supernatural beings, it is often they, the gods, who are portrayed as annoyed by the noises of humanity. One native explanation given for "the place of noises" at Haddam was the anger of a local divinity at the encroachment of the English settlers and their god. In his classic study of myth and ritual, *The New Golden Bough* (1890), Victorian anthropologist Sir James Frazer notes the role of "wild shouts" and "loud howls" in the village exorcisms of New Guinea and other places. It would seem that, just like human beings and animals, the ghosts and demons find noise repulsive.*

THE FIRST NOISE IN LITERATURE

Pride of place for the first recorded instance of noise annoyance—and for the oldest book in the world—usually goes to the Sumerian *Epic of Gilgamesh*, written in cuneiform on clay tablets and probably dating from sometime in the middle of the third millennium before the Common Era. In its climactic adventure the hero Gilgamesh goes on a quest to find a man named Utnapish-

* The modern world's counterpart to the eerie noises of ancient times may be the mysterious low-frequency "hums" noted by significant segments of local populations in particular geographical locations around the world. Perhaps the best-known instance is the Taos Hum in New Mexico, though there is also a Kokomo Hum in Indiana, a Bristol Hum in England, and an Auckland Hum in New Zealand. Investigators have put forth various explanations—including hypernormal sound sensitivity in certain members of the population and gravitational waves generated by the high-voltage electrical supply grid—while conspiracy theorists have advanced explanations more sinister.

tim, who together with his wife, kin, and animals was the sole survivor of the Great Flood. After the waters abated, the gods granted him immortality and settled him in a land distant from other human settlements. Utnapishtim begins his account of the Flood as follows:

> In those days the world teemed, the people multiplied, the world bellowed like a wild bull, and the great god was aroused by the clamour. Enlil heard the clamour and he said to the gods in council, "The uproar of mankind is intolerable and sleep is no longer possible by reason of the babel." So the gods agreed to exterminate mankind.

The annoyance of the gods undoubtedly reflects a human annoyance. The civilization that produced *The Epic of Gilgamesh* is considered the first urban culture in history, which means it also produced noises altogether new to the human ear. The city states of Ur, Uruk, and Sumer—hence "Sumerian"—are generally credited with the invention of both the wheel and the professional army, two unsurpassed contributions to the history of noise. Make it three, if the Sumerians were the first to construct apartment houses. We also know from the Sumerians' art that they were fond of musical instruments and ensembles of musicians.

Even the appearance of their mud-brick cities rising along the Tigris and Euphrates rivers must have seemed—no less than the clanging commotion arising from within their walls—like a "noisy" intrusion on the landscape. The biblical story of the Tower of Babel, an edifice intended to reach heaven but thwarted during its construction by the sudden phenomenon of linguistic diversity, is probably an allusion to the Sumerian ziggurats, artificial mountains erected to bring the Sumerian elite closer to their gods. You might think of ziggurats as the precursors of airports: sacrosanct, ambitious projects intended to launch their sponsors transcendently into the wild blue yonder. The sonic boom, Sumerian style.

Readers of *The Epic of Gilgamesh* will note striking similarities between its flood story and that of Noah and the Ark in the Bible. Most scholars agree that the older Sumerian story is the source. (The people of Judea spent time in exile in Babylon in the sixth century B.C.E. and wrote significant portions of the Hebrew Bible there.) Among the narrative changes that take place between the account in *Gilgamesh* and the version found in Genesis is the replacement of noise by general "wickedness" as the reason given for the Flood. "The Lord saw that the wickedness of humankind was great in the earth, and that every inclination of the thoughts of their hearts was evil continually."

It is a commonplace in biblical studies to posit an evolution from the capricious, polytheistic universe of the Sumerians, in which the gods destroy humankind because of annoyance at its noise, to the ethical monotheism of Genesis, where moral disgust brings on the Deluge. Perhaps the ethical difference is not so large as is sometimes assumed. I wonder if the Sumerians were using the auditory phenomenon of noise to indicate the same "wickedness" bemoaned by the writer of Genesis. Elsewhere in the Bible, the prophet Isaiah, hoping for better times than those in which he lives, speaks of a day when "[v]iolence shall no more be *heard* in your land" (emphasis added). With a civilization of warring cities on the ground below, the sound of violence may have formed much of the "noise" that was keeping Enlil and his fellow gods awake.

THE CRY OF WAR

Violence, at least in its organized forms, has always been noisy. Military historian Robert O'Connell writes that the history of weaponry from ancient times to the present day follows a "logic" that duplicates the personal qualities of early heroes like Gilgamesh and Ajax, who "generally embody certain stereotyped physical characteristics: They are big . . . tend to be fast, [and] . . . Finally, they are loud, indulging in horrific war cries calculated to terrify and thoroughly in keeping with the intimidating image the hero seeks to project." When the 2003 American invasion of Iraq was dubbed "Operation Shock and Awe" it echoed a strategy as old as the Trojan War.

In *The Iliad*, also known as *The Wrath of Achilles*, Homer gives us vivid descriptions of the hand-to-hand combat that characterized preindustrial war. It includes the type of war cry mentioned by O'Connell. In one passage the goddess Athena incorporates her voice into the cry of Achilles: "The great sound shocked the Trojans into tumult. . . . The hearts of men quailed, hearing that brazen voice."

Throughout history, any culture known for its prowess in war seems to have been known as well for its war cry. Writing in the nineteenth century of a traditional Scottish war cry (*Cairn-na-cuimhne*, "the cairn of remembrance") that had endured to his day among the people of Braemar, in Aberdeenshire, the English poet Samuel Taylor Coleridge observed: "Even now, after so many customs have been buried in oblivion, if this cry be raised, within that district, in any fair, or assembly of people, all the men collect for the purpose of protecting the injured individuals." Of course, the cry would have sounded very different to the injured individuals than to those foolish enough to mess with them.

In the same century as Coleridge recorded the war cry of Braemar, the famed "rebel yell" sounded on the battlefields of the American Civil War. It had the same tactical purposes as the cry of Achilles, even though booming cannons had been a regular feature of warfare for hundreds of years. It is said that no one living today knows how to reproduce the cry exactly; it is said that no one who heard it in the heat of battle ever forgot the sound.

The noise complaints of Sumerian gods notwithstanding, R. Murray Schafer feels that there is an "association" not only between noise and warfare but between noise and religion. "Both activities are eschatological [i.e., destructive of the world as it is], and undoubtedly an awareness of this fact lies behind the peculiar bending of the Latin word *bellum* (war) into the Low German and Old English *bell*(*e*) (meaning 'to make loud noise') before its final imprint on the instrument which gave Christianity its acoustic signal."

No doubt Schafer is thinking of the Western monotheistic faiths when he speaks of the "eschatological" noise of religion. The Muslim Qu'ran, for example, speaks of the Day of Judgment as the "Day of Noise and Clamour." The Christian equivalent is usually portrayed as being heralded by the blast of a trumpet. The Ten Commandments Moses receives on the summit of Mount Sinai are reported to have been delivered amid a roar of terrifying noise, and from a God known as "the Lord of hosts," one of whose weapons (a bow) could sometimes be seen in the heavens after a rain.

SINAI AND SABBATH

In what amounts to an interesting twist on the noise annoyance in *Gilgamesh*, the biblical book of Exodus portrays the Israelites as disturbed by the sound of their god. They haven't the power to quiet his voice, but they do appeal to Moses for his help in turning it down: "When all the people witnessed the thunder and lightning, the sound of the trumpet, and the mountains smoking, they were afraid and trembled and stood at a distance, and said to Moses, 'You speak to us, and we will listen; but do not let God speak to us, or we will die.'"

Apparently, the proper sonic volume for a human being is not altered even when the sound source is God. Loud is loud, and very loud is unendurable. The Israelites are not punished or rebuked for insisting that their auditory limits be respected. Their god is humbler than the potentates of the U.S. Federal Aviation Administration.

The Exodus story contains two ideas of profound relevance to the issue of noise, though neither of them pertains directly to sound. The first is the concept of "the neighbor" as the basic unit of ethical measurement: "You shall love your neighbor as yourself." *Neighbor* implies proximity of some kind, including aural proximity. A neighbor, we might go so far as to say, is anyone who lives within earshot. The notion that the first rule of neighborhood is reciprocity—doing unto your ears as I would have done unto mine—continues to inform present-day discussions of noise and civility. The motto of the Noise Pollution Clearinghouse is "Good neighbors keep their noise to themselves."

Also of relevance is the Hebraic idea of Sabbath, a day when no one, not even a slave or beast of burden, is permitted to work. Though the purpose of the Sabbath does not seem to have been noise abatement, any cessation of work is a cessation of noise. Given the American glorification of "productivity," and the stale aftertaste of puritanical "blue laws," it may be hard for a reader in the United States to think of "Sabbath" in anything other than "killjoy" terms. Jewish tradition sees it otherwise. (In books of rabbinical commentary like the medieval *Zohar*, for example, the Sabbath is recommended as an ideal time for spouses to engage in sexual intercourse.) The Sabbath is of special interest here for implying a temporal approach to noise control as opposed to the more common—and in some ways more problematic—spatial approach of zoning.

The history of noise abatement is to a large degree about dividing space into noisy and quiet areas. Even today the cutting edge of noise policy in the European Union involves the sonic mapping of densely populated areas and major transportation paths. Medieval records show land purchases and the emancipation of serfs in the vicinity of monasteries to create zones of quiet wherein contemplation would not be disturbed. In the noise initiatives of the late nineteenth and early twentieth centuries, schools and hospitals joined houses of worship as places deserving special consideration in regard to noise.

Although most of us would see that as a progressive step—especially if we happen to be patients in a hospital—zoning tends to create zones of privilege. The same can be said of contemporary noise standards that set the "acceptable noise level" for any new sound source according to a measurement of "ambient sound"—that is, the sound that already exists in a given place. So, for example, a noise code might say that a new installation cannot raise the sound level of a given location by more than 10 decibels above the existing ambient. The rationale makes sense up to a point: A loudly humming HVAC system will be more noticeable in a rural village than in the middle of Times Square. As someone who lives by choice in just such a village (population around 1,000), I'm happy to accept the argument.

The problem is that the argument implies that people who live near Times Square (perhaps *not* by choice, and perhaps without benefit of a retreat "in the country") somehow love peace and quiet less than I do. It also assumes that a soundscape like Time Square is sort of "wrecked" anyway, so who cares if we add a little more sonic pollution to the mix.

Temporal approaches to noise abatement are not necessarily more equitable than spatial ones. Still, if everybody, regardless of his or her status, is prohibited from making noise for a certain duration, then everybody within that time interval—that "Sabbath," if you will—gets the same aural benefit. This makes physiological sense, too, in that sound deafens not only in proportion to one's nearness to a loud source but also in terms of how long one is exposed to it.*

Recent efforts in New York City and elsewhere to make certain streets automobile-free on specified days might be seen as a secular extension of the Sabbath idea, one that combines temporal with spatial considerations. For a certain period, in a certain place, people give their cars, their scooters, and their ears a rest—while giving their legs and lungs a little workout. In the middle of Manhattan they have a taste of *shalom*.

Parallels to the Jewish Sabbath are found in cultures that do not use the term. Writing of the houses (and times) wherein the traditional Japanese tea ceremony is performed, and where there is "not a colour to disturb the tone of the room, not a sound to mar the rhythm of things . . . not a word to break the unity of the surroundings," Okakura Kakuzo says: "One may be in the midst of a city, and yet feel as if he were in the forest far away from the dust and din of civilization." If he gets there on a bike, so much the better.

STILL SMALL VOICE

One finds interesting references to quiet amid the awesome thunder and "joyful noise" of the Bible. (An acoustical engineer I spoke to said that the phrase "joyful noise" in "Make a joyful noise unto the Lord" [Psalm 66:1] is an oxymoron. "It should be joyful *sound*," he told me.) In what reads like a revision of Moses' Sinai epiphany, the prophet Elijah is witness to a clamor of meteorological noises—earthquake, wind, and fire—none of which contains the voice of God. Only in the "still small voice" that follows the storm is the divine communication heard. In the so-called Servant Songs that Christians take as prophetic of Jesus and that Handel would set to music, a later prophet, Isaiah, says that the chosen servant of God "will not cry or lift up his voice or make

* For example, in 1999 the World Health Organization recommended a threshold of 100 dB for "ceremonies, festivals and entertainment events," but with exposure times attached by way of qualification. If you dance at a 100-decibel fiesta no longer than four hours and attend comparably loud fiestas fewer than five times a year, the WHO feels your hearing will probably be okay. But if you're on the road six months of the year with a death-metal rock band doing two shows a night, maybe not.

it heard in the street." The same prophet says that "quietness and confidence will be your strength."

Quietness receives notable emphasis in the traditions of the East, especially those that teach seated meditation. Even when he was not seated, the Buddha is reported to have asked the monks accompanying him on his journeys to quiet their chitchat. "I shall go quietly in inhabited areas" is one of the rules he laid down for his followers. Drawing on a Vedic tradition even older than the Buddha, the nineteenth-century Hindu saint Sri Ramakrishna spoke of the attainment of *Samadhi*, or enlightenment, in terms of a progressive quieting down. He illustrates it with this parable:

> How long does one hear noise and uproar in a house where a big feast is being given? So long as the guests are not seated for the meal. As soon as food is served and people begin to eat, three quarters of the noise disappears. . . . When the dessert is served there is still less noise. But when the guests eat the last course, buttermilk, then one hears nothing but the sound "soop, sup." When the meal is over, the guests retire to sleep and all is quiet.

One can object that Ramakrishna is no more talking about acoustic silence than he is talking about buttermilk. The quietness of which he speaks is interior, a stilling of the mind, or as he goes on to say by way of explaining his parable: "The nearer you approach to God, the less you reason and argue." But such an objection presupposes categories of "inner" and "outer," literal and symbolic, that are likely to be more rigid than his were. For contemplatives like Ramakrishna and the fourteenth-century German mystic Meister Eckhart ("Nothing in all creation is so like God as stillness"), as for American transcendentalists like Emerson and Thoreau, the outer and inner worlds exist as complements. To test that idea, try to imagine Ramakrishna shouting into a cell phone or Thoreau contemplating Walden Pond astride a jet ski.

These examples of quietude do not contradict Schafer's observation about the noisy common ground of warfare and eschatological religion. So long as religion is based on some notion of revelation, both noise and silence will have a part in it. To a people wandering through desert wastes, a new revelation—something truly astounding—might well be heard as a thunderous noise. But to those who already live in the din of civilization, an authentically prophetic voice is as likely to be quiet as loud. Small wonder if Rosa Parks' quiet refusal to surrender her bus seat in 1955 should seem like the quintessentially rebellious act of her age, defiant not only of racism but of an entire culture's brash assumptions about what counts, who matters, and how best to rock the world.

ROMAN NOISE

Julius Caesar is credited with having authorized the first civic noise ordinance in 44 B.C.E. "Henceforward, no wheeled vehicles whatsoever will be allowed within the precincts of the city, from sunrise until the hour before dusk. . . . Those which shall have entered during the night, and are still within the city at dawn, must halt and stand empty until the appointed hour."

That the world's first noise law should have arisen in what was probably the largest and loudest city thus far known in the ancient world is no surprise. That it should have arisen from an empire that spread "peace" by the most violent and calamitous methods—as the Roman historian Tacitus quipped, "They make a wilderness and call it peace"—is more than a little ironic.

Beyond the innovation, and the irony, the most interesting thing about Caesar's ordinance is how counterintuitive it seems. It prohibits wheeled traffic *during the day*.

Why would Caesar have restricted wheeled traffic to the nighttime hours? One plausible answer is that his rule was not a noise ordinance at all, merely a traffic restriction—an attempt to reserve the daytime streets of Rome for pedestrian and, more important to the likes of Caesar, equestrian traffic. (The Roman plutocratic class was called "the Equestrian Order.") Still, that leaves the problem of a noisier nighttime, and the Roman night was certainly noisy: More than seventy years after Caesar's ordinance, the poet Juvenal is complaining that "the perpetual traffic of wagons" at night "is sufficient to wake the dead." Why not sufficient to change the law?

The best answer might be found by asking where Caesar himself would have slept. Certainly not in a ground-floor bedroom facing the street, perhaps not even within the city precincts. His sleeping quarters would have been higher off the ground or farther away from the main thoroughfares of merchant traffic than that of the average plebeian. If quietness was the aim of his ruling, he wanted it most when his own business—*his* bathhouse conversations, *his* senate oratory, *his* equivalent of a Forum "photo op"—would have been most affected. Caesar's law was consistent with the first principle of imperial economy: You pay, I benefit.

Roughly a hundred years after Julius Caesar, the Roman philosopher and playwright Seneca wrote what may be the first essay devoted entirely to noise. He based it on his experiences living above a gymnasium, the noises of which included all kinds of "grunting," "hissing," and "pummeling," to say nothing of "the hair remover, continually giving vent to his shrill and penetrating cry in order to advertise his presence." Depilatory techniques being what they were, it's conceivable that some of the hair remover's customers cried out as well. Seneca belonged to the Stoic school of philosophy, which taught that pleasure and pain were to be endured with equanimity and which gave us the word *stoical* to describe people who hold up well under strain. Seneca's approach to the noise from the gym followed his philosophy. Boasting that he took "no more notice [of] all this roar than . . . of waves or falling water," he claimed that to be distracted by noise was to succumb to one's own inner disquiet. "I cannot for the life of me see that quiet is as necessary to a person who has shut himself away to do some studying as it is usually thought to be." That quiet was by this time "usually thought to be" necessary to intellectual endeavor is indicated by a Roman law prohibiting coppersmiths from setting up shop on any street where a professor lived.

One wonders if Seneca would have been quite as stoic living above a modern gym, where music is sometimes cranked to 100 decibels in order to produce higher levels of adrenaline. Yet, one shouldn't underestimate him. Ordered to commit suicide by the emperor Nero, he cut his own wrists. When the blood ran too slowly, he opened veins in his ankles and behind his knees, and when that also proved unsatisfactory (he was old at the time and his circulation was evidently not the best), he swallowed hemlock (also maddeningly ineffective) and finally had himself suffocated in a vapor bath. He was tough; he was also honest in spite of his maxims. He closes his essay on noise by noting that he will shortly be moving from his apartment. "Why should I need to suffer the torture any longer than I want to?" Dedicated Stoic that he was, he still found the noise to be torture.

Although we live with sound sources and decibel levels beyond our ancestors' wildest dreams, the terms of our noise disputes are quite old. Seneca's advice is merely an ancient version of that familiar and often disingenuous claim that with just a bit of mental discipline *you* (at least Seneca had the decency to say *I*) can get used to noise. In fact, the research is pretty conclusive in showing that noise is not something people get used to, unless one considers hearing loss a form of acclimation. Over time noise annoyance tends to *increase*, even when the loudness and duration do not, and even when reported complaints start dropping off. "People have stopped complaining" never means people have stopped suffering.

That's not to suggest that getting used to a major noise is utterly impossible. With as much willpower as a diehard Stoic philosopher, and slightly more detachment than a Buddha, it should be a piece of cake.

While the gyms of Seneca's day were probably quieter than our own, his libraries may have been noisier. It seems to have been common practice in the ancient world to read texts aloud, and Roman libraries were often located in bathhouses, perhaps not the noisiest but surely not the quietest places in the city. (I'm assuming that hair removers plied their trade there as well.) Scrolls were kept in a central archive, with adjoining alcoves to which patrons could take the requested documents and mumble over them at their leisure. The hard marble surfaces would have accentuated every sound, as they can also do in modern libraries. But in contrast to the preternatural quiet of the Fifth Avenue Library's justly celebrated Reading Room,* one is likely to have heard a hive-like drone in a Roman library.

My hunch is that the low drone in a Roman library was a reassuring sound to its patrons. For one thing, a drone is generally less intrusive than a repeat-

* New York Public Library director David Ferriero told me that the architects who designed the library intentionally placed the Reading Room at the top of building, "as far from the noise and dirt of the street" as possible, and "as close to heaven as you can get." It is hard to imagine a more diverse—or quiet—population gathered in any space of comparable size. Heavenly indeed.

edly interrupted silence. So-called continuous noise is known to be less disturbing to people than impulsive noise—one reason why the constant whoosh of automobile traffic can be more endurable than the sound of airplanes taking off and landing, and probably a reason why some of us can concentrate better in a busy coffee shop than in a mostly silent library with a couple of chatty patrons (or librarians).

It's also possible that the murmur of other scholars had pleasant subjective associations. Luke Reynolds, a young teacher in Arizona, writes via noise stories.com about one of his favorite sounds: "The 'buzz' of my high school classroom. Students might be working in peer groups, editing, or creating a presentation. I look around, and they are all focused. The noise or 'buzz' emanates from the class, and rises like steam above all of us. . . . [I]t feels as though I have just turned over the engine of a brand-new automobile."

No doubt Mr. Reynolds (whose nominee for least favorite sound is the "painful silence" that followed his yelling at an unruly class of freshmen on a day when he'd had too little sleep the night before) would nod approvingly at Talmudic scholar Jacob Neusner's account of "the massive noise" of a traditional yeshiva. "The only moments of silence in a yeshiva come in the solitude of prayer: then you can hear a feather fall off the wing of an angel." Otherwise, I assume, you can hear the words of Talmud falling off the tongues of the scholars—wanted sounds in both cases.

FIENDS, BARBARIANS, AND LUNATICS

Neusner's yeshiva "noise" would doubtless have been quieter in the Middle Ages, at least in England after the year 1253. That's when a royal statute required Jewish services be said *submissa voce* (in a low voice) so as not to be audible to Christians. In the preceding year the Church had given official approval to the use of torture. By then the Inquisition was already two decades old. Except in the environs of an unmolested synagogue, these were probably not the best years for noise abatement. As cultural historian Hillel Schwartz reminds us, when we ponder the quieter medieval past we must not forget that, along with the ringing of church bells and the clang of blacksmiths' hammers, it included a fair amount of screaming.

Not surprisingly, the hell that loomed so large in the medieval imagination was conceived as a noisy place. In his *Inferno* Dante describes the sounds of the damned, doubtless an echo of those condemned by civic justice on earth:

Tongues mixed and mingled, horrible execration,
Shrill shrieks, hoarse groans, fierce yells and hideous blether
And clapping of hands thereto, without cessation

When fiends appeared on earth they were often reported to be accompanied by demonic noise.

With its sulfurous smell, deafening blast, and unsurpassed killing power, the advent of gunpowder struck many as a visitation from hell. Genghis Khan, a fiend in the eyes of those he conquered, may have been one of the first to make use of gunpowder as a weapon of war, not to hurl projectiles but to spread psychological terror.

Writing after guns had become standard issue in wartime, John Milton portrays gunnery in *Paradise Lost* (1667) as the invention of hell—and makes a trenchant connection between noise, force, and the will to obliterate anything that "stands adverse" to the agendas of the powerful. Here he has Satan exult in the future of arms development with his fellow devils:

These in thir dark Nativity the Deep
Shall yield us, pregnant with infernal flame,
Which into hollow Engines long and round
Thick ramm'd, at th' other bore with touch of fire
Dilated and infuriate shall send forth
From far with thund'ring noise among our foes
Such implements of mischief as shall dash
To pieces, and o'erwhelm whatever stands
Adverse, that they shall fear we have disarm'd
The Thunderer of his only dreaded bolt

Henceforth thunder will no longer be the prerogative of heaven alone.

If hell was imagined as noisy, in both the Middle Ages and well past the Renaissance, then noise was often heard as hellish. Visiting the London kennels where dogs were kept to bait bulls and bears in the ring (a popular "sport" in Elizabethan England), the playwright Thomas Dekker wrote in 1609 that "the very noyse of the place put me in mind of *Hel*." In medieval times the noise, as well as the heat, of castle kitchens also carried hellish associations. To fart was possibly to have made the sound of the Devil himself.

These archaic associations of noise, pain, and evil have an interest beyond that of historical curiosities. They remind us of a tendency that persists to our day: to find evil in the noise rather than in the pain that sometimes gives rise to it. For instance, it was against the law in Elizabethan England for a man to disturb the peace by beating his wife after 10:00 P.M. As late as 1830 the town fathers of New Orleans moved the jail to the back of the city after clergy complained about the "unpleasantness of being daily disturbed, during prayers in Jesus Christ's temple, by the crack of the whip and the screams of its victims." Recording the rape of Hispaniola in the mid-sixteenth century, Bartolomé de Las Casa, Dominican chronicler and critic of Spanish atrocities in the New World, tells of one commander who ordered his subordinate to dispatch the Indians he was slowly grilling alive because "the poor creatures' howls" were keeping the commander awake at night. Not to be daunted in his noble project, the officer drove stoppers into his victims' mouths and continued to play at being the Devil.

The explorations that brought Europeans into contact with aboriginal cultures gave further opportunity for another ancient noise trend: hearing "strange peoples" and their speech as barbarously noisy. The tendency is at least as old as the ancient Greeks. Their word *barbarian* has it origins in the sound that non-Greek speakers made in Greek ears: *bar-bar-bar*, or as we might say, *blah-blah-blah*. Not surprisingly, many Anglo-Saxons found the tongues of their subjugated Irish neighbors to be uncouth and unruly. According to the early seventeenth-century chronicler Fynes Moryson, the people of that island "are by nature very Clamourous, vpon euery occasion raysing the hobou (that is a doleful outcrye) which they take one from anothers mouthe till they putt the whole towne in tumult." Contemporary historian Bruce R. Smith tells us that early "English estimations of the Irish language classify it as one step removed from noise."

The shoe was on the other foot when the English colonists met the original inhabitants of Massachusetts, who by some accounts were quite appalled at the noise of their new neighbors. As one Englishman felt bound to observe in 1634, "I never heard yet of that *Indian* that was his neighbours homicide or vexation by his malepart, saucy, or uncivill tongue: laughter in them is not common, seldome exceeding a smile, never breaking out into such a lowd laughter, as doe many of our *English*."

One "Indian Sagomore," after "hearing an *English* woman scold with her husband, her quicke utterance exceeding his apprehension, her active lungs thundering in his eares," walked away believing the woman had uttered nonsensical noises, his best rendition of which was "Nannana Nannana Nannana Nan." This is not to suggest that white settlers heard no "noise" from "savage" throats and even from heterodox fellow Englishmen. Though the Shaker communities that took root in eighteenth- and nineteenth-century America were models of orderly social intercourse and industrious labor, their ecstatic worship was described by those who overheard it as suggestive of the cries of "witches" and of "Bedlam."

With that last comparison we complete the circle we began in hell. We should not leave without mentioning the infamous insane asylum that came to be a byword for noise. Founded in 1247, the Priory of St. Mary of Bethlehem became a repository for the mentally ill in 1375 and remains the oldest psychiatric hospital in the world. It bore the nickname of Bedlam at least as early as 1450. It was notorious for its noise—"so hideous, so great; that they [the inmates] are more able to drive a man that hath his wits rather out of them"—and for its cruel "treatments." Patients were routinely chained to floors and walls, whipped, and ducked in water, techniques that doubtless drove up the volume as well as the insanity.

JUST BEFORE WE STARTED SOUNDING MODERN

If one was lucky enough to avoid charges of heresy, treason, or madness, few of the sounds heard on the threshold of the modern era would have been oppressive. I am reminded of that whenever the power goes out in the nineteenth-century farmhouse where I live. The refrigerator, the furnace, the computer, the washing machine, the air purifier, the stereo—the fish tank filter in the days when our daughter kept fish—all die. And then, like a rush of heavenly music into the void: the songs of birds.

Often at such times I recall a passage from John Wain's biography of the writer and moralist Samuel Johnson, in which he attempts to imagine the visual world of England in the mid-eighteenth century.

> [I]t was a place in which ugliness was very rare; indeed, with the important exception of the ugliness that disease and disfigurement produce in human beings and animals, ugliness was unknown. . . . To us any object, from a city to a teaspoon, that is anything but hideous is immediately recognized as something special, probably the work of some world-famous artist. In his day there was probably no such thing as an ugly house, table, stool or chair in the whole kingdom. . . . [T]he main reason is known to all of us. It is that industrialism, by moving people away from the natural rhythms of hand and eye, and also from the materials which occur naturally in their region and to which they are attuned by habit and tradition, cannot help fostering ugliness at the same time as it fosters cheapness and convenience.

We would go too far if we said that no one in the eighteenth century heard an ugly sound. A famous eighteenth-century engraving by William Hogarth shows a musician standing by his open window, violin bow in hand, clasping his ears against the passing commotion of barking dogs, frolicking children, and street practitioners of his trade. The historian Liza Picard names among the urban noises of that time the "screech of iron tyres on cobbles and granite streets, crashing and bumping over the potholes and drains, horses' metal-shod hooves clattering, wooden axles squeaking, coachmen and carters shouting." Nevertheless, the majority of these noises were "organic" in nature, many of them the sounds of human and animal voices. And as these sounds were organic, they were likely to be familiar—associated with persons, beasts, and trades the hearer knew by sight as well as sound, and on which her own existence depended. Despite his reputation for misanthropy, Jonathan Swift, whose *Gulliver's Travels* caricatures humanity as a species of screeching, promiscuously defecating "yahoos," loved the street sounds of eighteenth-century London and Dublin.

> *The Smallcoal-Man was heard with Cadence deep,*
> *'Till drown'd in Shriller Notes of Chimney-Sweep.*
> *Duns at his Lordships Gate began to meet,*
> *And Brickdust Moll had Scream'd through half a Street.*

No doubt the rise of writing as a profession is partly what enables us to hear the eighteenth-century soundscape with greater clarity than any that had gone before. This development was tied to others of sonic consequence: the rise of manufacturing, urbanization, and a middle class that located its domestic ideal between the fanfare of the aristocratic palace and the screams of Brickdust Moll. Though the Industrial Revolution was not yet in full swing, and though organized anti-noise campaigns would not appear until the next century, the common features of noise complaints are fully in evidence by the eighteenth century. Tobias Smolletts' 1771 novel *The Expedition of Humphry Clinker* contains what has to be one of the most thoroughly fleshed-out noise disputes in English literature.

First we have a request for quiet at a noisy lodging house, justified by what would become a popular basis for complaints against noise: its adverse effect upon the sick. Matthew Bramble, an invalid, sends his nephew to ask another lodger to "make less noise, as there was a sick gentleman below."

The request is repeated to other lodgers at the house, one of whom, a "Creole colonel," makes the classic retort that "his horns had *a right* to sound on a common staircase; that there they should play for his diversion; and that those who did not like the noise, might look for lodgings elsewhere" (emphasis added).

There follows Bramble's colorful description of the noise itself, including a vestigial reference to the demonology of earlier times:

> [T]he devil, that presides over horrid sounds, hath given us such variations of discord—The trampling of porters, the creaking and crashing of trunks,

> the snarling of curs, the scolding of women, the squeaking and squalling of fiddles and hautboys out of tune, the bouncing of the Irish baronet overhead, and the bursting, belching, and brattling of the French-horns in the passage (not to mention the harmonious peal that still thunders from the Abbey steeple) succeeding one another without interruption, like the different parts of the same concert, have given me such an idea of what a poor invalid has to expect in this temple, dedicated to Silence and Repose, that I shall certainly shift my quarters to-morrow.

Perhaps the best part of all is the reaction of Bramble's sister: "This intimation was by no means agreeable to Mrs. Tabitha, whose ears were not quite so delicate as those of her brother—She said it would be great folly to move from such agreeable lodgings, the moment they were comfortably settled. *She wondered he should be such an enemy to music and mirth*. She heard no noise but of his own making" (emphasis added). Sound familiar?*

Like other noise complaints, Bramble's may not be free of nonacoustical prejudices. The bouncing baronet is Irish, the surly colonel is a Creole, the horns are French, and the slaves who play them are "negroes." The "scolding" comes from women, since men never scold. At the same time, Bramble makes no remark explicitly linking noise to the "lower classes," the "rabble." That sort of insinuation is at least as old as the Middle Ages: The noise of the English Peasant Revolt of 1381 was described by its upper-class chroniclers in such a way as to suggest that the rebels were barely human. The link between social class and audible commotion informed certain nineteenth- and twentieth-century diatribes against noise. It also informs, with contrasting sentiments, a number of academic sound histories in our postmodern period. The bourgeois reformer decries underclass noise as a want of good manners; his descendent, the bourgeois cultural historian, justifies it as subversive of the

* When I visited in London with Val Weedon, co-founder of the UK Noise Association (also a rock band promoter and a longtime devotee of the '60s group "The Faces"), she told me how the English tabloids had branded her "Britain's Number-One Party Pooper." As a counterweight to the proverbial Goody Two Shoes, I hereby nominate Mrs. Tabitha: the person who hears every noise complaint as an attack on "music and mirth."

reigning order. Both would seem to know all about slums; neither seems ever to have heard of slumming.

In that regard consider Samuel Johnson's poem "London," written roughly thirty years before Mr. Bramble's litany of complaints about the loud boarding house, and the equivalent of a hit single in its day. In one passage Johnson describes the city at night, when the sounds of "music and mirth" seldom boded well for the solitary traveler:

> *Some frolic drunkard, reeling from the feast,*
> *Provokes a broil, and stabs you for a jest.*
> *Yet even these heroes, mischievously gay,*
> *Lords of the street, and terrors of the way;*
> *Flushed as they are with folly, youth, and wine,*
> *Their prudent insults to the poor confine;*
> *Afar they mark the flambeau's bright approach,*
> *And shun the shining train, and golden coach.*

Johnson was not exaggerating these details for rhetorical effect. He wandered the meaner streets of London and knew them well. Poorly lit and scarcely patrolled, they were the haunts of rowdy gangs who filled the night with violence and menacing noise. A London gang who called themselves the Mohocks were infamous for performing various outrages on hapless pedestrians, including boring out their eyes and standing women on their heads in order to "commit various indecencies and barbarities" on their persons. As Johnson notes in his poem, these outrages were "prudently" confined to the urban poor. Those travelers rich enough to have "flambeaus" to light their way and "golden coaches" to carry them were left alone.

But the perpetrators of the crimes were themselves poor, right? In fact, the Mohocks included a bishop's son and a baronet, the latter of whom was eventually arrested. Johnson's reference to such thugs as "Lords of the street" was not a mere figure of speech.

The nighttime world of eighteenth-century London invites us to reconsider at least three common assumptions about the relationship between poverty and noise. First, the assumption that the noise of poorer neighborhoods is always made by the people who live there. Where have baronets always gone when they wanted to raise a little Cain? For that matter, where have train tracks and interstate highways usually been sited?

Second, the assumption—the very callous and condescending assumption—that everyone who lives in a poorer neighborhood is happy with the level of its noise.* It's what "those" people do. "It's their culture." Their ears are different. Anyone who's worked a community-action hotline knows better.

Third, and perhaps most pernicious, the assumption that any member of a poorer neighborhood with the financial means to make noise and the physical muscle to back it up is the most authentic representative of that community. We might call this the "take me to your leader" school of social thought. If a middle-aged white man catches our attention with a loud motorcycle, if a teenaged black kid catches our attention with a loud boom car, then they are, respectively, the archetypal symbol of blue-collar rebellion and the righteous voice of "da hood." Never mind the people who've lived and worked for years in those same communities, all of whom need their sleep, and some of whom can barely afford to pay their water bill let alone go out and buy customized straightpipes or woofer-enhanced speakers. They don't count.

Before the pipes and the speakers, though, other inventions would have to be heard, many making their debut in the age of Johnson, Smollett, and Swift: the steam engine (1712), the power loom (1785), the hydraulic press (1796). A new volcano was about to erupt.

---

* Based on surveys in nineteen urban "open spaces" in Europe and China, University of Sheffield researchers Lei Yu and Jian Kang have concluded that "the long-term sound experience at home could be an important factor to influence the sound level evaluation in urban open spaces." How so? "People from noisier homes"—which presumably might include homes in overcrowded, poorly constructed, multifamily housing—"could show *less sound tolerance* in urban open spaces" (emphasis added).

CHAPTER 5

What Laura Heard

The Industrial Age to the Present

> *Laura and Mary held tight to each other's hand. . . . They had never seen a machine before. They had never heard such a racket.*
>
> —LAURA INGALLS WILDER, *LITTLE HOUSE IN THE BIG WOODS*

I can hardly read a page of Laura Ingalls Wilder's Little House series without picturing, along with whatever prairie landscape or pioneer craft she is bringing to life, a pair of small feet in terrycloth pajamas wiggling near the edge of my couch. My first acquaintance with the books came when my wife and I read them out loud to our daughter; they figured vividly in her first games of pretend, so that in the passage quoted above I see her as a third little girl huddled together with the Ingalls sisters in the presence of that awesome, clattering machine. She is less apprehensive than Laura and Mary are, because loud machines have always been part of her world.

The racket-making machine is a grain thresher that Pa Ingalls has hired at harvest time. Powered by eight horses, not by coal and steam, the thresher is like a bridge between one world and the next. The machine enables Pa to thresh his grain much faster than he could by hand, but it also takes him one step closer to the industrial age that will make him and his kind obsolete. Understandably, Pa himself has no such misgivings. "That machine's a great invention!" he says to his wife, Caroline, when the threshing is done. "Other folks can stick to old-fashioned ways if they want to, but I'm all for progress. It's a great age we're living in."

My daughter's is not the only childhood I find in Wilder's pages, reflected in the saucer eyes of those Ingalls girls clasping hands by the clamorous gears. I also find the childhood of my own gadget-driven age, reflected in Pa's wholehearted embrace of what Wilder in her chapter title calls "The Wonderful Machine."

SUCH A RACKET

The changes that the Industrial Revolution would eventually bring to the air, water, soil, and biological diversity of the planet were foreshadowed by more immediate changes in the soundscape. Noise not only increased with industrialization; it took on different forms. The "flat line" sound of the industrial hum, unvarying and seemingly endless, was something altogether new in the world.

Soundscape scholar Barry Truax speaks of the change in terms of what he calls "acoustic ecology."* He writes that "few high intensity or continuous sounds exist in the preindustrialized world. Therefore, more 'smaller' sounds can be heard, more detail can be discerned in those that are heard, and sounds coming from a greater distance form a significant part of the soundscape. In terms of acoustic ecology, one might say that more 'populations' of sound exist, and fewer 'species' are threatened with extinction."

The first industrialized cities must have produced a strange cacophony, infinitely more varied and abrupt in its advent than what a Sumerian villager would have heard laboring on the first walls of Ur. The Sumerian would have heard an increase of mostly familiar sounds, more voices, more wheels, more animals, more plowshares beaten into swords. But his counterpart in eighteenth- and nineteenth-century London would have heard, in addition to the older sounds of iron wheels on cobblestones, church bells, and roosters, the

---

* Truax, together with his colleague and mentor R. Murray Schafer, helped to found the World Soundscape Project (out of Simon Fraser University in Vancouver), which since the late 1970s has attempted to catalog, and where possible to record, the silenced and endangered sounds of the planet.

newer sounds of steam whistles, thundering belts and pulleys, pumping pistons, and slamming metal presses. As one moved closer to these newer sounds—not only in an incremental, historical sense, but in the moments it took to walk from the street to the factory floor—the older sounds would have been swallowed up, like the cries of horses jettisoned into a hurricane sea.

One of the weaker sound species threatened by industrial machinery was the work song. In our age of recorded music and professional singers we tend to forget that, for millennia, song and work went hand in hand. In medieval Wales, for instance, a plowing team consisted of two men: one to hold the handles of the plow and another to walk by the head of the ox, singing to it. On the factory floor songs withdrew within the singer—one mill worker reported having "a hymn to sing to herself, unheard within the deep solitude of unceasing sound"—while choral singing vanished entirely. "Labor was orchestrated by the number of revolutions per minute," Lewis Mumford wrote in *Technics and Civilization*, "rather than the rhythm of song or chant or tattoo."

Conversation also became difficult, discouraged alike by the din of the machines and the brutal regimes of the bosses. A British investigative committee in 1832 noted that workers caught talking were routinely "beaten with the strap." Some of the words drowned out by the din of machinery must have included warnings. There is probably no way to know how many accidents were attributable to noise, but we know there were plenty of accidents. Mary "Mother" Jones, the famous labor leader, saw a young girl scalped alive when her long hair got caught in a machine.

The loss of hearing was obviously another casualty, one that affects the industrialized world to this day. "Boilermakers' disease" became a common euphemism for deafness. The effects of industrial hearing loss were recognized as early as 1713, but attempts at prevention were slow to take hold—sometimes due to resistance from the workers themselves. As late as 1959 a Dutch factory worker remarked: "A nail boy wearing earplugs will never become a man."

I have wondered if occupationally induced hearing loss accounts for some of the "positive relish for noise of a distinctly emphatic kind" that some social historians associate with early working-class life; one of them suggests a parallel "to the popular appetite for strong tastes—pickles and 'rusty' bacon." In other words, I have wondered if, all pickles aside, the "relish" was mostly due to loss of sensation. When I first moved to northeastern Vermont, I wondered why some of the men here seemed to talk so loud—if this was some class or macho thing—until my speech pathologist wife pointed out that most of these men earned their bread with chain saws and got their meat with big-bore rifles.

Reaction to the noise of industrialization was slow in coming. When it did come, it appeared first among literary writers, who chronicled the din in their books, though in their remarks off the page they were as likely to be appalled by the transient sounds of urban life (organ grinders, whips, etc.) as by the sounds of industrial machinery.* The reasons for this delay are various. For one thing, industrial workers faced dangers and had demands (the abolition of child labor) more pressing than noise; for another, their first exposure to industrial noise came when their collective voice was least likely to be heard and, if heard, most likely to be silenced with a strap.

Machinery has a certain beauty, too, not unrelated to the speed and number of its motions and thus to the level of its noise. A French observer in the nineteenth century noted that the typical American "has a perfect passion for railroads, he loves them . . . as a lover loves his mistress." A large, noisy machine could also grant its operator a certain status; the extra decibels of a bigger, more complicated machine might go hand in hand with higher wages.

In the end, the acceptance of industrial din may have come from the agrarian culture that preceded it. The first industrial workers were often not even a generation removed from peasants, and if anything characterizes peasant

---

* Karl Marx writes in *Capital*: "As soon as the working class, stunned at first by the noise and turmoil of the new system of production, recovered in some measure its senses, its resistance began." The resistance is not to noise per se, though noise may reasonably be construed as one of the factors informing initiatives like the restriction of child labor. Elsewhere in *Capital*, Marx says that "factory work exhausts the nervous system to the uttermost," which may also imply the effects of noise, though his focus is mainly on the way that factory work "does away with the many-sided play of the muscles, and confiscates every atom of freedom, both in bodily and intellectual activity." It is probably tenuous at best to infer the importance assigned to noise by the number of times it is mentioned in contemporary texts. A scarcity of references *could* mean that noise was not a major concern; it could also mean that its oppressiveness was so obvious as to scarcely merit mention.

life, it is a belief in fate. Possibly it is this fatalism that Marx had in mind when he spoke of the Industrial Revolution as having liberated workers "from the idiocy of rural life." Perhaps he spoke too soon. What the medieval plowmen would have attributed to the motions of the stars or the immutable will of God, his industrial descendents still attribute to the irresistible march of technological progress or the omnipotent demands of "the global marketplace." New Yorkers appalled by the screech of the elevated trains in 1878 were told, as if by the village soothsayer, that it "has to be." Almost a hundred years later, in 1960, when members of the British Noise Abatement Society roused aviation minister Duncan Sandys from bed in his "silk dressing gown, blue pajamas and red slippers" to tell him that the noise of Heathrow Airport "prevented them from sleeping and made life 'unbearable,'" he spoke in the same fatalistic mode: "[W]e must resign ourselves to the fact that noise and power go hand in hand."

"BELEAGUERED BY DISCORDANT HOSTS"

It comes as no surprise that nineteenth-century writers such as Charles Dickens and Emile Zola provide some of the first trenchant testimonies to industrial noise, not only because bearing witness is what writers do but also because "writers" as we now think of them were another product of the machine age. Machines both printed their books and built the bourgeois classes who read them. A nineteenth-century writer (like a twenty-first-century writer) who criticized industrial noise was simultaneously "doing his job" and biting the hand that fed him.

This dependence, along with our tendency to complain most about the noises that affect us most directly, explains why the majority of complaints from nineteenth-century writers are aimed at street noises rather than factory sounds. Traveling through Italy in 1874, the Victorian art critic John Ruskin was "driven wild" by the "crashing discordance of the church bells" in Venice and later, in Fiesole, by drunken revelers in the night and by the "accursed omnibus" riding by "with noises like breaking wind" during the day.

Ten years earlier the English Parliament had debated the vexing issue of street musicians, many of whom were Italian organ grinders. One house lord (as likely to hear an organ grinder outside his window as to roast his own beef) maintained that "street music" was a necessary pleasure for the urban poor; others countered that the musicians were little more than petty extortionists. In a letter of support for a "Bill for the Suppression of Street Noises," Charles Dickens and twenty-six other signatories, including Alfred Lord Tennyson and Thomas Carlyle, maintained that "no sooner does it become known to those producers of horrible sounds that any of your correspondents have particular need of quiet in their own houses, than the said houses are beleaguered by discordant hosts seeking to be bought off."

The British campaign against street music was colored by class prejudice and anti-immigrant sentiment. Period cartoons depict the Italians as a dirty, surly bunch, slightly less kempt than their monkeys and nearly as hirsute—though it is probably safe to assume that anti-Italian prejudice was not what motivated Mayor Fiorello LaGuardia in his efforts to ban organ grinders from the streets of New York nearly a century later. Mostly at issue in the Victorian controversy was the self-proclaimed right of a new class of home-based artists and intellectuals to work in peace without the interruption of noise from the street.

On the continent, the German philosopher Arthur Schopenhauer (1788–1860) was also a great hater of street noises, the cracking of whips in particular. (His countryman Goethe had been troubled by the barking of dogs.) Like his peers in England, Schopenhauer felt that noise interfered with his thinking and that his thinking entitled him to special consideration. "There can be no harm in drawing the attention of the mob to the fact that the classes above them work with their heads, for any kind of headwork is mortal anguish to the man in the street"—was his idea of reaching across the aisle.

Had these writers been able to universalize their hatred of noise, and the dignity of human work, their legacy to noise abatement might have been of more value. Still, I find it hard not to sympathize with their complaints. Dickens' letter to Parliament closes with mention of "the frightful noises in despite of which your correspondents have to gain their bread." Bourgeois though he was, Dickens labored with the feverish intensity of an immigrant, never forgetting the bottle-blacking factory in which he'd worked as a child. Such memories were another product of the industrial age. Who knows what psychic influence they might have had on the way in which emergent middle-class people like Dickens heard the noises of the "riffraff," or on his reasons for calling those noises "frightful"?

CARLYLE'S SANCTUARY

For a noise story that shows both the tenor of its times and the potential futility of noise suppression in any time, you could hardly do better than Thomas Carlyle's failed attempt, beginning in 1853, to make a soundproof study.

When Thomas and Jane Carlyle moved to 5 Cheyne Row (now number 24) in the Chelsea district of London, it was not the upscale neighborhood we find today. "Chelsea is unfashionable," Thomas wrote to Jane after some scouting around, but "the houses are cheap and excellent." They could rent the one on Cheyne Row for £35 a year, be close to the center of London, yet still a comfortable distance from its bustle and noise. Other artists and writers of the time lived in Chelsea for the same reasons. From the backyard of the house "nothing was visible but leafy clumps, green fields, and red high-peaked roofs glimmering through them. . . . Of London nothing visible but Westminster Abbey and the dome of St. Paul's." Though Jane had concerns about the dampness off the nearby Thames River, she judged the house the best of their prospects—"commodious, with closets to satisfy any Bluebeard"—and the couple settled there in 1834. It was there, Thomas would write after Jane had died, that "we spent our two and thirty years of hard battle against fate."

A good part of the Carlyles' battle was with noise. Like Ruskin and Schopenhauer, Thomas was acutely sensitive to distracting sounds. "All summer I have been more or less annoyed with noises," he wrote to his sister in 1853, "even accidental ones, which get free access through my open windows." Among them were the sounds of pianos, parrots, dogs, fireworks from a nearby park, and the usual extortionist organ grinders. Carlyle reached the breaking point when, just as he was starting work on his monumental biography of Frederick the Great, his next-door neighbors filled their yard with a flock of "demon foul" that crowed and squabbled the whole day long.

With the arrival of the so-called demon foul, Thomas determined to embark on a project he and Jane had discussed for ten years: to raise their house another story by constructing a soundproof study at the top of it. After choosing a contractor—"a practical Liverpool railway man" who "went to the chief Builders here . . . and told them my sad case, 'a literary man,' &c."—Carlyle drew up a set of meticulous specifications, including provisions for a skylight, closets back and front with double doors, a ventilation system, and insulating air chambers in the ceilings and walls.

Predictably, the noise of the construction drove him nuts—and soon out of the house—though Jane, who directed much of the work, seemed rather to enjoy it. "The tumult has been even greater since *Mr* C went than it was before. . . . But now that I feel the noise and dirt and discord with my own senses only and not thro *his* as well, it is amazing how little I care about it. Nay in superintending all these men I begin to find myself in the career open to my particular talents." Almost as predictably, the work took longer than expected, went over estimate, and was shoddily done.

In the end Thomas was to pronounce the work "a flattering delusion." Though the chamber successfully muffled or eliminated some noises, others became more audible in the resulting quiet: distant railway whistles, bells, and "evils that he knew not of" drifting up from the lower rooms. The ventilators didn't work as planned, so that the skylight had to be opened in warm weather—not the quietest season of the year. "The silent room is the noisiest in the house," Jane observed. "Mr C is very much out of sorts." Nevertheless, Mr. C continued to work on his *History of Frederick the Great* and finished it after thirteen years of arduous labor.

Of the Carlyles and their house, Virginia Woolf wrote: "One hour spent in 5 Cheyne Row will tell us more about them and their lives than we can learn from all biographies." That was also my impression when my wife and I got to spend some time there. The middle floors had that cozy, plush, but funereal quality found in many Victorian houses; one immediately understands the period's fascination with the spirit world. That we went at dusk, several days before Halloween, may have contributed to the effect.

Two rooms stood in sharp contrast to the rest, the soundproof study being one. It seemed ghost-proof if not soundproof, brighter than the other rooms because of the skylight, without shadows or illusion. Starkly masculine, it could have been a cabin on the upper decks of the Mayflower. The floorboards creaked noticeably underfoot. Carlyle has never been a hero of mine, but I can name no better place of pilgrimage for anyone who hates noise—a shrine and a warning oracle as well. You can imagine a motto inscribed over the door: *To be obsessed with quiet is never to be possessed of it.**

The room I found most affecting, however, was at the very bottom of the house. Here were the kitchen and scullery, as well as the narrow iron bedstead occupied by Fanny, the sole servant of the couple Virginia Woolf describes as "two of the most nervous and exacting people of their time." Like the study, the room was austere and free of mystique. I imagined the master at the very top of the house, and Fanny eye-level to the street in the basement, where she would have heard both the outside noises and the tramping of her employers overhead, Jane's incessant coughing, the sometimes bitter words between husband and wife, the bell that summoned her at all hours of the day and night. Her kitchen offered a symbolic reminder: Whenever we examine a noise dispute, and especially when we frame it in terms of class conflict (privileged complainer versus the boisterous masses, etc.), there is often some disadvantaged, invisible Fanny whom we ignore. Was Fanny any less a member of the "lower orders" than the organ grinder? Was her inner life any less important than that of the man in the sanctum under the roof? Did Fanny never need to hear herself think?

But there can be poetic justice even when the social kind is lacking. Though Carlyle would never completely escape the noise of Cheyne Row, Fanny did. It seems that she ran off with one of the workmen who built the soundproof study.

GETTYSBURG

At roughly the same time as Carlyle was pursuing his "hard battle against fate" and "demon foul," the Northern and Southern regions of the United States were engaged in their hardest battle against each other. "It is inconceivable to most visitors to Gettysburg," writes Civil War historian Kent Gramm, "how

* It is interesting to note that although Carlyle was strongly affected by noise, some within his own circle, the poet Tennyson for example, found him loud and overbearing.

loud the battle was. . . . The noise was beyond anything most human beings, except those who have experienced modern combat, have ever heard. If we contemplate the choking, blinding smoke that covered the battlefield in times of intense combat with a furious noise so loud one can barely think, let alone attempt to communicate, we can begin to understand the conditions under which the fighting men labored." The fighting men themselves were often at a loss to describe what they had heard. "Who can describe such a conflict as is raging around us!" wrote Union General John Gibbon, just before Pickett's Charge. "To say that it was like a summer storm with the crash of thunder, the glare of lightning, the shrieking of the wind, and the clatter of hail-stones would be weak."

The medical records of Union Army soldiers show that a third returned home with hearing loss, which for the first time in U.S. history was acknowledged by the government as a service-related disability.* Though the term *shell shock* came into use during the First World War, it is clear that the Civil War produced disabilities for which that term would have applied. A doctor who examined former Union infantryman Dixon Irwin, whose bizarre behaviors indicated what we today would call "flashbacks," reported that Irwin's "eye had a peculiar appearance as a man who is frightened, and he spoke of the damn big guns. Whenever he spoke of the cause of his trouble he said it was the constant roar of the guns in the service. . . . He was wild and very excitable and imagined persons were after him and upon the firing of a gun he was frantic."

War had always been loud relative to the sounds of peace, and loudness had always proclaimed dominance, but the deafening warfare of the industrial age was able to dominate over peacetime itself—at least in the ears and nervous systems of returning soldiers who struggled to readjust there. This is not to say that noise was the worst thing about modern combat. It is only to say that as modern combat grew worse, it inevitably grew noisier. War and noise proclaimed each other's truth.

Meanwhile, slaveholders had been praising their way of life—that is, their ultra-dominance over other human beings—in terms of quietness. Pointing with disparagement to the bustling industrial North, patrician Southerners

* The General Law of 1862 and the Disability Act of 1890 were instrumental in bringing this about.

congratulated themselves for what one of them called "the quiet and retirement of plantation life." During a visit to New York in the early 1850s, a white South Carolinian named William M. Bobo remarked that "[a]ny one who walks the streets of New-York with eyes and ears open, sees and hears many strange and horrid things." He was quick to attribute those sights and sounds to "Northern Institutions."

Of course he could have stayed home and heard plenty that was strange and horrid. As Louis Hughes wrote in his memoir *Thirty Years a Slave*: "Everything was in a bustle—always there was slashing and whipping. . . . It was awful to hear the cracking of that whip . . . so loud and sharp was the noise." It is possible to silence the oppressed but not to oppress them *silently*. Subjugation must always make a sound.

SHUSH!

While the theater of war was to grow progressively louder from the nineteenth century on, American theater audiences for the performing arts underwent a change in the opposite direction. Partly owing to the cultural exaltation of the performing artist, partly owing to the rise of the upper middle class and the accompanying division of stage entertainments into "high" versus popular (or "vulgar") art, middle-class audiences after the Civil War began to conduct themselves in a more "refined" manner than they had in the past. Earlier in the century mostly male audiences of mixed class had been free to carry on during performances, much as continental aristocrats had done in the days when court musicians and composers wore livery. In the new dispensation, audiences were less mixed as to class, more mixed as to gender, and—in the case of the high art crowd—more concerned with proper behavior. Thirty-six new guides to etiquette appeared in the United Sates in the 1830s, thirty-eight in the 1840s; the rate doubled in the 1870s. A prevailing theme of these guides was that "refined" people did not call attention to themselves, whether on the street or in the theater.

Exemplary of this development and its European parallels was the imperious Chicago musical conductor Theodore Thomas (1835–1905), who once stopped a performance of Beethoven's Eighth after a talkative audience member had made several unsuccessful and noisy attempts to light a cigar. "Go on, sir! Don't mind us!" declared Thomas. "We can all wait until you light your cigar." During another performance in New York in 1867, after the audience had hissed and howled at an "unconventional" sampling of Liszt, he again stopped the music in order to allow the rowdies to leave. "Whoever wishes to listen without making a noise may do so. I ask all others to go out. I will carry out my purpose if I have to stand here until two o'clock in the morning." Apparently that was not necessary.

Like the anti-noise sentiments of Carlyle and his circle, Thomas's standards carried an odor of elitism. According to his wife he came to the conclusion that "neither children nor what are called 'wage-workers' were sufficiently advanced intellectually to be able to appreciate the class of music which was his specialty." And, as with Carlyle and his circle, it is all too easy for modern commentators to criticize the elitism while preserving the same elitist distinctions, taking as typical the loudest members of the working-class audience while crediting the middle classes with better manners than they may actually have possessed. After describing how one determined vaudeville owner had "marshaled the forces of middle-class refinement against the guerrilla fighters of male working-class culture" who in their "battle against gentility . . . hooted at the stage acts, shouted obscenities at the female performers, and did their best to break up the show," historian of manners John F. Kasson writes:

> Nonetheless, in the decades surrounding the turn of the century there still remained theaters controlled by working-class immigrants that retained the boisterous informality and conviviality that had characterized antebellum American theaters generally. The Yiddish theater of New York's Lower East Side in particular aroused intense, even fanatical involvement.

While I cannot fault Kasson's excellent history, I take exception to his lumping together of very different kinds of boisterousness.* I would argue that the

---

* I certainly don't fault the superb conclusion to his book *Rudeness & Civility: Manners in Nineteenth Century Urban America*: "[T]he larger challenges of civility in American urban and public life generally persist—and while they are not solvable simply by the achievement of a full and humane democratic social order, one wonders how anything less can ever suffice." Amen.

"boisterous" audiences of the Yiddish theater and the show-stopping dictatorship of Theodore Thomas had more in common with each other than either had with the "guerrilla fighters of male working-class culture." The key to that similarity is in Kasson's phrase "fanatical involvement." The only thing the "guerillas" in the vaudeville audience were fanatically involved with was themselves. Along the same lines I would argue that the ecstatic worship of an African Pentecostal church has more in common with the liturgical solemnities of a high Anglican mass than it has with the outbursts of someone who attended either service merely to mock it.

In this I take my cue from Johan Huizinga's classic study of play, *Homo Ludens,* in which he says that nothing is more inimical to the spirit of play—a category in which he includes both ritual and drama—than a "spoilsport." What else would we call someone who disrupted a vaudeville show? Of a different order entirely is the Yiddish theatergoer who was reportedly so moved by a performance of *The Jewish King Lear* that he ran down the aisle shouting, "To hell with your stingy daughter, Yankl. She has a stone, not a heart. Spit on her, Yankl, and come with me. My yidene [Jewish wife] will feed you." *That* man knew how *to play,* and on the highest level; only a spoilsport would say he was making noise.

TWO CHAMPIONS OF A QUIETER WORLD

Perhaps no figure looms so large in the anti-noise agitation that marked the turn of the century as Julia Barnett Rice. Her successful campaign against the gratuitous "social" tooting of tugboat captains on the Hudson River (almost 3,000 blasts counted in a single night) led to the passage in 1907 of the landmark Bennet Act. Enforceable in harbors across the United States, it was the only national piece of anti-noise legislation passed during a period of mounting opposition to noise.*

* Historian Raymond Smilor gives an entire catalog of turn-of-the-century city noise ordinances in Baltimore (against drum corps), Boston (against bell-ringing on the street), Buffalo (for regulating auctions), Kansas (against gongs), Portland, Oregon (restricting the hours of piano playing), St. Louis (against bells on animals), and San Francisco (against any clatter "having a tendency to frighten horses").

Like the campaign against organ grinders in Victorian England, much of this opposition was directed against "street" noises—from trumpet-blowing peddlers, for example—and tended to work down the social ladder.* Rice's peculiar genius, not untypical of other women reformers of her time, was to look beyond those noises that annoyed her personally. In this she stands apart from figures like Dickens, Carlyle, and Schopenhauer, who, like many writers (including one I know quite well), just wanted to be left in peace. Yes, Rice herself lived on the Hudson River—in quite a mansion, apparently—and was annoyed by the whistles. But when she formed the Society for the Suppression of Unnecessary Noise in 1906, she turned her efforts toward preventing noise around the city's schools, hospitals, and mental asylums. The resulting "quiet zones" that we now take for granted around most of these institutions were her legacy. Even in her campaign against the boat whistles, she showed herself greater than her own annoyance, carefully soliciting the opinions of boat captains in order to determine to what extent the noise was unnecessary. To appreciate her fully, one has to imagine a campaign against motorcycle noise that polled the members of biker clubs or an anti–"boom car" movement that began by taking a survey of young people on Facebook.

Though he had once been a riverboat captain, and though his character Huck Finn "lights out for the Territory" to escape just such civilizing females as Julia Barnett Rice, Mark Twain agreed to serve as honorary head of the Children's Branch of her society. New York children were soon composing their own pledges not to make noise near hospitals. "I offer up this sacrifice," wrote one earnest child, "so as to comfort the sick near hospital and any place I know where sick persons are, and to prevent all sorts of noise that are not necessary."

It was in fact while visiting schools to promote her hospital campaign that Rice realized how much they, too, were vulnerable to noise. "It is no exaggeration to say that noise robs class and teachers of 25 percent of their time," she noted. In addition to illustrating the reformist spirit of her age, Rice's campaign serves as a good example of the cyclical nature of reform itself. When,

* Chicago peddlers rioted in 1912 to protest noise bans, which they perceived—rightly—as intended to drive them from the streets altogether.

in the 1970s, New York researcher Arline Bronzaft found marked reading delays among children on the train-track side of a New York City school, it caused a small uproar. How quickly we forget.

Rice's influence crossed the Atlantic, prompting the German philosopher Theodor Lessing to form the *Deutscher Lärmschutzverband* (German Association for Protection from Noise) in 1908. Lessing also published a newsletter, *Das Recht auf Stille* (The Right to Quiet), which included advertisements for doorstops, earplugs, acoustical insulation, and quiet hotels.

Lessing is probably one of the most curious figures in noise history, though as an anti-noise campaigner he was far less successful than Rice. In contrast to those who saw noise as a "lower-class" disorder, Lessing heard it as the sound of a tension underlying modern life itself, a desire on the part of hyperrationalized man to return to "unconsciousness and oblivion." And unlike those for whom "the other" (the Italian organ grinder, for instance) was always too noisy, Lessing humbly looked to "the other" for an alternative to noise. Western culture, he felt, was egotistical and immature, whereas the cultures of the Far East had achieved higher degrees of both quietness and social integration.

A curious mix of noise-hater, socialist, romantic, feminist, and pessimist, Lessing was as complex as the issue of noise itself. He was one thing more, a Jew. In 1933, almost twenty years after his noise-abatement society had disbanded with the onset of World War I, he was assassinated in exile by two Czechoslovakian Nazis. Not quite a martyr for the cause of noise abatement, he was nevertheless a victim of those forces he had perceived to be at the lower frequencies of the modern din.

HUMBLE CUSHIONS, CUSHY HEGEMONY

In a museum for homely scientific devices, not far from Archimedes' hip bath, Newton's apple, and Benjamin Franklin's kite, there would need to be a place for Wallace Sabine's seat cushions. With the aid of these unlikely objects Sabine was able to define the concept of "reverberation time" and earn his title as the "Father of Modern Architectural Acoustics." That also makes him the Father of Modern Soundproof Design. Carlyle would have loved him.

In 1895, when he was a 27-year-old physics professor at Harvard, Sabine was given the improbable task of trying to improve the acoustics of the newly constructed Fogg Lecture Hall at the university. Sabine had no particular expertise in sound and the senior faculty had no idea how to approach the problem. That's probably why the job fell to Sabine.

On nights after the hall was no longer in use, Sabine took sound measurements with a pipe organ and a stopwatch, using Oriental rugs from the Fogg, seat cushions from a nearby theater, and people he positioned in the hall, to alter the sound absorption. (He determined that an average human body was equal to six seat cushions in reducing reverberation.) He was eventually able to show how the "echo effect" at the Fogg could be decreased by the installation of sound-absorbing materials. He would put these discoveries to use in designing Boston's renowned Symphony Hall, first opened in 1900 and still considered one of the finest in the world. Millions of people lead quieter lives—at least in the interiors of their buildings—thanks to the young professor whom I will always picture as a pair of worsted trouser legs moving under a stack of seat cushions "by the light of the silvery moon."

The application of scientific expertise to the problem of noise was, like science itself, a mixed blessing. On the positive side, science provided an objective language for discussing noise problems and a practical means of addressing them. The invention of the audiometer in 1925, for example, and the popular adoption of the decibel in 1928 bolstered the faith expressed in a 1930 issue of the *Saturday Evening Post* that "the fight against wasteful racket is out of the hands of cranks and theorists and is being directed by trained technical minds."

Unfortunately, some of the most intractable noise problems involve cultural attitudes and social inequalities not easily addressed by "trained technical minds"—or readily acknowledged by those with the wherewithal to hire them. Increasingly noise has tended to concentrate in locations where trained technical minds are least likely to pay rent. On top of that, the trained technical mind swiftly gained ascendance over the untrained human ear. With new instruments and formulae to rely on, scientists quickly reduced noise to a measurable phenomenon and human hearing to an inexact and thus "inferior" form of measurement. What this meant in practical terms was that winning a community noise dispute, or determining the acoustic character of a community "under development," belonged to that party with the greatest means to buy expertise.

The situation need not be so bleak. The psycho-acoustic experiments of Japanese scientist Koji Nagahata have led him to conclude that "even ordinary citizens can accurately recall the loudness of sounds with which they are very familiar." His findings corroborate those of others in the same field, including researchers examining how well people can recall the loudness of a given sound after a significant time lapse. If science has often been used to wrest credibility from the human ear, it might also be used to corroborate the ear's witness.

A CONTEST FOR INVENTORS

Nothing would undercut the contributions of trained technical minds to the problem of noise so much as their inexhaustible ability to create new noises. If necessity is the mother of invention, invention has proved to be a very fertile mother of noise. We might go so far as to imagine a contest to determine the invention or the inventor who deserves first prize—say, a golden loudspeaker—for filling the world with the most commotion. The claims of the principal contenders occur within a few years of one another: The Wright Brothers made their historic flight at Kitty Hawk in 1903; Henry Ford introduced his automotive assembly line in 1908. Transportation noise, from highway traffic and from aircraft, accounts for the largest percentage of environmental noise complaints today. Among neighborhood noise complaints, loud music is perhaps the most common, so Marconi's radio* (1895) and Edison's phonograph (1877) ought to be in the running, too. Add to these a whole catalogue of industrial machines, from rock crushers to hydraulic drills, though none are as widespread as cars and planes.

---

* Credit for inventing the radio is often given to Marconi but remains contested. Jagdish Chandra Bose, Oliver Lodge, Alexander Popov, and Nikola Tesla all have claims to the honor. Even Ford's reputation as the inventor of automobile mass production is challenged by Oldsmobile's assembly line of 1901. Credit for the noisiest inventions often goes to the "noisiest" inventor. In fact, you may recall that "reputation" was formerly a meaning of the English word *noise*.

Ironically, the winner may be standing mutely in the wings. Though it makes only the faintest sound, the incandescent light bulb, patented by Thomas Edison in 1879, may have done more than any other invention to make the world a noisier place. The electric light did for the colonization of the soundscape what "guns, germs, and steel" did for the colonization of the globe—it opened up new territory. The dark night, like the "Dark Continent," was there for the taking. So was the carbon fuel to power the lights and produce further noise in the process.

In his essay "Denying the Holy Dark," John M. Staudenmaier writes of the world before artificial lighting: "Activities that need good light—where sharp tools are wielded or sharply defined boundaries maintained, purposeful activities designed to achieve specific goals, in short that which we call work—all this subsided in the dim light of evening. Absent the press of work, people typically took themselves safely to home and were left with time in the evening for less urgent and more sensual matters: storytelling, sex, prayer, sleep, dreaming"—all things that nighttime noise can interrupt. It is ironic that what we have come to regard as the liberating, anarchic energies of "nightlife" might best be described as an extension of the "purposeful" dynamism of the industrial day.

NOISE ACQUIRES PROPHETS

"It is time for electric lamps with a thousand points of light to brutally cut and tear your mysterious, enchanting and seductive shadow!" So wrote the Italian Futurist Fillipo Tommaso Marinetti in 1910. "Time and space died yesterday," he'd proclaimed the year before. His 1909 *Futurist Manifesto* set the tone of a short-lived but influential movement of painters, sculptors, and composers that glorified industrial dynamism, including noise. The use of noise as an instrument of glorification is probably as old as the first male gorilla who pounded his chest, but the glorification of noise *as noise*—this was something new. Other objects of Futurist enthusiasm had older pedigrees.

"We will glorify war," Marinetti goes on, "—the world's only hygiene—militarism, patriotism, the destructive gesture of freedom-bringers . . . and scorn for women. We will destroy the museums, libraries, and academies of every kind, will fight moralism, feminism, every opportunistic or utilitarian cowardice.* We will sing of great crowds excited by work, by pleasure, by riot . . . we will sing the vibrant nightly fervor of arsenals and shipyards . . . deep-chested locomotives . . . planes whose propellers chatter in the wind."

The Futurists were fond of issuing manifestos; the most relevant here is Luigi Russolo's *Art of Noises*, published in 1913. It is a curious document. Referring to the concert halls of his day as "hospitals for anemic sounds," Russolo calls for a music that will embrace and exalt the noise of the machine age, delivering to its audience "a generous distribution of resonant slaps to the face." Yet, many of the noises he celebrates—"the whistle of the wind," "the flapping of curtains and flags," "the generous, solemn, white breathing of a nocturnal city"—amount to little more than acoustic casualties of the industrial clamor he claims to love.

A year later Russolo gave a concert in Milan featuring specially made and named instruments of his own invention, including a "howler," a "gurgler," a "croaker," and a "roarer." Unimpressed, his audience began making some noise of its own. One might suppose this would have pleased the Futurists—why not a "cat-caller" and a "boo-maker" to add to the mix?—but instead Russolo's artistic cronies jumped into the seats and began punching people, eleven of whom wound up in the hospital.

I mention Russolo and the Futurists mainly because their preoccupations support some of the points I've been making about the historic relationship between noise and violence, between the arrogance of power and contempt for the weak. In general, though, I have a hard time classifying their artistic works, or any work of art however cacophonous, as "noise." In this refusal I take exception not only to Russolo and his pugilistic friends but also to the German physicist Hermann Helmholtz's often-cited distinction between a musical tone, "which strikes the ear as a perfectly undisturbed, uniform sound which remains unaltered as long as it exists," and noise, in which "many various sensations of musical tone are irregularly mixed up and as it were tumbled about in confusion" (1877).

* Marinetti's catalog of scorn calls to mind Nietzsche's assertion that "the human being who has become free . . . spits on the contemptible type of well-being dreamed of by shopkeepers, Christians, cows, females, Englishmen, and other democrats."

The essential difference between music and noise is neither acoustic nor aesthetic but ethical. In the case of deafening amplification, we would need to add physiological. With that one exception, nothing produced in a concert hall or a recording studio for a willing audience, all of whom are free to walk out or switch off, can be noise in the fullest sense. In contrast, anything amplified to a deafening volume or imposed on me against my will is noise no matter how "pure" its tone or "classic" its pedigree. Lou Reed's *Metal Machine Music* performed at the Gramercy is not noise; Gregorian Chant piercing my bathroom wall is.* Noise comes stillborn into the orchestra pit; it breathes and wails only in the balcony, on the street, surging down the factory aisle—or diving at us from the sky.

Within a few years after Russolo's concert, the German Luftwaffe would equip its bombers with devices that enhanced their noise as they descended upon libraries, museums, academies of every kind, moralists, feminists, and children. The Futurist future had arrived. The Nazis themselves, however, were not notably fond of dissonant music. Official tastes of the Reich inclined toward romantic (and of course, Germanic) bombast—Wagnerian opera and the symphonies of Strauss, a preference that would influence a sweeping postwar reaction against tonality among postwar composers.†

But the noise of military machinery and triumphal spectacle, the masterful bark of the loudspeaker—these the Nazis loved and used, most notably at the Nuremberg rallies. "The standard bearers stood in a row before Adolf Hitler, whose booming voice went across the total silence of the enormous gathering." Then "unending shouts of 'Heil' joined with the music and the sounds of the fanfares and the beating of the drums." I suppose that to the participants none of this was noise, because none of it was unwanted—though we can

* A gray area in this distinction is when the sound values of a musical performance impede the ability of individual musicians to "communicate" musically with other musicians, or else interfere with an audience's ability to hear certain parts of the performance (the words of a song, for instance)—scenarios we will explore in a later chapter.

† Much of my information about twentieth-century music comes from Alex Ross's wonderful history of the subject, *The Rest Is Noise* (2007). Among a host of insightful observations, he says this: "The automatic equation of radical style and liberal politics and of conservative style with reactionary politics is a historical myth that does little justice to an agonizingly ambiguous historical reality."

hope that it sickened at least one heart hiding under a brown shirt. Of this we can be sure: Some of those who heard the noise of Nuremburg were terrified, and they were meant to be.

SILENT FLASH

The war that might be said to have begun with a loudspeaker came to an end with what its surviving witnesses remembered as a silent flash. As recounted in John Hersey's *Hiroshima*, Dr. Masakazu Fujii was sitting on his porch in his underwear, reading his newspaper, when "[h]e saw the flash. To him—faced away from the center and looking at his paper—it seemed a brilliant yellow. Startled, he began to rise to his feet. In that moment (he was 1,550 yards from the center), the hospital leaned behind his rising and, with a terrible ripping noise, toppled into the river." Another survivor, Dr. Terufumi Sasaki, recalled something like "a gigantic photographic flash" illuminating the corridors of his hospital just before the blast demolished the building. The Reverend Mr. Kiyoshi Tanimoto "had heard no planes" and would wonder later "how such extensive damage could have been dealt out of a silent sky."

Nine days later a loudspeaker was set up in the ruins of the Hiroshima railway station, and the emperor's voice announced that Japan had surrendered.

With the atomic age, humankind succeeded not only in splitting the atom but in splitting—if only for a split second—the age-old connection between apocalypse and noise. Nuclear power would be lauded as quiet and clean. The spent uranium rods, radioactive for 10,000 years, are as silent as tumors. Discrete beyond measure, a neutron bomb would even leave the enemy's buildings intact.

The silent flash over Hiroshima and Nagasaki, foretelling "clean bombs" and quiet annihilation, exposes the lie that the world can be made more beautiful simply by reducing its noise, that if we could just silence Blake's "dark Satanic mills" the devil would have no place to go. This is merely the flipside of the nonsense that touts noise as a revolutionary act. A quieter world, especially if it were also a more carbon-neutral world, would certainly be more sustainable than the one we know now. That it might also be more humane is conceivable, but by no means guaranteed.

THE BOOM

Less than a year after atom bombs were dropped on Hiroshima and Nagasaki, the *Atlantic Monthly* published a full-length feature article titled "The Science of Noise." With nothing to go on but the text, we could easily date the piece by its references to "air raids that luckily never came off"—including "a siren which emitted the loudest sustained shriek ever achieved, more than 25,000 watts of acoustic violence"—and to naval aviation engineers who "made it possible to talk to an entire town from an altitude of 10,000 feet," a development with "current utility for ordering the Japanese about."

We could also date the piece by its blithe dismissal of the dangers of noise. "More imagined than real are the horrible consequences—insanity, nervousness, nausea, inefficiency, confusion, fatigue and diverse upsets—so graphically depicted by well-meaning din phobists and other self-appointed guardians of the public good." A few paragraphs and puffs of his pipe later, the author assures us in that unmistakable newsreel voice: "Man is a resistant organism with amazing talents for adaptation to nature's buffetings, and it takes more than a thunderous noise to lay him low." The author does acknowledge "an estimated forty thousand servicemen" who came home with damaged hearing, but adds: "Many of these had subnormal hearing before they joined the ranks and should never have been drafted."

It's easy to picture the author and his peers in all their Father-Knows-Best glory, white lab coats buttoned over white shirts and ties, extolling the bright chromium future with condescending smiles for the modern housewife and the budding astronaut poised for blast-off on her knee. Then, the moon turns full, Jupiter aligns with Mars, their clean-shaven faces grow lupine ruffs of wild hair, their black eyeglass frames morph to wire granny rounds, psychedelic auras surround their heads, and their ties, wide as bibs, explode in floral profusion. Everything is changing—everything, that is, but the fundamental faith in the benign effects of mind-blowing sound.

The "baby boom" years would have boomed even without the babies. Especially in America, the middle of the twentieth century saw the greatest increase in noise since the advent of the Industrial Revolution. "The military-industrial complex," of which President Eisenhower felt obliged to warn the nation, was gearing up again, as were all the modes of carbon-fueled transport. Inaugurated in 1956, the U.S. interstate highway system would eventually take up 65,000 miles, radically increasing truck traffic and spelling the demise of the railroads. The two-car family would become the national norm. By 1958, the first year that American commercial jets arrived at Heathrow Airport, more people were crossing the Atlantic by air than by boat.

On suddenly ubiquitous television sets advertisers promoted consumption as a patriotic duty. In 1956 Robert Sarnoff of the National Broadcasting Company attributed postwar prosperity to the fact that "advertising has created an American frame of mind that makes people want more things, better things and newer things." Add "louder things," including a panoply of household appliances and gas-powered lawn tools, seldom engineered to their full potential for quietness because consumers then, as now, associated noise with power. Though Muzak had its origins as far back as 1934, the basic Muzak doctrine that every silence is a vacuum abhorred by nature, God, and man found the postwar mission fields of restaurants, office buildings, and funeral parlors ripe for conversion. In 1965 the Beatles were drowned out by their Shea Stadium fans, but what seemed to point to a new era was merely waving good-bye to the old; nobody drowned out The Who.

Without a doubt rock and roll changed music radically and beautifully. It made a lot of people happy, some people deaf, and a few people rich. I'd feel bereft without it. But in the end it was more a culmination than a subversion of the postwar culture that produced it—a glorification of sheer volume, both of sound and of sales, a boon for technocrats, and a quasi-fascist exaltation of the *Ubermensch*, with his slavish groupies, his thuggish retinue, and his ability to overpower every voice but his own—unless, seized by a fit of *noblesse oblige*, he holds the microphone to the audience and deigns it the favor of parroting his refrains.

It is difficult for people who believe they have been liberated by loudness to find anything but oppression in quiet. American politics—from the street theater of the 1960s to the growth of "shock jock" radio four decades later—and American education would both be affected by the sonic pieties of The Boom. In his 1979 book *The Culture of Narcissism*, Christopher Lasch noted that many high school teachers "actually fear quiet and restraint in their classrooms, justifying their failure to enforce order on the grounds that imposition of silence interferes with spontaneous expression and creates unnecessary fears. 'A quiet classroom may be an awfully fearful situation for someone,' said one teacher, whose classroom grew so noisy that the students themselves clamored for quiet." According to one observer cited by Lasch, the purpose of such a classroom was to teach children "their first lessons in how to live in the 'friendly' and 'relaxed' climates of the contemporary bureaucracies of business and government."

In such an atmosphere it is no surprise that noise would increasingly become a common source of metaphor, just as agriculture provided the figurative language of previous times. We now complain of "the static" in our lives; we demand "more bang for the buck." When David Brock wrote his 2004 book about American right-wing propaganda, he called it *The Republican Noise Machine*. Not everyone agreed with his politics, of course, but no one needed a nanosecond to grasp his figure of speech.

CHAFING AT THE LIMITS

Historian Emily Thompson is undoubtedly correct in pointing to the concept of "noise pollution" as an outgrowth of the environmentalist mindset that emerged in the 1970s. Hers is but another way of saying that the postwar belief in "limitless potential" was eventually seen to have limits. At least for a while, and at least by some. "We have learned," President Jimmy Carter declared, "that more is not necessarily 'better,' that even our great nation has recognized limits"—none being so great, it turned out, as the patience of the American electorate with any leader who dared say such a thing.

Still, the recognition of noise as a pollutant, a potential health risk, and a serious public annoyance began to grow. In 1969 the International Music Council of UNESCO unanimously passed a resolution that denounced "the intolerable infringement of individual freedom" and upheld "the right of everyone to silence, because of the abusive use, in private and public places, of recorded or broadcast music." In the Noise Control Act of 1972 the U.S. Congress declared "that it is the policy of the United States to promote an environment for all Americans free from noise that jeopardizes their health or welfare." Two years later the U.S. Environmental Protection Agency issued its "Levels Document," which disputed the acceptable day-night average sound level (DNL) of 65 dB recently adopted by the Federal Aviation Administration. In other countries, too, public opposition to noise increased, with demonstrations at Heathrow Airport that have continued to this day, and the Dutch "Let's Be Gentle with Each Other" campaign against neighbor noise in the late 1970s. Within roughly the same year came Canadian R. Murray Schafer's *The Tuning of the World* and, in France, Jacques Attali's *Noise*, which would put the study of sound on the map of academic discourse and in the awareness of people well beyond the academic sphere.

Loud as the 1970s were, I have talked to young acoustical engineers who speak of those years as a sort of golden age, a time when the "older guys" were cutting new ground, especially in reducing industrial and transportation noise, and funds were readily available to meet those goals. That changed in the United States when Ronald Reagan cut funding for the Office of Noise Abatement and Control in 1982. No president or Congress since has seen fit to reverse his decision.

The Reagan move continues to exert an influence not only on the scope of noise research in the United States but perhaps on the American conception of noise itself. Especially in academic discourse, noise seems to have been rehabilitated. Once viewed as an environmental pollutant that affected us all, noise has increasingly come to be seen as a cultural signifier that identifies one's "tribe" and its supposed inclinations. Some people prefer to listen to Beethoven on the Bose, some people prefer to go deaf on the bus. By extension, some people like to eat steak every day and some people like to eat day-old bread. You can call this "diversity" if it makes you feel better, but in terms of the overall Reagan agenda, *victory* might be the better word.

TERROR AND TORTURE

With the exception of a few war veterans—survivors of Pearl Harbor, for example—most living Americans, including even war veterans themselves, have never heard the sound of hostile aircraft. The distant drone of approaching jets, the strobe-like sound of helicopter blades, does not produce the same reaction for us as it would for our civilian counterparts in Afghanistan or Iraq. That altered with the terrorist attacks of 2001.

In the summer of 2005 the *New York Times* announced the release of more than 12,000 pages of oral history gathered from firefighters and medical responders in the months after the 9/11 attacks. They testify to the terror—and to the noise—of that terrible day. "I heard this metallic roar," "a rumble that I will never forget," "groaning and grinding," "I thought there had been an explosion or a bomb," "popping sounds . . . going both up and down and then all around the building," "frigging noise," "like being in a tunnel with the train coming at you," "screaming from the top of my lungs, I must have been about ten feet away from her and she couldn't hear me because the building was so loud."

One perhaps unintended effect of the 9/11 attacks was that people experienced the sound of a sky without aircraft. Several times now I have heard someone comment, always in the positive, on the uncanny celestial quiet of September 12, 2001. The observation always takes me aback; I suppose that in the knee-jerk days immediately following the attacks it would have been deemed "pro-terrorist" or at least insufficiently pro-U.S.A. I happen to think otherwise, but the comment does connect to the terrorists at least insofar as suggesting their motivations, their desire for a world in which the noise of Western culture does not bear down on them from on high. I doubt we will ever be able to "listen" to our enemies or cause them to listen to us until we can hear our own noise with their ears.

Meanwhile, we have found occasion to make them hear it with a vengeance. During the 2004 siege of Fallujah, the 361st PsyOps company "prepared the battlefield" by blasting the city with heavy-metal music, repeating a tactic used against Panamanian president Manuel Noriega in 1989. One of the "no touch torture" techniques employed in the U.S. "war on terror" has included subjecting detainees to prolonged sessions of loud music.* According to Binyam Mohamed, a British resident tortured "by or on behalf of the U.S. in Morocco, Afghanistan, and Guantanamo," his interrogators hung him up in a "pitch black room" where "[t]here was loud music, [Eminem's] 'Slim

* "No touch torture" was discussed in CIA documents as far back as 1963.

Shady' and Dr. Dre for 20 days." Other songs reportedly used to break down prisoners include Metallica's "Enter Sandman" and—a favorite—Barney the Purple Dinosaur's "I Love You."

The use of noise for torture is not new. What feels new is the use of the torturer's own cultural artifacts to inflict the pain. The "playlist" at Guantanamo may well have passed up the chain of command from the soldiers themselves, who presumably listen to some of the same music, at something approaching the same volume, in their own barracks or darkened bedrooms back home. Aside from the obvious messages implied by the choice of certain songs—Springsteen's "Born in the U.S.A." kicking your sorry Bedouin butt—something very interesting is going on here. To get your head around it, you have to imagine, for example, French colonial officers torturing suspected Algerian terrorists with nonstop cancan performances and lumberjack portions of *pot-au-feu*. When a country appropriates its most popular art forms in the service of torturing its enemies, is it not admitting repulsion at its own culture? There is something sad and defeatist in all this. Were I a suspected Muslim terrorist undergoing torture, I would hang on to that thought to steel my resolve. I imagine some do.

THE AGE OF TINNITUS

The noisy mind games we play with our enemies blur with the mind games we permit the culture to play with us. In his 2007 book *Musicophilia*, neuroscientist Oliver Sacks discusses the phenomenon of "earworms" (from the German *Ohrwurm*)—"catchy tunes that may, in fact, be nothing more than advertisements for toothpaste but are, neurologically, completely irresistible." In their most extreme form earworms can mirror the compulsive neurological tics of autism or Tourette's syndrome; for people who have those conditions to begin with, an earworm can be especially maddening. Sacks himself has experienced an upscale version of the malady:

> I have lately been enjoying mental replays of Beethoven's Third and Fourth Piano Concertos, as recorded by Leon Fleisher in the 1960s. These "replays"

> tend to last ten or fifteen minutes and to consist of entire movements. They come, unbidden but always welcome, two or three times a day. But on one very tense and insomniac night, they changed character, so that I heard only a single rapid run on the piano . . . lasting ten or fifteen seconds and repeated hundreds of times. It was as if the music was now trapped in a sort of loop, a tight neural circuit from which it could not escape. Towards morning, mercifully, the looping ceased, and I was able to enjoy entire movements again.

Earworms—melodies that come from outside our bodies but take up unbidden residence in our brains—typify some of the ironies of noise in the twenty-first century. First, there is the irony that a consumer culture, which glorifies and magnifies our "personal choices," allowing us to program our listening devices in virtually every conceivable manner, should render us ultimately helpless to control what we hear in our own heads. We are told we can "Have it your way," even as the nonstop musical score of the consumer marketplace has its way with us.

Second is the irony that the old historical distinction between music and noise, repeatedly challenged by musical composers throughout the twentieth century, has been reestablished more firmly than before—only in reverse. What we call "music," often completely divorced from performance and scarcely heard with more than half our attention, has effectively become background noise; while the sound artist, hoping to be *heard* in some meaningful way, uses what had hitherto been called "noise" to make music. Or at least the closest possible approximation to music: something we consciously *choose* to hear.

Lately I've enjoyed an exchange with a sound artist out of Newcastle, England, who goes by the name of Meinkinder. The son of "a traditional working-class family (my dad was a factory worker, my mum a school assistant)," he has always loved "all kinds of music" and has played in a number of bands. Influenced by such albums as Sonic Youth's *Bad Moon Rising* and Reed's *Metal Machine Music*, and possibly by "stark industrial landscapes," he began to experiment with different sound compositions, some of which he has kindly shared with me. In keeping with his political beliefs—he once belonged to the UK Marxist organization Militant Tendency—he makes his work available for free, "eschewing the conventions of organized record labels and distributors." Reviews of his first official release, "The Bloodshed Will Redeem Us" (a title he took from a poster outside a church), give a rough, if not entirely generous, idea of his work, referring to it as a "furious sixteen-minute blood-drenched vocal and noise maltreatment" that "ends in shattered shards of teeth and a flickering winged thing heralding a trembling pulse and dodgy voiceover that verges on the uncomfortable. Nasty."

Another critic describes the piece as "the kind of music you might want to use at home to get rid of unwanted company." I doubt the artist would put it to such use, however. "Like anyone else," he says, "I get highly irritated if the guy in the apartment upstairs is clanking around in the early hours." He likes to spend his weekends "walking about the Northumberland countryside and coastline . . . immersed in the beauty of nature," a noise artist very much after my own noise-hating heart.

As for my impressions of the work itself—which includes a series of twenty-one tracks named for each of the children killed in the 1993 Waco siege—let's just say that I can't imagine playing these compositions while I eat dinner. I suspect Meinkinder would say that's the point. I can't imagine his pieces turning into earworms, either—perhaps one reason why noise art, for all its occasional "nastiness," does not seem to have been appropriated as an interrogation tool. The only thing I can imagine doing with the CD Meinkinder sent me is intentionally listening to it—or not.

The seemingly endless multiplication of electronic devices, recreational vehicles, auditory come-ons, bleeped instructions, and Pavlovian reinforcements threatens to make our cultural condition one of collective tinnitus. In order to survive with any sanity, we either shut ourselves down or mask the buzz with other noises, prompting those who wish to get our attention to ramp up their volume even more. Not infrequently "a noise that could wake the dead" is meant to do just that: to wake people sensuously "dead" from noise.

At the same time, our suppressed irritation at this aural assault, coupled with the social identities we associate with various gadgets, can create its own noise in our heads: a curmudgeonly earworm that fixates on certain devices to the exclusion of others, tempting us to hear with our biases instead of our ears. The day before my wife and I toured Carlyle's house we went to Ye Olde Cheshire Cheese, on Fleet Street in the St. Paul's district of London, a tavern that has been serving customers ale and steak-and-kidney-pie for centuries. There is a table in the back corner, near the fireplace, where you can sit in Charles Dickens's customary seat or, if you dare, at the head of the table under the imposing portrait of Samuel Johnson.

Shortly after we had gotten our drinks, a well-dressed, impeccably groomed young man sat down at the far end of our long table and began talking into his cell phone. Imagine, a cell phone in the dining room of Ye Olde Cheshire Cheese! Pa Ingalls would probably have been delighted, lover of progress that he was, but what would Dr. Johnson have said? No sooner had I wondered than I heard the ghost of Dr. Johnson making reply:

"You misapprehend me, sir, and your neighbor as well, for his manner is not rude, and quiet conversation I can never hold odious. It is a man's discretion, not his devices, that merits our praise or censure. I'd more lief eat my supper with a score of gentlemen accoutered such as he than with one prating companion who talked me dead on his disapprobation for cell phones."

Smiling at me across the table, my wife put it more simply. "He's just having dinner with a friend."

CHAPTER 6

Their World Too

Noise Today

It's my world too.

—FRITS PLATTE, FRIESLAND, THE NETHERLANDS

In Beirut neighbors of noisy nightclubs have started pelting patrons with eggs and water bombs. In Thailand residents around Subarnabhumi Airport are engaged in an ongoing dispute with authorities over the noise of jets. A recent editorial in the *China Daily* decries the ceaseless din of creating *jingzhuang*, "elaborately decorated," apartments. (Making wood and stone furniture "on the premises" is noted as especially disturbing.) The Environmental Protection Agency of Ghana, concerned about the amplified noise coming from churches, mosques, and nightclubs, has begun observing International Noise Awareness Day. The Environmental Health Service of Moscow, where oil wealth has radically increased traffic and new construction, has declared 70 percent of the city "a zone of noise discomfort." Native peoples in the Arctic Circle, one of the most naturally quiet regions on earth, show marked rates of hearing loss attributable to the extensive use of snowmobiles. One out of six Australians suffers the same malady; by the year 2050 the rate is projected to be one in four.

Noise has gone global. Even the word *global* has a noisy, self-important ring to it, loud with superlatives and grand designs, a din of jumbo jets and electronic twitter, the staccato fire of insurgency and counterinsurgency, filling every silence on earth.

History has finally brought the peoples of the world within earshot. Ralph Waldo Emerson's famous line about "the shot heard round the world" can refer to almost any shot now. The merest whistle of my tea kettle, announcing a boiled cup or two of water and an increment of carbon emissions necessary to bring that water to a boil (and an excess beyond that, since the water boils before the whistle sounds), speaks to my ability to disturb neighbors as far away as Bangladesh. It speaks volumes about my disproportionate "volume" on this earth. As a man in the northern Netherlands protested to me—and he could have been speaking for any one of billions of people in any one of hundreds of countries, though he was speaking mainly for himself and his partner and about the combined noises of neighborhood commotion and NATO aviation exercises that might soon drive them from their low-income apartment—"It's my world too, you know."

SAD-EARED LADY OF THE LOWLANDS

I met Frits Platte and his partner Annette Reinboud online before I met them in person at their home in a small Fries town near the Waddenzee. Annette, who says she is acutely noise sensitive due to childhood sickness and the effects of heroin use earlier in her life, belongs to Stichting BAM, a Dutch anti-noise organization devoted to such causes as removing piped-in music from commercial establishments. Like so many of the noise-hating "killjoys" I've met in my travels, Annette is a lover of music, with a special affection for the Rolling Stones. She also has a sharp sense of irony.

"Everybody's looking for silence," she told me. "And they go to workshops. They go to silent retreats in India. And they take a plane and they fly over my house. Wow! Silence!"

It's possible I had flown over her house too, though I was not looking for silence at that point so much as for a brief glimpse of a culture with an interesting and complicated relationship to noise.

The Netherlands is the most densely populated country in Europe, with close to 17 million people squeezed into an area roughly the size of Maine. One out of three Dutch inhabitants complains of neighbor noise; three-quarters of

Dutch homes are exposed to noise levels above 50 A-weighted* decibels. Even animal noises get thrown into the mix; the country boasts 12 million pigs for its 17 million people, raising concerns not only about odor but about the "information" communicated by the animals' squealing. (The pigs are packed together worse than the people.) Like any densely populated place, the Netherlands provides a plausible glimpse of the future, when there are likely to be many more of us, and when a post-petroleum economy might leave us with fewer options for getting around, or at least with more reasons for staying put.

The Netherlands also has two geographical features that favor noise propagation and, at the same time, force the Dutch to weigh their noise issues alongside other concerns. First, the country is flat, much of it reclaimed from the sea and much of it at or below sea level. Second, water is everywhere in the Netherlands, on the land, around the land, in the land—and very much in memory. As recently as 1953 heavy storms broke through North Sea dykes, killing 2,000 people and displacing 70,000 more. The country is crisscrossed by some 3,135 miles of rivers and canals, equal to the length of one Mississippi River with 800 miles left over. An abundance of water also favors the propagation of sound.†

The combination of flatness and proximity to water complicates as well as exacerbates certain problems of noise. On the one hand, a wind turbine erected on this kind of landscape is likely to produce more noise for its densely packed population than the same tower sited offshore or on remote and mountainous terrain. On the other hand, the Dutch landscape and location make the country especially vulnerable if sea levels continue to rise due to global climate change. Our driver and traveling companion for the journey that took us to meet Annette and Frits was a Dutch researcher named Frits van den Berg, who has spent a great deal of time examining wind tower noise, the public perception of it, and the dismissal of that perception as a "new Dutch disease."

---

* A-weighted decibel readings give emphasis to the middle and higher frequencies in order to produce figures more indicative of what the human ear actually hears. For more information see "Common Terms Used in Discussions of Noise" at the back of the book.

† This seems to be true underground as well as over it. Because of the saturated land, the low-frequency sound of ocean sonar—which has raised concerns about effects on ocean mammals—is reported to penetrate Dutch ground for as far inland as 20 kilometers.

Several cultural traits also add interest to Dutch noise issues, not the least being a tradition of social equality, which noise historian Karin Bijsterveld, of the University of Maastricht, calls "the starting point" for any discussion of the issue. Others I talked to agreed, one of them noting that the Netherlands is reputedly "a bad place to be a celebrity," with little public patience for "stars" who put on airs. I found less agreement on why this should be so, though van den Berg felt that the ongoing historical battle with the sea might have been a factor. As early as 50 C.E. the Roman historian Pliny spoke of the natives as "a wretched race . . . inhabiting either the more elevated spots or artificial mounds. . . . When the waves cover the surrounding area they are like so many mariners on board a ship." Building these mounds (which are still visible in the rise of the oldest Fries towns), and the dikes and canals that came later, demanded extraordinary effort and cooperation—and would have been tough on any celebrity's manicure. However it originated, this emphasis on equality disposes the Dutch to be concerned about the unfair distribution of noise effects, especially in regard to children. Fred Woudenberg, head of the Amsterdam Municipal Health Service, told me that the distribution of transportation noise in the Netherlands does not show the same low-income correlation found in other countries, though inequities do show up in the greater abilities of more affluent people to find quiet.

An emphasis on equality can also enter into the dynamics of individual noise disputes. I found it interesting that Bijsterveld herself, who is not shy about asking people on trains if they've seen the signs prohibiting cell phone use, seemed to have difficulty accepting my contention that one neighbor's right to make noise was not equal to another neighbor's right to quiet if the noise penetrates that second neighbor's walls. "In the Netherlands the starting point is that everyone is equal," she insisted. Presumably it is the starting point for anyone who goes next door to complain about the noise—or winds up changing his or her mind about going.

The Dutch are also celebrated for their tolerance, a trait often ascribed to the general tendency of trading nations (theirs was the world's leading commercial power in the seventeenth century) to accept cultural differences as a standard rule of doing business. Like all national reputations, this one bears scrutiny: The broadcasts coming from minarets have met some of the same resistance in the Netherlands as in other Western countries, and Woudenberg went so far as to say that even the sounds of children at play, traditionally a welcome part of any country's soundscape, have an ominous timbre in some Dutch ears because more and more of the children are Turkish and Moroccan. Recent years have seen a rise in right-wing backlash against immigration. Still, in its laws at least, the Netherlands remains one of the more liberal democracies in the world. Marijuana is sold legally in the tea shops of Amsterdam; brothels are licensed and sex workers have their own union.* Physician-assisted suicide is an option for certain patients. Though she is prudently leery of making national generalizations, Bijsterveld sees these phenomena—along with the "Let's Be Gentle with Each Other" anti-noise campaign of the 1970s (itself a gentle approach, raising awareness as opposed to laying down any law)—as typically Dutch, given the tendency of the populace to start by "accepting the defined behavior." People will smoke dope, people will make noise, people with terminal diseases will attempt to end their lives. "We accept that, and then we start thinking about how to deal with it."

Her comment may explain Dutch efforts to link noise effects to specific rates of mortality and financial expenditure. After all, what human behaviors are so emphatically "defined" as avoiding death and saving a buck—or, in this case, a euro? It is estimated that within ten years noise could be the "number-one burden of disease" in the Netherlands. One attempt at quantifying this makes use of a measurement called the DALY, or disability-adjusted life year. As Fred Woudenberg explained, a DALY is equal to a year of lost life, with various disabling conditions assigned a numerical weight; so that, for instance, two years of living with cancer is equated to one year's loss of life. You can

* When I got back to the States I e-mailed the sex workers' union, Rode Draad ("Red Thread") to see if noise was an issue for them. The so-called Red Light district struck me as one of the noisiest in Amsterdam. The reply is worth quoting: "Sexworkers are as much affected by noise as other people in the entertainment industry, in shops and at home. In gentlemen's clubs the noise is less than in for instance discos. But there is one kind of noise we find particularly irksome. It is noise made by people who think they are so special that they are set aside from other workers. And by people who think they live in different neighbourhoods than the rest of the population. . . . The stigmatization creates a lot of noise. Because of that, we sometimes also make noise. We don't underestimate the impact of noisy working places. But really, it is not different from other workplaces. If there are complaints sexworkers can lodge a complaint with the labour inspectorate."

also calculate what that year is "worth" in terms of the average money amounts that people are willing to pay in order to lower their risk of dying. Obviously these will vary from country to country, which gives the alarming but unintended impression that a year of life in an EU country, which now goes for about 70,000 to 80,000 euros, is "worth" more than a year of life in a place like Sierra Leone. According to official calculations published by the National Institute for Health and the Environment, noise costs the Netherlands about 38,000 DALYs per year or close to 3 million euros. Woudenberg believes, however, that the "real number," one based on other effects besides the criterion of severe sleep disturbance used in the analysis, would be much higher, perhaps as high as 150,000 DALYS or 10.5 million euros.

I hope that at least some of my readers will find such a scheme utterly preposterous and demeaning—only to reflect further about how much it differs from their own evaluations of noise and its effects. For instance, many of us (including me, apparently) accept the deleterious health effects of aircraft noise as a necessary trade-off for more "important" gains. What gains? Money, perhaps? Woudenberg's point was less strident than mine. He suggested that by allotting half the money a country already spends *because* of noise effects (say, 5 million out of 10 million euros) in order to *eliminate* those noise effects, the country would reap savings equal to the other half it did not spend, and reap some additional quality of life into the bargain.

Long before the Dutch had learned to count DALYs, they had discovered a way to make their flat terrain work against disease and noise both. They had taken up the bicycle. One sees bicycles everywhere, not only in Amsterdam, where clusters of them are chained to every railing and bike lanes are marked along the streets, but all across the countryside. (During the controversial 1966 wedding of Princess Beatrix and former German Army officer Claus von Amsberg, protestors jeered the groom with "Give us back our bikes!"—a reference to their confiscation during the last world war.) I would see many sights that impressed me during our brief stay, but none of them, not even the vision of Van Gogh's "Starry Night" a few breathtaking feet from my face, seemed so compelling and even sensuous as the sight of throngs of men and women, some in suits and dresses, others less formally attired, some toting babies and some long past childbearing years, but all fit and graceful, wheeling in anarchic harmony through the shady lanes of the Vondelpark. Call me an elitist—in my belief that some ways of getting through a life are superior to others, I *am* an elitist—but all I could see and hear in my mind was someone spilling out of an SUV and shouting to his children above the stereo speakers to pipe down and stay put as he huffs miserably toward an ATM machine for a fresh wad of DALY-jolly cash. Think of it as the *goal* in *global.* Nothing in Karin Bijsterveld's brilliant book seemed to nail her subject quite so well as the last glimpse my wife and I had of her, bicycling up the ancient streets of Maastricht, under the lamplight and into the quiet dark.

BACK TO REALITY, BACK TO THE ROAD

For all those bicycles, 87 percent of Dutch travel is by automobile, with the average person traveling 0.76 times a day for a distance between 1 and 2.5 kilometers. The Netherlands is part of the broader European trend of rising automobile ownership (a 22 percent increase equal to 52 million additional cars between 1995 and 2006) and rising levels of traffic noise. Of an estimated 120 million Europeans "extremely annoyed by noise," 67 million live with daily noise levels above the day-night average of 55 decibels recommended by the European Union, with traffic noise counting as the single greatest cause. (Road noise has three main sources: engine and drive train, including exhaust; tire friction on the pavement; and wind. Rattles in the vehicle's body can also be a factor.) Cars and pavements are both quieter than they were several decades ago—trucks driving on European roads are 90 percent quieter than they were in 1970—but traffic noise overall continues to rise.

So do the health and social costs, which are hardly restricted to the Netherlands. A 2001 German study estimated that as many as 3,900 myocardial infarction cases in that country could be attributed to road traffic noise. A 2009 study led by Goran Pershagen of Stockholm concluded that people exposed to traffic noise exceeding 50 decibels—that is, 5 decibels lower than the recommended EU day-night average—have a 40 percent increase in heart attack risk.

We have seen similar stats before. After a while they can sound like just another form of noise, a static of numbers that ultimately delivers too faint a signal. Cars make noise; noise makes people sick. Well, so does unemployment, so does not having a way to get your kids to school. At times I've had that same response. Nevertheless, one is occasionally able to pick up a peculiar signal amidst all the statistical din. A pie chart for "Passenger Transport Volume by Journey Purpose in Germany" in a 2008 European Environment Agency (EEA) report called *Traffic at the Crossroads* is a good example. According to the chart, getting to work accounts for a mere 15 percent of passenger trips; escort of children and others, for 9 percent. Leisure activities account for 31 percent, which in combination with the 19 percent for shopping takes up exactly half the pie. Supposedly this breakdown "is very similar for other Western European countries."* So—if the health effects associated with traffic noise can strike us as exaggerated, they are surely no more exaggerated than our sense of transportation "necessities."

This is not to imply that Europeans are a frivolous lot when it comes to driving patterns.† The typically smaller percentages for work and school travel in Germany might well be accounted for by the fact that three-quarters of Europe's population live in cities, where public transportation satisfies most workaday needs, thus raising the percentages for a category like leisure. And even if car owners vowed to restrict their driving to transportation to and from work, the larger economic effects of this austerity would eventually leave many of their neighbors with no work to drive to. If anything, the numbers probably support University of Groningen professor C. Vlek's statement that noise is yet another "baked in" environmental effect of "the technology used and the physical infrastructure in which it is based." In other words, we can

---

* In urban areas of the United States, trips to work equal the combined percentages for shopping and personal business. Public transportation accounts for less than 2 percent of total passenger miles.

† The Swiss have a car-sharing "Mobility Cooperative" whose 70,000 members account for 2 percent of total passenger transport in the nation and tens of thousands of tons of saved carbon. Members can reserve a car by Internet or phone when they need one; otherwise they use public transportation covered by an annual flat fee. The goal of the company is to have a car available on every street corner.

take noise out of our civilization with about the same ease as we can extract an egg from a cake.

It may be that aircraft noise, another major source of noise, not only in the European Union but around the world, is every bit as "baked in" as automobile traffic. I do wonder, though, if this has less to do with the nature of the cake than the dogma of the bakers. If I had five dollars or three euros for each of the people I've heard qualify a statement about the harmful effects of aircraft noise by saying that they liked flying and wouldn't for all the world want to do anything to stop it—as if they faced the rack and the screw for so much as implying otherwise—I could afford my own private jet. In any case, aircraft noise has proven to be every bit as noxious and intractable as that of other forms of transport. The EEA report cited above notes that over the past fifty years "the certified noise levels of aircraft have been reduced by over 30 EPNdB (certified perceived noise in decibels), which corresponds to an eight-fold decrease in loudness." That is no mean achievement. At the same time, the current growth rates of European air travel exceed even those of the United States, which the FAA has projected to reach 1 billion "enplanements" by 2021. In short, European air noise continues to rise.

The "intermittent" quality of air traffic noise can also make it more disruptive than the steady whoosh of road traffic. The widely discussed RANCH (Road Traffic and Aircraft Noise Exposure and Children's Cognition and Health) study, which looked at 9- to 10-year-old children living around three major EU airports—Schiphol in Amsterdam, Barajas in Madrid, and Heathrow in London—published findings in 2005 indicating that nighttime aircraft noise exposure negatively affects both reading comprehension and recognition memory. *Now don't get me wrong, I fly too*—but ever since flying to London and speaking to two of the researchers who conducted the study, one of whom also took pains to assure me that she liked to fly, I have had to ask how many of my various reasons for "enplanement" have trumped the importance of a 10-year-old child learning to read. Middle-aged man writing a book, 10-year-old kid trying to read one. And we haven't even touched on carbon.

At least in terms of acknowledging and studying the problem of noise, the European Union probably leads the world, with individual nations sometimes running ahead of the pack.* Ten major airports in France, for example, levy fines on planes that exceed certain noise levels, increasing the amount charged according to six different categories of loudness and multiplying the fine by 10 in cases of nighttime departures. The revenues (about 25–55 million euros a year) are used exclusively for soundproofing dwellings near the airport—a good example of the principle, honored more often in the breach than in the observance, that "the polluter should pay." Though airports in other EU nations (the UK and Spain, for instance) have adopted similar measures, the French are so far unique in having an independent agency—Autorité de Contrôle des Nuisances Sonores Aéroportuaires (ACNUSA)—that oversees the process and has the authorization to ground the aircraft of any airline that refuses to pay its fines. It is likely, however, that some airlines circumvent the deterrent by routing their noisier planes through airports with less stringent standards.

Recognizing that a union of nations linked by international systems of transit calls for a unified approach to noise, the EU released its European Noise Directive in 2002, requiring the completion of strategic noise maps of major railways, roadways, and airports for every "agglomeration" of 250,000 or more people by 2007, with updates every five years. The maps for the next round in 2012 will also include smaller agglomerations. Recommendations based on the mapping were published in 2009.

Shortly before the completion of the first phase of the directive, I visited with Wendy Hartnell and Antonia Acuna, who have been overseeing the process in the United Kingdom, which has 20 percent of the EU's designated agglomerations.† They showed me the maps, now available to the public as the directive stipulates, with a sober satisfaction. They had tackled a complex technical problem knowing that, in Hartnell's words, "where we are with noise now is where we were with air quality thirty years ago" and were justly pleased

* Readers wishing to explore EU noise policy in greater depth than I am able to do here will probably want to begin with the often-cited *Green Paper* of 1996, "the first step in the development of a noise policy with the aim that no person should be exposed to noise levels which endanger health and quality of life." The noise maps mentioned later in this chapter can be found online at www.noisemapping.org.

† The noise maps for England can be seen at http://defraweb/environment/quality/noise/mapping. Results for the first round of noise mapping in the EU are at http://noise.eionet.europa.eu/index.html.

with the work they had done. Nevertheless, she admitted, the maps "are just a tool to help us come up with these action plans, and that's where it all starts getting tricky." At the risk of insinuating a pejorative nuance of *tricky,* I asked what role on-site measurements had played in the creation of the maps. I was surprised to learn that they were based entirely on computer modeling. "It's a fairly high-level view of what the noise problems may be," Hartnell said. "The further you zoom in, the more difficult it is to say that it actually reflects reality."

I have heard critics say that this is true of European noise policy in general: It appears quite progressive from a distance, but the further you zoom in, the more the rigors of "the process" seem to take precedence over any actual reduction of the noise. That may be, but in all fairness, any "zoomed-in" perspective will also reveal the "baked-in" nature of the problem.

NEW HIGHWAYS IN EIRE

Against this background the Irish have for the past ten years been engaged in building the country's first, and one of Europe's last, national highway systems. Slated for completion in 2010, the goal of the project is to link Dublin to the south, west, and north of the island in the hope that some of the prosperity of the country's economic boom will travel with it. The Irish have many more cars than they did even a decade ago, a key factor in the 165 percent increase in their country's greenhouse emissions between 1990 and 2006, by far the largest jump for any European country. A 2008 study conducted by Trinity and University Colleges in Dublin found that 90 percent of residents in the city center endured nighttime noise levels above the 45-decibel day-night average recommended by the World Health Organization, a finding confirmed for me by an older gentleman dressed in a tweed jacket and tie for a showing at the Irish Film Institute in the lively Temple Bar district, where men his age seem almost as unusual as ties.* Recalling a time when "bicycles were everywhere," he said he found Dublin "louder *and brasher*" than the city

* Ireland is a youthful country, with a greater proportion of its population under the age of 25 than any other member of the European Union. How this affects or might affect noise levels in the country I do not know, though it probably is a factor, in type and tone if not in measured decibels.

he lived in as a child. Other longtime residents told me the same thing, most of them citing increased traffic, some citing the boom in immigration, and one of them (another older gentleman, albeit considerably less dressed-up than the first) attributing it to Ireland's tendency "to go to extremes with everything," including economic development and especially in its political enthusiasms, which he described as "making icons out of shite bags."

Call it what you will, the Irish government is hoping to avoid some of the extremes, including noise, of other highway systems elsewhere in Europe. According to Vincent O'Malley, an environmental manager with the Irish National Roads Department, many of those systems "were in place before noise became an issue." Based on close to twenty different environmental criteria, including traffic and highway construction noise, the Irish plan is to construct and route the system in such a way as to minimize impacts on wildlife, residential areas, historic and cultural sites, and agricultural lands—a challenging list given a country of green landscapes, numerous ancient monuments, and generally small agricultural holdings (some no larger than fifty hectares), where "severance" of one part of a farm from another can be devastating. In some cases O'Malley feels the new routes will actually direct traffic noise away from areas currently affected by it. In other cases he hopes to achieve the same result through engineering, including the use of sound barriers and quieter pavements. The pains being taken are perhaps best illustrated by the rerouting of one section of road to reduce impacts on snails. There are currently 1,100 passes through the new roadbeds to allow badgers to pass safely under the highway instead of becoming roadkill on top of it.

Such measures have raised the predictable accusation that the government cares more for snails and badgers (and bats, too, who get new houses whenever construction disturbs old roosts) than for human residents granted no such waivers. There have also been strong objections to the project on environmental grounds. Although a government that cares for snails is probably going to be more cognizant of noise than one that doesn't, the fact that the government has given itself fifteen years past the completion date of the project to meet all of its design goals does give one pause. As a longtime resident of another region renowned for its emerald greenness (*Vermont* is from the French for "green mountains"), I can't help but regard the Irish plans with mixed emotions. The history of my state can be divided pretty neatly into Before-Interstate and After, with the latter phase marked by a loss of family farms, a compensatory plague of Holstein tchotchkes, and the steady immigration (via Interstate) of "flatlanders" like me. A man from a small village in County Mayo on the far western side of Ireland, O'Malley probably feels the pros and cons of a national highway system as acutely as anyone. Nevertheless, he believes with visible conviction that the project can be truly "innovative" and even a model for other countries, proof that it's possible to cut down the travel time to Donegal without cutting down the badgers and honking the nuns out of the cloisters into the bargain. The Irish experiment will certainly be a project to watch for anyone interested in one of the world's most persistent noise problems. Who knows but that it will raise the bar for transportation systems and national icons alike.

WAKE-UP CALL

If the unification of Europe carries the hope that the war on noise will be waged on multiple fronts and at a higher level of commitment, it also carries the danger of every big plan—namely, that the smaller sounds and silences of local custom will be overwhelmed by a bigger noise. The Irish national highway system is at least potentially an example of that trend. The passing of the Spanish siesta may be another.

What Nobel laureate Camilo José Cela once referred as "yoga, Iberian style" has been observed in Spain for at least 400 years. The word *siesta* comes from the Latin *sestus* for "sixth" and refers to the sixth hour of the day, when the sun is hottest. The custom seems to have originated among fieldworkers in rural areas and spread to the cities. Though we might think of it as unique to Spain and Spanish-speaking countries, a form of the custom appears in other Mediterranean countries as well. Especially in the islands, Greeks observe the *mikro ipno* between 3:00 and 5:00 P.M., during which time making phone calls and especially riding motorcycles is discouraged, just as construction noise has been proscribed by "silence proclamations" covering the same hours in the Andalusia region of southern Spain. Similar taboos apply in Syria, which has an additional hour of prohibition beginning at 2:00.

It is interesting and by no means irrelevant to note that all three of these countries are renowned for raucous yearly festivals: the Carnival at Santa Cruz, which has been called "the biggest party held in Europe," the festival of San Isidro in Madrid, the Basra and Silk Road Festivals in Syria, the pre-Lenten carnivals held throughout Greece. One suspects there's much more than a feminine rhyme to the relationship between *siesta* and *fiesta*. It may be that people who know how to share quiet also know how to share merriment. Be that as it may, show me "a city that never sleeps" and I will show you a city where there are no collective festivals—thousands of complaints annually about "party noise," but no all-embracing party.

Lately the siesta has come under fire. Even as Spain's noise-abatement standards catch up to those of other European countries—Madrid has long been considered one of the noisiest cities in Europe, and national limits were not instituted until 1993—there is increasing pressure from multinational firms and reformers within the country to abolish the siesta and put Spaniards on a "standard" work schedule. In this event, the longer lunches and later closing times of the traditional schedule, which often result in people eating dinner as late as 10:00 P.M., would give way to prevailing European norms.

It is silly to be nostalgic for the loss of a custom that is not one's own and that may be good for the majority of its possessors to lose. I have no firsthand knowledge of how the siesta works in its native context, who pays for the privilege, and who reaps the biggest benefit. Apparently it has always been more of "a man's thing" than a woman's. Certain features of the custom, such as the late hours for meals, seem to date from no earlier than the Spanish Civil War, when straitened circumstances forced many people to hold two jobs. Even as I write, a man has been gored to death during the running of the bulls in Pamplona, a gentle intimation that perhaps some customs need to die.

Even so, a world in which Spaniards no longer take siestas is a world in which Spaniards will be a little more like everybody else. As is already true of their major cities, their town squares will be noisy at all hours of the day, and the noise will announce a further step toward cultural homogenization. White noise, so called because it contains all sound frequencies just as white light contains all the colors of the rainbow, may be our best auditory metaphor for such a world, one in which the bandwidths of all cultures theoretically have a place, while the overriding "hiss" of global capitalism, media conglomerates, and motorized traffic is all we actually hear. I am reminded of a fable in Russell Hoban's ironically futuristic novel *Riddley Walker*, about an owl who tries to swallow up all the sounds of the world, after which he plans to swallow up the silence, too. I think of that owl as noise, though I'm told he likes it if you call him Global.

THE BEST (AND THE WORST) SOUNDS OF JAPAN

The owl in Hoban's fable almost succeeds. Just as he's about to swallow all the sounds in the world, a blind boy realizes what he's up to and listens them back into existence. The boy has fellow travelers in the soundscape movement that followed the publication of R. Murray Schafer's *Soundscape: The Tuning of the World* and has continued in the work of the World Soundscape Project of Simon Fraser University in Vancouver and in the studies of Hildegaard Westerkamp, also of Canada, Brigitte Schulte-Fortkamp in Germany, Bennett Brooks in the United States, and Keiko Torigoe in Japan.

Now white haired but still vigorous, Schafer explained the origins of his soundscape work to me in a little car filled with his pipe smoke, parked on a muddy track in a remote wilderness compound in Ontario, not far from the lake where he stages his sonically rich and completely "unplugged" outdoor theater productions. It was raining as he spoke and I couldn't help wondering how much more he heard in that rainfall than I did.

"I came to noise because I went to Vancouver," he told me. "Buildings in Vancouver, like all buildings on the West Coast, are less insulated than buildings in the east. So, they're less insulated for noise, too. And I immediately found increased noise. It bothered me a great deal. Also, that was in the 1960s, an exceptionally noisy decade. Commercial jets came in the 1960s. For the first time music was being pumped out to 100 decibels and more in rock concerts. Cities were expanding. It was the era of what Detroit called the muscle cars. Noise was being celebrated, you might say, in a number of ways and it bothered me immensely. So, I tried to teach a course in noise pollution at the university. And it was not successful for the simple reason that it was all negative. 'So the world is noisy,' the students said. 'What can you do about it?'

And that's when the word *soundscape* occurred to me. Let's not concentrate on noises. Let's concentrate on the entire acoustic atmosphere," including its history and people's subjective responses to the sounds they hear.

Later in our conversation Schafer spoke of his work and that of his colleagues as "a way of fighting against noise by celebrating the beautiful." His words reminded me of what Albert Camus had said in *The Rebel,* that a rebel is a person "who says no, but whose refusal does not imply a renunciation. He is also a man who says yes" to some other value, the way that an uprising of slaves says "yes" to freedom. In Camus's mind this is what separates a true rebel from a mere nihilist. It is also what separates a sound-loving noise-hater like Schafer from a mere crank.

The soundscape approach has found some of its most passionate devotees in Japan, owing in no small part to a woman named Keiko Torigoe, who studied in Canada and translated some of Schafer's books into Japanese. Probably the idea resonated strongly with certain features of the Japanese aesthetic tradition; somehow it feels more like an idea that came home than like a foreign import.

In 1997 the Japanese Environmental Protection Agency launched a program called *One Hundred Soundscapes of Japan: Preserving Our Heritage*. Chosen from 738 applications, 249 submitted by individuals, the Best Hundred Sounds comprise a list of acoustical landmarks, a cross between a Seven Wonders and a Greatest Hits. They include such audible treasures as the "Drift Ice in Ohotsku" (No. 1), which during the months of January through April "creates strange, out of this world sound," the "Cranes of Idemizu" (No. 97), "making a great symphony with their singing and beating of wings," the "Sound of Wood-Engraving from the Cobble-Stone-Lined Town of Inami" (No. 54), the "Strong Yet Elegant Old Steam Locomotive of Yamaguchi Rail-line" (No. 78), and the "Cicada Chorus of Yama-Dera [literally 'mountain temple']" (No. 4), where the seventeenth-century poet Basho composed a famous haiku:

Lonely silence,
a single cicada's cry
sinking into stone

At first glance the relevance of such a program to the issue of noise pollution seems tangential at best. What does a cicada have to do with a 747, and what chance does it have against one? Perhaps a better chance than we think. If one of the Best Hundred Sounds is in your neighborhood, if you are proud that it is, and especially if you derive some of your income from tourists coming to hear that sound, you are likely to be zealous for its preservation. In effect the Japanese have managed to do for a select list of sounds what other countries routinely do for big noises: identify them with the commercial interests and patriotic loyalties of the people. Only Finland boasts a similar program.

Japan has traditionally been a country of sharp acoustical contrasts, some puzzling to a stranger's ear: the serenity of the traditional teahouse and the deafening thunder of no-less-traditional Taiko drumming (sound pressures as high as 120 decibels); the etiquette that frowns on "riotous laughter" yet virtually requires it in certain situations, so that according to one seasoned observer "the laughter at the average Japanese bar is so loud that about the only people able to have a normal conversation are Wagnerian tenors out on the town." Westerners wanting a more relativistic sense of noise could do worse than to visit Japan.

Perhaps the most curious noise they would hear is that of loudspeakers, some of it coming from *uyokusha,* the right-wing propaganda trucks that routinely circulate through the cities and towns before elections. A holdover from the Second World War and the years of reconstruction that followed, the widespread use of loudspeakers for public announcements began to arouse opposition as early as the late 1950s, with a number of municipal bylaws enacted against them. Still they persist. In a 2009 issue of *Kansai Scene Magazine,* writer Mary Miyamoto describes a "recent day trip to the sleepy village of Yagyu, in Nara prefecture" that "was punctuated by an ear-splitting siren blast at noon to tell us it was lunchtime" and "by speakers on the outside of the bus [that] emitted the same 32 bars of Beethoven's 'Fur Elise' for the duration" of the journey. Visiting Iriomote Jima on Okinawa, Miyamoto was awakened every day by loudspeakers that "run the length of the island wishing the residents (mostly cows) a good morning."

One is tempted to interpret these details as another instance of noisy Western influence—with a Beethoven score, no less—imposing itself on a traditionally "quiet" Eastern culture. Yet, Miyamoto feels that a Confucian emphasis on community and order, possibly added to the nostalgia of an older generation, makes some Japanese reluctant to give up the loudspeakers. It's also possible that stoic resignation of the kind John Hersey observed in Hiroshima after the atomic bombing plays a part. Japanese themselves will disagree over the explanations, just as they do over the loudspeakers. Suffice it to say that the world remains a curious place, full of complex cultures and beautiful sounds, and Hitler's pet appliance continues barking over it.

SACRED SQUAWKING

Japan is not the only country plagued by loudspeakers. Nor is their use in other countries confined to political and commercial announcements. Surveying noise around the world, one is struck by the number of noise complaints arising from religious observances. The Metropolitan Assembly of Accra, in Ghana, has singled out the amplified noise of Christian and Islamic houses of worship for particular condemnation. Before city authorities took the matter in hand in the late 1990s, the World Health Organization reported loudspeaker noise at the Durga Pooja festival in Calcutta reaching 112 decibels, barely a match for the firecrackers—a standard feature of Hindu festivals—with decibel levels as high as 150. A Buddhist writer sitting in his hermitage on a hill in Kandy, Sri Lanka, on a Saturday night lists the following audible loudspeaker noises: "Buddhist chanting from three temples, Sinhala Christian folk music from a church (which had been going on non-stop the whole day), and prayer calls from several mosques."

It makes sense that religion should seem noisier at a time when religious fervor is on the rise, both as an outcome of globalization and as a bitter reaction to it. Islam and especially Christianity are both growing at exponential rates. As Schafer and others have noted, religion has always appropriated for itself a right to make "sacred" noise (Holy Roman Emperor Joseph II called steeple bells "the artillery of the church")—a right Schafer associates with religion's "apocalyptic" impulse. At this point in our history, though, and especially where religions like Christianity and Islam are in violent conflict, one suspects that noise has more to do with staking out turf than proclaiming the end of the world. It's not hard to imagine the loudspeakers performing much the same function as megaphones at a sporting event: Our team has more fans than your team, more spirit, more power.

Buddhist contributions to this pandemonium seem especially ironic given the founder's reputation as a lover of quiet. In Burma a Buddhist text called the *Patthāna* is "blared for hours and days from loudspeakers." In Sri Lanka the chanting of "protective" prayers can be broadcast for the entire night. How different from the Buddha, who is supposed to have rebuked monks for going "among the houses making a loud, great sound" and to have expelled some visiting monks from one of his monasteries for talking "like fishermen hawking fish." Says one Buddhist text: "Different are these wanderers of other sects, who, having assembled and come together, are noisy, making loud and great sounds.... And different is the Fortunate One who uses remote dwellings in forests, woods, and groves, which are quiet, free from loud voices, deserted, secluded from people, conducive to seclusion."

But the irony extends well beyond Buddhism, and really beyond religion. No less than it destroys silence, amplification tends to destroy intimacy. Either it destroys intimacy by drowning out conversation, or else it counterfeits intimacy by making physical proximity irrelevant to social intercourse. This is glaringly at odds with the classic, nearly universal religious tradition of a teacher imparting wisdom to pupils who sit, literally, at his or her feet. Yeshivas can be noisy places, but I have not encountered any anecdotes of rabbis teaching Talmud with megaphones. A religion of microphones and loudspeakers is a religion of leaders and followers, which is not the same thing as a religion of teachers and disciples. The goal of the latter is to raise the disciple to the level of the teacher; the goal of the former is to keep the followers "informed," and in formation.*

Even for the nonreligious, it will be interesting to watch—and to listen—as religious bodies negotiate the choice between using amplification to proclaim

* I have probably simplified the religious aspects of the problem for the sake of my argument. It could be argued that the teacher-disciple model I've extolled is intrinsically elitist, and that the loudspeaker has the effect of "democratizing" religious practice, putting the laborer in the rice paddy on an equal footing with the adepts in the ashram. To that I would counter that putting the guru in the rice paddy would be a more genuine expression of the goal. Don't shout at me from another room; come into my room.

their message and using quietness to provide an alternative message. (For Buddhists, that would also come down to the original message.) At what volume do chanting monks become indistinguishable from electric guitars? The writer Pico Iyer, who went to Japan many years ago to study in a Zen monastery, writes: "It is no surprise that silence is an anagram of license. And it is only right that Quakers all but worship silence, for it is the place where everyone finds his God, however he may express it. Silence is an ecumenical state, beyond the doctrines created by the mind. If everyone has a spiritual story to tell of his life, everyone has a spiritual silence to preserve."

Everyone has a right, in other words, to believe that "It's my world too." But the degree of faith required depends increasingly on where you happen to live.

NOISE IN THE DEVELOPING WORLD

During the pandemic outbreak of swine flu in 2009, Mexico City's 20 million inhabitants noted an embarrassingly pleasant side effect. In addition to bluer skies (air-quality index up to "bad" from "unbearably bad") and reduced crime rates (muggings down 37 percent in a single week), the stricken city was quieter. Pedestrians and dog walkers intrepid enough to don face masks and leave their homes found the streets "emptied of the exhaust-belching, horn-blowing Hummers, SUVs, moving vans and trucks that normally crowd them." Other sounds reasserted themselves; an Associated Press report begins with "leaves rustling in the wind." Hand-in-hand with the greater quietness was what some residents observed as a greater civility. Adolfo Perez Garcia, who runs a small corner snack stand, put it this way: "Now at least we can walk in the streets—albeit with masks—without slamming into each other. We're less vulgar and mean to each other. And we're not telling each other off like we usually do."

Anyone who happened to be in Manhattan in the months after the September 11th attacks might recall a similar sensation. Though any sane person would prefer being jostled and honked at to dying of flu or jumping from a burning skyscraper, more than a few people wondered: "What's different, and why does it seem so refreshing?" Perhaps the clearest sign that a civilization has gone over the top is when it requires a catastrophe to make its streets seem human again.

Perhaps, too, you have to live in a place like Mexico City to get the full effect. The din of the developing world is like nothing most of us in richer countries will ever experience. More and more cities of Latin America, for example, are reaching Mexico City's 20 million mark, with increasing noise levels, not only from industry and traffic but also from "music and propaganda" broadcast on the streets. The World Health Organization reports that nighttime noise levels in Buenos Aires are barely lower than those of the day.

Farther east, the bustling cities of India's economic boom routinely contend in news stories for the dubious honor of "noisiest in the country" (Delhi usually wins), which in the case of India could well mean noisiest in the world. Residents of Mumbai (former Bombay), with a population of 14 million, are exposed to a constant 80–85 decibels, almost double the limits recommended for city noise by the World Health Organization. The pile drivers used in the constant construction of high-rise buildings can be as loud as jets at takeoff, and even so seemingly innocuous a noise source as a three-wheeled motorized rickshaw, modified for increased speed, can achieve 100 decibels. In order to better navigate Mumbai's bustling narrow streets, some motorists have removed their rearview mirrors for easier passage, compensating for the loss of visibility with additional horn honking. According to retired Mumbai doctor and noise activist Yeshwant Oke, "People feel agitated and angry, impotent to some extent. Indians are very docile. They would rather suffer than have enmity with the neighbors. But lately patience is wearing thin, and more and more people are complaining to get relief." More and more of them are also losing their hearing. A survey now more than ten years old found that 35 percent of the population in four major India cities had bilateral sensory neural hearing loss.

More recent surveys in Karachi, Pakistan, yield similar results, suggesting that the numbers in the Indian survey have probably risen substantially by now. One study showed that 33.3 percent of rickshaw drivers, 56.9 percent of shopkeepers in noisy bazaars, and 82.8 percent of traffic constables suffered noise-induced hearing loss. The ear's capacity to restore hearing after limited exposures is confounded by the typical Karachi working day. According to a 2009 report, most Karachi drivers work ten- to twelve-hour shifts; 62 percent of them work seven days a week. No less revealing, or sad, are the data showing that 84 percent of those surveyed were aware of the ill effects of noise while 96 percent of the sample reported wearing no hearing protection. As for the relative availability of adequate hearing protection in Karachi—or whether it is even advisable for a traffic constable to wear it!—the report does not say. On a more positive note, Pakistanis have learned, and a warming world might lead us to learn from them, that noise and heat can be combated with the same strategies. Building with hollow blocks of porous material, for example, keeps homes both cooler and quieter. Some residents arrange close ranks of clay pots on their roofs as another way of achieving the same double benefit.

Also on the positive side are certain court rulings out of India that might leave plaintiffs in a noisy American city envious. Overall India presents "a limited regulatory environment" regarding noise, owing partly to its giddy rate of development and perhaps also to its colonial past. The landmark Factories Act of 1948, for example, makes no provision for the protection of workers from noise; the Aircrafts Act of 1934 makes no mention of noise either. Articles in the Indian Constitution guaranteeing freedom of speech and freedom of religious expression have both been used in defense of loudspeakers and firecrackers. Against this background, the courts have had some notable things to say. In 1997 the High Court of Calcutta ruled to restrict the manufacture, storage, and selling of fireworks on the grounds that the right to live peacefully, the right to sleep at night, and the right of leisure are "all necessary ingredients of the right to life." In a 1993 ruling involving a Christian denomination's use of loudspeakers, the high court of Kerala held that free speech and religious expression applied to human beings and not to mechanical devices. While the precedent set by such a ruling feels a bit dicey—couldn't the government smash a printing press on the same grounds?—the basic principle rings true, especially for anyone who's ever been tormented by the inalienable rights of six D-sized batteries and a "repeat" button. The same might be said for a 2000 ruling by the Indian Supreme Court, which archly stated that no religion requires its adherents to preach or pray using electronic amplification.

My favorite, though, came a year later in *Sayeed Maqsood Ali v. The State of Madhya Pradesh* (India's largest state) in which the court held that even an individual can "maintain a writ petition against noise pollution." The individual in this case was a cardiac patient; the *dharamsala* (hostel) in his neighborhood was ordered to reduce its noise. Like the honeycomb of clay pots on top of Pakistani houses, the Indian rulings stand as examples of good sense and human ingenuity operating in the relative scarcity of what more "advanced" countries like to think of as their more sophisticated means.

A NEW KIND OF NOISE?

It is tempting to think that people in the developing world are merely playing our old LPs, grooving to the golden oldies of the Industrial Revolution, that what they're hearing now is essentially what Europeans and North Americans first began hearing in the eighteenth and nineteenth centuries. In other words, it's tempting to succumb to the big brother, little brother version of history. We had Miss Noisy for third grade—and what a witch *she* was; now it's China's turn. But these days Miss Noisy is teaching from a different book.

For one thing, most of the developed nations had the advantage of receiving their noise sources incrementally. The steam engine was followed by the gasoline engine, then the jet engine and the amplifier, each sound layered upon a progressively louder ambient, with some trade-offs along the way. (We retired the steam engine, for one thing, and the noise of iron-shod hooves on cobblestone streets gave way to the lighter acoustic footprint of trolley cars.) Think of it as the difference between walking gradually into an icy lake and being thrown fully clothed off a dock. The image is especially apt for the millions of displaced farmers who move from rural areas to urban centers, a migration that currently accounts for 95 percent of the urban population growth in the developing world. These people wind up not only in a radically noisier place than the one they lived in before—but in the *noisiest districts of that noisier place.*

A World Health Organization report on "the developing sector" of South Africa gives what can be taken as a generic picture of these districts:

> This sector of the population has the greatest exposure to high noise levels, both at home and in the workplace. Overall, they are relatively poor and cannot afford to live in quiet areas, or afford large plots or solid building materials. A large component of this sector resides in squatter communities where buildings are made of any material available, from plastic to corrugated sheets and wood. The buildings are right next to each other and there is almost no noise attenuation between residences.
>
> People in this category usually live close to major access routes into the cities, because they make use of public transportation and taxis to get to their places of work. Often, too, they live close to their places of work, which are usually big industries with relatively high levels of noise pollution. These people usually work in high noise areas, and because of their lack of awareness of the effects of high noise levels often do not make use of available hearing protection equipment. Because of lack of funds, these people also cannot get out of high noise areas and go to recreational areas for relaxation and lower noise levels.

The report notes that people in "the rural sector" also hear noise, most of it from agricultural machinery, but an "advantage they . . . have is that they return to homes in quiet surroundings and their hearing has a chance to recover."

A second difference between developing countries and older industrialized nations is that people in the latter have always had a greater likelihood of hearing machinery in its best working order and at its highest available level of efficiency. They have had few if any hand-me-downs. That will not be the case in a country plunged more recently and precipitously into industrial development. The air fleets of the developed world, for example, have been phasing out louder aircraft; so-called Chapter 1 and Chapter 2 planes (roughly equivalent to Stage 1 and Stage 2 in the United States) had been removed from most airports by early in this century. Chapter 4 commercial jets, appreciably quieter than many of their Chapter 1–3 predecessors, have recently begun to appear on U.S. and European runways.* Ever wonder what happens to the old planes? Some of them wind up in poorer countries. In 1999, five years after its first representative multi-party democratic elections, the government of South Africa produced a "White Paper on National Policy on Airports and Airspace Management," which called for the "non-addition" of Chapter 2 aircraft beginning in 2001 and a phase-out of those aircraft to begin in 2003. But the measures were stalled, for predictable economic reasons, with the result that Chapter 2 aircraft continued to be added to air fleets in that country.

Of course, we do not need to "think globally" in order to recognize that recycling in a context of economic inequality means that the poor inevitably wind up in the role of bottom feeders for the rich. If I buy a quieter, more fuel-efficient lawnmower, what happens to the noisier one? Probably it will be hoisted into the trunk of a sagging, badly muffled gas guzzler that pulls up to my garage sale. (Granted, the difference between the new lawnmower and the one I'm getting rid of is probably a matter of degree within the same technological "chapter," not one or two chapters apart.)

Another difference is that older industrialized nations tended to receive their noise sources in some "logical" order—that is, in a descending order of utility. The tractor would have come before the racing car. By the time people heard the racing car, they would have reaped the benefits of the tractor, including the financial means and physical well-being to go to the municipal offices and raise hell about the racetrack. This sequence is not always a given in poorer countries. I have young friends who've worked in remote Paraguayan villages where there are TVs but no sanitary latrines, compact-disc players but no healthy way of baking bread.

Finally, the developed nations have the psychological reassurance of knowing that their development has been largely indigenous rather than imposed.

* The International Civil Aviation Organization (ICAO), a United Nations body, first set aircraft noise limits in 1971. In 1990 the ICAO resolved to phase out Chapter 2 aircraft commencing in 1995 and ending in 2002, at which time these aircraft were permanently banned from European airports. Chapter 3 aircraft requirements were enforced from then until 2006, at which time Chapter 4 came into effect.

The interstate highway system in the United States, which allows for such extravagances as using more energy to ship a quantity of food than the food itself contains, was not adopted with a strong nudge from the World Bank or as a desperate measure toward paying off a crushing burden of national debt to Tanzania. As studies have repeatedly shown, noise annoyance exists in inverse proportion to one's ability to control it. And, as any perusal of popular media will also show, a central dogma of "globalism" is that it *can't* be controlled; it can only be embraced to one's benefit or resisted to one's harm. Either crank up the volume of your economy, or risk being drowned out by your party-hearty neighbors across the way.

A person would need to know much more than I do about history, anthropology, and psychoacoustics to judge the merits of my point, but it seems reasonable to suggest that people in the so-called developing world are hearing a new kind of noise. It seems arrogant to claim that we in the first industrialized nations know what it is, or that our ancestors heard the same thing back in the day.

Probably, though, it would be just as arrogant to presume that all of this noise counts as such for the people who hear it. Writing recently after a trip to Nicaragua, my friend Gary Greenberg offered me some trenchant reflections on the sounds and silence of a developing country. Waking up at 3 in the morning in a cabin with no electricity after 10 P.M., in a remote place with "no freeways or ice machines, no radios or records," he was both taken by the "gorgeous" aspect of the night, in which the only light came from "a bright-orange crescent moon" and "the only sound was the wind"—and unsettled by it too, with "the wind drumming the terrifying news of my insignificance against my shack and into my head. After ten minutes or so, I couldn't stand this anymore. I got out of bed and, unthinking, reached for my camera. I went out on the deck and took a picture of the moon. . . . I think I might have shot the darkness if the moon hadn't been there, because I wasn't interested so much in preserving the moment as I was in hearing the familiar beeps of my camera, its synthesized shutter sound . . . to seek consolation in noise, in other words." Well, you might say, that's how a man who lives in Connecticut seeks consolation. But Greenberg was listening beyond himself:

> Indeed, it seemed to me that the Nicaraguans, at least some of them, craved our noise. On a bus picking its way through the potholes and washouts of a mountain road, the music, Latinized American pop, was as terrible as it was loud (and this was funny: the bus, a retired American school bus, still had the markings from its life driving West Virginia students to their ABCs, including a sign that said, "THIS KCS BUS MAY BE EQUIPPED WITH AUDIO/VISUAL SURVEILLANCE, GPS TRACKING, AND STUDENT TRACKING," around which were pictures of the Virgin Mary and Jesus, in the lurid Latin American style, and I thought it was a perfect juxtaposition of our two belief systems and their respective presiding deities and their justice.) But don't worry. I'm not going to go all Tom Friedman on you. Just noticing how complex and fascinating a way noise is into the ambiguities of what we have wrought.

We would do well to follow Greenberg's example of attending to the ambiguities of noise, especially when we hear it across cultures. Noise can be as relative as any other value. That said, the human ear respects relativity only up to a point. It insists on one absolute truth, which is that prolonged exposure to sounds above 85 decibels renders it less able to appreciate Latinized American pop and numinous Nicaraguan nights—an insistence that makes the ear an invaluable guide through the diverse commotions of our "global" age.

START SPREADING THE NEWS

I end this survey of global noise that began in Old Amsterdam by turning to the city that used to be called New Amsterdam, one that stands in many people's minds as a byword for noise, "the city that never sleeps." It's up to you, New York, New York.

In 2007 the New York City council unanimously passed a thoroughgoing revision of its antiquated, thirty-year-old noise code.* Initiated by the Bloomberg administration and largely overseen by an intrepid lawyer-engineer

* The section of the new code dealing with community noise can be found at http://nyc.gov/html/dep/pdf/law05113.pdf; the section dealing with construction noise, at http://nyc.gov/html/dep/pdf/noise_constr_rule.pdf.

named Charlie Shamoon, the revision was a five-year labor, including various drafts, legal challenges to those drafts, input from virtually every type of business in the city, and an eighteen-month period of outreach to inform people about the law before it went into effect. The code set new and stricter standards for nightclubs, construction sites, boom boxes, motorcycles, barking dogs (barking after ten straight minutes is a no-no), and that American version of the Japanese *uyokusha* truck known as Mr. Softee. (In a compromise over the original plan to shut them up altogether, the ice cream trucks are permitted to blare their music in transit, when the sound is likely to be no more than a brief annoyance, but must turn it off when the trucks are parked—and surrounded by children.) Among the law's innovations are specific recommendations for reducing certain kinds of noise (better maintenance of air-conditioning units, thicker plywood sheeting around construction sites)—and even the recent invention of a sound-insulating jacket for that iconic urban noisemaker, the jackhammer. "It's rare," Shamoon told me with obvious pride, "that people writing a law would try to invent things, but we had these partners out there," one of whom developed the jacket.

Shamoon is pleased with other innovations in the law and with the fact that it is now being considered as a model for other U.S. cities. He's especially happy with the much-discussed "plainly audible" standard set for "transient" noise sources like motorcycles, which are usually many blocks away by the time a police officer can take a reading with a noise meter. For critics this restoration of the unaided human ear as a credible detector of noise is far too subjective; for Shamoon it is an attempt to inject an element of common sense—and personal control—into the law, an objective he said he might have lacked had he been an engineer only and been overly preoccupied with the logarithmic minutiae of decibels. "With motorcycles, if I can hear you plainly from 200 feet away, you're in violation. Since noise diminishes over distance, I know that if I can hear you from 200 feet away, and I was standing right next to you, you'd be blowing my brains out." Along the same lines he notes that "a kid with a boom box can't buy a $2,000 sound meter, but he can have a friend run down the sidewalk 25 feet"—the distance at which the device should not be plainly audible—"to determine if he can hear it."

The law doesn't do away with meters entirely. In the case of stationary businesses like nightclubs, where fines can be steep, Shamoon feels that "we owe it to them to take out a meter." He also feels that everyone is owed an initial warning. "It's not a game of gotcha." And it is not, to his mind anyway, an attempt to silence anyone's right of free expression. Recalling a 1999 case involving a street musician who was "plainly audible" inside Broadway theaters, Shamoon spoke of a "duality" of rights. "He had a right to express himself with his music. But the people on the street also had rights." Les Blomberg, who advised Shamoon on writing the law, often puts that duality this way: "The right of your fist ends at the tip of my nose."

Clearly the task of reducing noise in New York City, where noise-related complaints to the governor's hotline average in excess of 1,000 calls a day, is as Mayor Bloomberg called it "Herculean." I spoke to Shamoon in a Starbucks where the latte machine prevented our being "plainly audible" to each other across a tiny table. The law does not deal with "neighbor noise," historically one of the most persistent, intractable, and potentially deadly forms of noise that exist. A year after the law went into effect, I asked two cops on Fifth Avenue whether they noticed any difference in the level of city noise and immediately felt like a naïf, as if I'd just made them an offer for the Brooklyn Bridge. "See, where's *he* going?" said one in disgust, as a cab went rocketing down the avenue, blaring its horn and making no better progress for it. Still, for what it's worth, my own out-of-town ears have noticed a difference. Along with other regular visitors, including Rutgers professor Eric Zwerling, who assisted Shamoon in writing the code, I hear less honking, find myself less often quickening my pace past a construction site. Of course, it's possible that the law has done more by raising awareness than by levying fines, a possibility that would not disappoint Charlie Shamoon. On a recent visit to the city I was impressed by a handwritten sign tacked outside a German Bierhaus in the West Village, less daunting than the corner signs announcing $350 fines for unnecessary honking, but perhaps closer to the intent of the law: "Dear Patrons, When leaving our restaurant or if you are smoking outside, *please* try not to be loud or create noise. Our neighbors thank you."

THE NOISE WE NEED MORE OF

I had met Charlie Shamoon for the first time at a private reception to celebrate the inauguration of the revised New York noise code and the ten-year anniversary of the Noise Pollution Clearinghouse. It was held on Manhattan's Upper West Side at the home of Judith and Bill Moyers, whose son John had served as board director of NPC for the past eight years. In keeping with the occasion and the milestone it marked, my wife and I indulged in the extravagance of a hired car, not quite a limo but big and black, to take us in our dress-up clothes to the fete.

If there's a heaven for noise people, I imagine it will look something like what we found when we stepped off the elevator and into the Moyers' elegant living room. There was Arline Bronzaft, grand dame of the New York noise scene, now an advisor on noise cases that come through the city's hotline and a classroom guest with her new children's book, *Listen to the Raindrops.* We were introduced to Nick Miller of the prestigious Boston acoustical firm HMMH, who's advised the National Parks on overflight noise and on its Natural Sounds Program and who would prove an invaluable help with this book; to the environmental economist Charles Komanoff, whose publications cover such diverse topics as wind energy and jet skis, and whose passions include the rights of urban bicyclists and the musical excellencies of McCoy Tyner; to Doctor Peter Rabinowitz of Yale University, who has done extensive work with that most self-defeating of acoustical phenomena, hospital noise; to Janet Moss and William Lang, who work on international noise policy and were gearing up for a conference in Istanbul. I didn't see her, but perhaps we brushed against the ghostly elbows of Julia Barnett Rice, who campaigned against tugboat whistles at the turn of the century and whom Charlie Shamoon admires and also imitated when he solicited input from contractors on how to regulate construction noise.

We got to meet Bill Moyers as well, a living rebuttal to the notion that "quiet" best describes a lapdog. Gracious but always off to the side, appearing at intervals to chat with his admirers, then slipping out of sight, Moyers has been relentless in chronicling the economic inequalities of American society and the influence of moneyed interests on both the major parties of our political system. He was tough on Bush and he hasn't lost his head over Obama. It moved me not a little to see Les Blomberg getting an award as a guest in his house. I sensed that many of those present felt the same thing, not only for Les but for themselves. Their hard work, their "weak" issue, had made it in New York, and therefore, if the song is right, could make it anywhere. I guess we'll have to see.

Nice as the evening was, the most memorable part of it for me was the chauffeured ride over the river to Manhattan and back again. Torrential rains had been falling for days and many parts of New Jersey were flooded. The driver's GPS and the numerous detour signs routed us through Paterson streets I'd not seen since I was a kid, when Paterson was our suburban catchword for black, brown, poor. Too little had changed since then. The newest-looking things I saw were the coils of razor wire strung round the lots of small auto repair shops and tire stores. Heads turned to our passing car, and I felt a silly urge to roll down the window and call out, "This isn't how we usually travel, you know."

Returning through Paterson in the darkness with a new and piteously lost driver, we got to see more of the waterlogged city, the washed-out bridges, the iconic Passaic Falls roaring like an open floodgate in *The Epic of Gilgamesh*. This was the world, as it was in the beginning and promises to be again—not the enchanted, exclusive gathering of an hour before, but this: water everywhere and poor people flooded out of their homes. I would get another glimpse of that world later in the year when I visited New Orleans and saw, after two years of cleanup, the remaining devastation from Hurricane Katrina, the looted houses, the toxic, tin-box "FEMA trailers" ($80,000 each) that were making people sick. I was shown the place where one of the Army Corps of Engineers' shoddily built levees had given way and a tanker boat had ridden the deluge into a residential neighborhood, a rude harbinger of the "crony firms" that would soon be retrieving the dead at $12,500 a corpse. I was told that the boat had made "a terrible noise" coming through.

Having surveyed our noisy world, what can we conclude but that there are many ways in which it could stand to be quieter—and at least one in which it needs to grow noisier still. I mean the "unwanted sound" of protest. No hope for the world is complete without it. As I write, Iranians have taken to the streets to contest the legitimacy of their presidential election; Indians in Peru are blocking roads to prevent the encroachment of oil companies onto their lands. In China there are close to 75,000 demonstrations a year by people protesting their displacement by a "market" version of the Great Leap Forward. Apparently these people do not believe that by playing their stereos loud and honking their car horns late at night—all things that many of them lack the material means to do anyway—they are subverting the powers that be.* I don't believe it either. If at times I have seemed to take a reactionary, peevish view of noise, it is partly because I despise the shell game that substitutes "transgressive" displays of acting out for acts of disciplined resistance. The forces steamrolling the world and crushing the weak, the master schemes concocted to benefit the master minds, all phenomena that are thousands of years old and first announced themselves with trumpets and bells, and later with cannon fire and loudspeakers, flattened cities and broken sound barriers, shock and awe and prisoners trussed up in the dark as heavy metal music is blasted into their cells, will not be put to rout by cranking up our favorite tunes or buying motorcycles or buying anything else. They will succumb because we resist. Some of that resistance will make noise, but as soon as we begin to believe that every petty noise is a gesture of resistance, the Big Boys have as good as won.

In the end, after all the physicists, musicologists, and social theorists have had their say, there are only two kinds of human noise in the world: the noise that says "The world is mine" and the noise that says "It's my world too." We need to quiet the first and make more of the second. We need to hear the whole world inside the *too*.

---

* Jacques Attali writes in *Noise*: "Thus we may risk the hypothesis that the use-level of automobile horns in the city is related to its political and subversive potential, and the establishment of control over it is indicative of a credible reinforcement of political power at the expense of the subversive elements." This passage does not do justice to Attali's brilliant analysis, which is hardly ever this silly.

PART III

Lighter Footsteps: A Broad Perspective

CHAPTER 7

Loud America

... beneath the nearly invincible and despairing noise,
the sound of many tongues, all struggling for dominance.
—JAMES BALDWIN, *NO NAME IN THE STREET*

The other day as I was driving down the main drag of Lyndonville, Vermont, near my home, a large sleek pickup truck with what appeared to be a middle-aged man at the wheel roared out of a side street and cut in front of me. I caught only the quickest glimpse of him before he became invisible behind a window-sized decal of the American flag. Blazoned in large black letters along its stripes were the words "Drive em like you stole em!"

Why a man this old and well-off (it was no cheap truck) wanted to drive like a car thief, or why he imagined this was how any competent car thief would drive, or why even a thief would want to drive a pickup truck if he had nothing to pick up with it (this one was as shiny and unused as a ceremonial sword), intriguing as these questions are, did not interest me so much as what the motto had to do with the flag. Presumably there was some connection between the two, and presumably we could all be counted on to see it, though I wonder if as recently as fifty years ago the average American would have seen much of anything except possibly a case of flag desecration. I suspect even Jesse James would have scratched his head for a minute over this one.

Yet if, as our ads and leaders keep telling us, America means exceeding the limits, cranking it up, letting her rip, galloping through the chicken yards of the world with feathers flying and guns a-blazing, then the connection is more obvious than I'm making it sound. You "drive it like you stole it" because if you're a true-born American and want something badly enough, you will steal it, and have stolen it—from the Indians back in the day, from your great-grandchildren in the sweltering, bankrupt days to come. Parse it any way you want, to "drive em like you stole em," you have to make some noise. You have to be willing to stick your prerogatives in somebody else's ears or face, as the driver was doing with his customized flag, which seemed to stand for nothing so much as his middle finger.

Welcome to Loud America. In Sturgis, South Dakota, home of the world's largest motorcycle rally, sits the Loud American Roadhouse. Like the truck that cut me off, the name of the bar got me thinking about what *American* has come to mean in our time and whether *loud* ought to be included with *free* and *brave* in our national anthem. There are reasons for thinking that it should: the fact that the United States has more cars on the road* and planes in the sky† than any other country, the fact that we have been in no fewer than fourteen major military engagements since the Second World War, even the less spectacular fact that "America's biggest exports are television programs and movies." Our rock concerts, professional sports events, political conventions, and "mega-church" services—increasingly indistinguishable from one another and getting louder all the time—serve as models for the "civilized" world. That adds up to a lot of noise.

To say it adds up to pure evil, however, is going too far. Especially if we're talking about the sounds of a vibrant society, or what one political scientist refers to as the "rasp" of democratic discourse,‡ most of us are going to want

---

* Americans currently own around 250 million of the world's 806 million cars and light trucks, with 90 percent of the working population driving to work, 76 percent driving alone. As of 2005 the average household owned 2.6 televisions; as of 2007 U.S. citizens owned 270 million of the world's 875 millions guns.

† The Federal Aviation Administration recorded 757.4 million "enplanements" for 2008.

‡ Jonathan Schonsheck identifies "three broad categories of incivility" in his essay "Rudeness, Rasp, and Repudiation," which he defines as follows: "Very generally, *rudeness* is essentially impoliteness. *Rasp* is the friction of jostling political, moral, religious, and ethnic groups that is inevitable in any multicultural 'liberal democracy'—a system, or theory, or philosophy of government that cherishes the values of toleration and mutual respect. Not everyone, however, subscribes to toleration and mutual respect; the *repudiation* of these values generates the third, and most serious, category of incivility."

some boisterous sound. It has been suggested that noise "bears witness to the democratic dimension of public space," that in the American immigrant experience, for example, the ability to make noise on the street meant that one was no longer "a guest" of his or her adoptive nation but a full-fledged member of the family. In the old country you stepped furtively out of your door, crept meekly out of your ghetto—not so in the Land of the Free. "I sound my barbaric yawp over the roofs of the world," wrote Walt Whitman, celebrating both himself and the country he loved. Perhaps the fellow in the pickup truck is trying to say the same thing.

But is it the same thing? Was it democratic vitality that James Baldwin heard in "the nearly invincible and despairing noise" of New York City and in "the sound of many tongues all struggling for dominance"—or something else? Baldwin had sharp ears, and we should like to know.

LET'S START WITH A LITTLE MUSIC

I've told myself that when I finish writing this book, I'm going to mark the occasion with a pilgrimage to the Saint John Will-I-Am Coltrane African Orthodox Church on Fillmore Street in San Francisco, where the jazz giant John Coltrane is revered as a messenger of God, and his masterpiece *A Love Supreme* has an honored place in the worship. Though my devotion to the man is less than that of the faithful there, I frequently mark my Sundays with the same piece of music, especially when I don't make it to church.

Other Sundays, when I do go to church, and my wife and I are driving home, discussing the week ahead and playing what used to be called "the devil's music" as we go, she might turn and say, "What do you want to hear next? Bob Dylan, I bet." Usually she's right, and usually she knows which Dylan I want, anything from *Time Out of Mind* on, anything in that weathered, graywacke voice, so companionable to my middle-aged ears.

These two American artists, one white and one black, one still going strong and the other dead after the briefest career, one who took folk into rock and brought it back again transformed and the other who took jazz out of this world, are connected in my mind for having come to a similar crossroads at almost the same moment. Briefly put, Dylan lost some of his most loyal fans and Coltrane lost two members of his legendary quartet, along with some of *his* most loyal fans—both men's losses owing to what their detractors dismissed as noise. Some of these detractors made noise by way of emphasizing the point, most famously when Dylan was booed at the Newport Folk Festival and at subsequent concerts in the United States and England.

All this happened between the years 1965 and 1966, when I would have been around 12 years old, vaguely aware of Dylan and totally ignorant of Coltrane, and thus unable to make any comparisons between the two. But once I'd learned of these events, they seemed to typify a good deal of what else was happening in America at the time—and a good deal of what it has come to mean, for better and for worse, to live in the place Oscar Wilde once called "the loudest country that ever existed."

CHASIN' THE TRANE

What would Wilde have made of John Coltrane, I wonder. His fellow Englishman and poetical descendent Philip Larkin couldn't stand him. Coltrane made jazz "ugly on purpose," Larkin wrote in one review, especially in his later work, "when he was fairly screeching at you like a pair of demonically possessed bagpipes." More sympathetic listeners heard what biographer Eric Nisensen calls an "almost Mozartean celebration of life," a quality often attributed to such mid-career classics as *My Favorite Things* and *A Love Supreme*. Coltrane was well aware of the conflicting passions he aroused. In an interview in 1962, five years before his death and a few years before his free-jazz stage was in full "screech," he said:

> Some people say "your music sounds angry," or "tortured," or "spiritual," or "overpowering" or something; you get all kinds of things, you know. Some say they feel elated, and so you never know where it's going to go. All a musician can do is to get closer to the sources of nature, and so feel that he is in communion with the natural laws.*

* Among those who felt "elated" was a young musician named Frank Lowe, who said, "John Coltrane helped keep me alive when I was in Vietnam. I heard him everywhere I traveled before I shipped out. I brought some Coltrane records with me to Vietnam and listened to them constantly. His music was like life to me, and death was just down the road a few hundred yards away."

Coltrane was as mystical as his words suggest. After a religious experience in 1957, he began seeking what his wife Alice Coltrane called "a universal sound." The search would take him to African and Asian musical traditions, the theories of Albert Einstein, LSD, and ever freer forms of musical self-expression, especially in live performances. As he neared the end of his forty-year life, some in his audience—and two members in his band—walked away from what they felt had become mere noise.

Coltrane himself was not a noisy person. In fact, he seemed to personify the best of what we mean by quiet. Jazz critic Nat Hentoff called the physically imposing Coltrane "one of the gentlest people I've ever known." Several acquaintances, including Hentoff, said they could never recall his having said a single malicious word to or about another person. Once a woman in a jazz club scolded him for playing "loud, crazy music," this after having tried to hook his arm with an umbrella during the performance. He spoke to her kindly. Even in the way he dressed—mostly black or gray suits—he was quiet. A great one for practice, he would spend the first hour or so of his morning sessions on the road by fingering the keys of his instrument without blowing any notes so as not to wake the rest of the hotel. Lesser musicians who asked to sit in with him were never turned away; one producer quipped that had he signed on every talent Coltrane urged him to, he'd have had to issue at least 400 new contracts. Loud egos are rarely so generous. Nor are they known for caring much about the damages left in their wake. Coltrane's longtime—and "good time"—bassist and drummer, Jimmy Garrison and Elvin Jones respectively, once stepped out between sets for a bit of carousing that included breaking a few plate-glass windows. They came back, "laughing and giggling," only to find Coltrane waiting for them in the doorway with tears in his eyes.

On stage he blew notes that ruptured the blood vessels in his nose.

Perhaps the most notable part of Coltrane's quiet ethos was his capacity for listening. "It was part of his nature," writes musical scholar and composer Bill Cole, who knew him personally, "to listen to everyone he played with."* Some

---

* Coltrane played with, and in some cases studied under, such figures as Thelonious Monk, Duke Ellington, Miles Davis, and Coleman Hawkins. Those who listened avidly to him included composers Steve Reich, Keith Jarrett, and Ravi Shankar, after whom the Coltranes seem to have named one of their sons.

of this was Coltrane; some of this was simply jazz. Dialogue has always been at its heart. Everybody says so, of course, but the cliché becomes more interesting if you credit the theory that the development of jazz was influenced by the noise of urban commotion, a plausible idea so long as it doesn't obscure the artistry of the music and the depth of its African roots. The genius of jazz is that it took the noises that threatened conversation and refigured them as a form of conversation. It took the honking horn and humanized it. Say the word *humanism* and I hear the names Desiderius and Duke, as in Erasmus and Ellington. Wynton Marsalis describes the achievement in more political terms:

> In American life you have all of these different agendas. You have conflict all the time and we're attempting to achieve harmony through conflict. Which seems strange to say . . . but it's like an argument that you have with the intent to work something out, not an argument that you have with the intent to argue. And that's what jazz music is. . . . You have yourself, your individual expression, and then you have how you negotiate that expression within the context of that group. And it's exactly like democracy.

What Marsalis says is especially resonant if we think of the conflicts taking place in America between the year when Coltrane's famous quartet first came together (1962) and the year it came apart (1966),* and if we consider why it came apart.

Pianist McCoy Tyner joined Coltrane's quartet at the age of 21, though the two men had first played together when Tyner was as young as 17. Coltrane had hired him for a brief engagement, along with bassist Jimmy Garrison, during the same week that he decided to give up alcohol. Sobriety has seldom had a stronger recommendation. Jazz historian Gary Giddens says that not until McCoy Tyner did Coltrane "find his true soul mate." A devout Muslim and a former student of Thelonious Monk, who Tyner said had taught him to hear the "magic in sound," Tyner brought a mix of musical hu-

* Those four years saw the assassination of John F. Kennedy, the murder of Medgar Evers, the bombing of a Birmingham church that killed four African American girls, all in 1963, the election of Lyndon Johnson and the passage of the Gulf of Tonkin Resolution in 1964, the murder of Malcolm X in 1965, and the founding of the Black Panther Party in 1966.

mility and virtuosity that proved the perfect complement to Coltrane's inspired flights. More than once I've been surprised to discover that musical passages I regarded as sublime examples of "Coltrane" were equally or entirely sublime examples of Tyner.

Tyner left the group, and drummer Elvin Jones with him, after Coltrane radically expanded it—and the boundaries of its sound—by adding a second drummer and as many as two extra saxophonists, all of them explosive in their playing styles.* Tyner explained his departure by saying that he wished to form his own trio in order "to continue to grow," but on another occasion he spoke more bluntly: "I didn't see myself making any kind of contribution to that music. At times I couldn't hear what anybody was doing! All I could hear was a lot of noise." Jones's explanation was almost identical: "At times I couldn't hear what I was doing—matter of fact, I couldn't hear what anybody was doing. All I could hear was a lot of noise."

Neither artist had bad words for Coltrane. Jones went so far as to say of the new music: "Well, of course it's far out, because this is a tremendous mind that's involved, you know. You wouldn't expect Einstein to be playing jacks, would you?" Tyner, too, expressed his respect for Coltrane's right "to do what he wants to do as a creative artist." For both men—for Jones, who would later say, "I believe he was an angel," and for Tyner, who saw Coltrane as proof that "God still speaks to man"—there was no loss of faith. But while Coltrane was trying to hear God, Tyner and Jones were finding it ever more difficult to hear Coltrane. The desire to do so, no less than the failure, is what interests me most about their story and the whole loud American scene.

"IF I HAD AN AX . . . "

No one ever heard, and I doubt anyone could ever imagine, John Coltrane telling his band mates to "play fucking loud" as Dylan told his backup band, The Hawks, in Manchester, England, in 1966. Then again, Dylan might have

* Among Coltrane's bold and various innovations it ought to be mentioned that he replaced Tyner—whose departure he is said to have felt like "the loss of a son"—with his wife Alice Coltrane, this at a time when Miles Davis, to cite one notorious example, refused to let any woman so much as come backstage.

had more provocation. True, Coltrane had been pelted with a few tomatoes while touring Europe in the early '60s, and there was that woman with the umbrella, but no one called him "Judas!" as a member of the audience did to Dylan in Manchester. And even an inebriated Philip Larkin might have comported himself with more dignity than the audience at the Newport Folk Festival the year before.

The fiasco at Newport has been called "the most storied event in the history of popular music." Briefly put, Bob Dylan, who as a 22-year-old sensation had injected new life into the festival by performing alongside Joan Baez and Peter, Paul, and Mary in 1963, stepped onto the stage again in 1965—preceded by an introduction in which Peter Yarrow called him "the face of folk music to the large American public," though on that night Dylan had ominously abandoned his proletarian work shirt for one with polka dots—and with backup from the Butterfield Blues Band played a raucous sixteen-minute electric set of three songs, including his recently released hit single, "Like a Rolling Stone." Accounts vary as to who booed, how many booed, and why they booed (the faulty sound system has been blamed), but it seems safe to say there were boos. "Goddamnit, it's terrible!" Pete Seeger was heard shouting. "You can't understand the words! If I had an ax I'd cut the cable right now!" In subsequent mythologies Seeger would acquire an ax and come within one fatal stroke of severing the lifeline to America's bright rockin' future.

Like other inheritors of a conquest, we find it hard to understand the passions of the vanquished. Why such a fuss about amplifiers? Historian Greil Marcus notes that, for some of the "folkies" at Newport, "rock 'n' roll was pandering to the crowd, cheapening everything that was good in yourself by selling yourself to the highest bidder, putting advertising slogans on your back if that's what it took." Oscar Brand, active in an earlier folk revival, put it more explicitly: "The electric guitar represented capitalism."* These days it is taken

* I was interested to learn that performers at Newport all worked for the same $50-a-day scale, so that "Bob Dylan did it for the same money as prisoners from a Texas chain gang." A week after Newport, the Butterfield Blues Band would get "an unheard-of $100 a night to play Club 47—an amount that forced club manager Jim Rooney to clear the coffeehouse after each set like a standard nightclub rather than allow people to loiter all night over a fifty-cent mug of coffee. The Cambridge folk scene was dead."

for gospel that the folkies who booed were self-righteous and ill-mannered, which they were, and deaf to the future, which they also were, if only in the sense that their misgivings about amplified music would prove correct beyond their wildest dreams. Surely the most prescient left-leaning folkie in Newport could not have imagined rock music as a military interrogation tool, as a trove of theme songs for major political parties and pro sports teams, as an essential part of any pitch to buy a soft drink or an SUV. (For that matter, could the folkies have imagined even Bob "Judas" Dylan doing a lingerie ad?) Inasmuch as the Newport fracas can be framed as a noise dispute, the court must rule for Dylan. The booers were the true noisemakers, a fact Dylan himself wryly pointed out: "I can't put anybody down for coming and booing; after all, they paid to get in. They could have been maybe a little quieter and not so persistent, though." Dylan was the wronged party at Newport, and for me as for most of my contemporaries, Dylan remains the hero. But if there was any prophet at Newport in 1965, he was a man wishing he had an ax.

Like Coltrane's, Dylan's rebellion, as much against his own celebrity as against the self-appointed guardians of "the folk," can be seen as an act of artistic self-preservation. You can't play "Blowin' in the Wind" or "My Favorite Things" forever, or as C. S. Lewis put it, you can't be a good egg all your life. You must either hatch or rot, and Dylan refused to rot. He said as much when he was asked why he'd decided to go electric. "I was doing fine, you know, singing and playing my [acoustic] guitar. It was a sure thing, don't you understand, it was a sure thing. I knew what the audience was gonna do, how they would react. It was very automatic." Close your eyes for a moment too long and you'll open them to find that you're a nostalgia act for a TV fundraiser. There may be no recourse, especially if you're being booed into that kind of a corner, but to "play fucking loud."

On an even more basic level the Newport story is the American story of declared independence, of reinvention, of breaking away from the tribe, the British, "these little town blues" that sent Frank Sinatra crooning over the George Washington Bridge. On that level we know who we're rooting for at the Newport Folk Festival and, if we know where Coltrane took his last and most controversial band, who we're rooting for at the Newport Jazz Festival too.

Where the stories get more complicated, which is to say where they get even more profoundly American, is in the tension they reveal between the hero's noisy breakout—together with the noisy booing it inspires—and the desire of those people straining to hear him out. In that light the most poignant characters in the stories I've told are Tyner, Jones, and Seeger, the people who couldn't hear what they very much wanted to hear. *You have a right to go where you want,* they seem to say, and Tyner just about does say, *but do you have to leave me behind?* More than "Like a Rolling Stone" or *A Love Supreme, that* is our American anthem. In its classic version it is the immigrant father talking to his American-born son, the mother who never went to college wondering what her daughter could possibly mean by "women's studies," the designer of the Brillo Pad carton wondering what makes Andy Warhol such an artist for stacking one on top of another. More lately, and I would say more loudly, it's a liberal old lady calling in to a shock-jock radio show truly wanting to understand what all these white men are so angry about, or a conservative old gentleman trying to get a word in edgewise with someone like me. Something's going on here and Mr. Jones doesn't know what it is—obviously—but even if Mr. Jones would *like* to know what it is, nobody's going to tell him.

You can think of it as a matter of *jamming*. The word referred to two kinds of sound in those years when my hipper classmates began wearing polka dot shirts. One sense, still in use today, is the Wynton Marsalis sense of musicians improvising together, negotiating a shared acoustic space.* The other was the name of a loud vacuum cleaner–like noise that came through the old shortwave receiver my father hooked up for me to fool around with before I swapped it for the more fashionable diversions of a portable transistor. That kind of jamming, he told me, was the Russians blocking our radio broadcasts from reaching behind the Iron Curtain. The people who booed Dylan at Newport, the people at later concerts who adopted a tactic of coordinated clapping to throw him off his notes, were clearly jamming in the Soviet sense, a fact any of the old Stalinists among them would no doubt have been pleased to hear. But the shots are not always that easy to call. Often we are left with the

* The denotation dates at least as far back as the 1920s, though like other jazz terms it may possibly have its origins in the West African Wolof language.

question, and I think of it as "The Tyner-Seeger Question," though obviously I'm thinking of it mainly in a political rather than a musical sense:* Are we jamming like jazz musicians or are we jamming like the KGB? Is the noise we're making the rasp of democratic discourse or a repudiation of the discourse itself? Are people even able to hear?

There's more to that question than the platitude that "nobody benefits if there's no conversation." In fact, some people benefit splendidly. As Alexis de Tocqueville noted in his classic *Democracy in America*: "Despotism, which by its nature is suspicious, sees in the separation among men the surest guarantee of its continuance, and it usually makes every effort to keep them separate. . . . [A] despot easily forgives his subjects for not loving him provided they do not love one another." In other words, a despot can tolerate any kind of "subversive noise" as long as what it subverts is human solidarity.

When are we jammin' and when are we jamming? There is no easy answer, but I think it helps to know the question. And I suppose it helps to remember what we stand to lose whenever someone as gifted and committed as McCoy Tyner can no longer hear his own piano or imagine himself "making any kind of contribution" to the music we play.

JAILHOUSE ROCK

Perhaps the most extreme version of the failed American conversation—or as the warden in *Cool Hand Luke* famously calls it, the "failure to communicate"—is prison. As surely as Johnny Cash is an American icon no less than Bob Dylan, the names Folsom, Alcatraz, and Sing-Sing sound as resonantly American to our ears as Motown, Mount Vernon, or Disneyworld. Ironically the Land of the Free is also the land of incarceration, with Baldwin's "invincible and despairing noise" on the rise in both places.

As of 2009, 2.3 million Americans are behind bars, which amounts to nearly one in every hundred citizens. With 5 percent of the world's population, the United States has 25 percent of the world's prisoners, 44 percent of

* No less an authority than Plato might have raised an eyebrow at any distinction between *musical* and *political*. In a famous passage in his *Republic* he says that "the modes of music are never disturbed without unsettling the most fundamental political and social conventions."

them black. At current rates of incarceration one out of every three black men in the United States will at some time in his life have served time in a state or federal prison. Incarceration may be our most ambitious social program since the New Deal, a way of housing our poorest people of color and employing our poorest whites.

Though the United States has quadrupled its number of incarcerations over the past thirty years—an instance of "cranking up the volume" that includes the rhetoric of the "war on drugs"—the nation has not significantly expanded its prison space. This overcrowding, together with the acoustical nightmare of hard cement and metal surfaces, makes prisons round-the-clock chambers of reverberating noise. "You don't need an alarm clock in prison," according to a recent survival guide. When the lights come on in the morning, "you wake to the noise of 500 or more men urinating, flushing toilets, coughing, stirring, rustling about their cells for their clothing." At night, "strange noises rise from the steel tiers. One man is praying to Jesus, another to Allah, and a third to God knows whom. Other men weep or moan. . . . The most frightening sounds are those of men crying for their children." In such an environment even the silences can be unnerving. "It's when it gets quiet that you know something's either going to happen or something just did," writes Ralph "Sonny" Barger in his autobiography *Hell's Angel*. "Every jail I've been in sounds exactly the same: a combination of machinery running, people talking, noise traveling through the ventilation system, twenty-four hours a day, seven days a week. When that stops or the pitch changes, and the human voices are gone but the machinery is still running, look out."

The punishing effects of prison noise, some of them "baked in" to the environment, others intentionally meted out, can linger long after a sentence has been served. My noisestories.com correspondent Patrick Lincoln was sentenced to six months in prison for his role in a "School of Americas Watch" vigil at Fort Benning, Georgia, to protest U.S. military intervention in Latin America. "I've always been sensitive to noise, especially while trying to sleep," he writes, "but this experience aggravated my sensitivity to the point where I now have to sleep with earplugs and a hat over my eyes or I can't fall asleep." He continues:

> I was in a unit with 50–80 other guys and most of us had to use earplugs (stolen from the welding shop at the prison and given away for free) to drown out the snoring, radios, guards coming by at all hours, the buzz from the lights always on outside, etc. After talking back to a guard I was punished by being assigned to work at the prison's powerhouse, which produced an incredible noise all night long.

Abdulrahman Zeitoun, who is neither an outlaw biker nor a civil resistor, but a New Orleans businessman who had the misfortune of being mistaken for a terrorist as he paddled his canoe through the flooded city streets rescuing people and animals stranded by Hurricane Katrina, was handcuffed upright next to a loud prison generator for three days. The arrest turned out to be "an accident"; the technique was not.

As for Patrick Lincoln's sensitivity, which he regards as "my remembrance of a place where stimulation was used to pacify . . . and where human ingenuity and solidarity stood up in the face of oppression," he should probably be careful lest he be diagnosed as a paranoid schizophrenic. Ted Kaczynski's "oversensitivity to sound," based in part on his written and altogether cogent complaints about the noise of his Sacramento lockup, was cited as evidence in his court-ordered psychiatric evaluation. I don't mean to suggest that the "Unabomber" is not, or is, a paranoid schizophrenic, only that anyone who regards complaints about prison noise as somehow anomalous needs to spend a few nights there.*

Some would argue that if convicted murderers find things a bit too noisy in the hoosegow, that's just a bit too bad for them. Such a view, in addition to

---

* Where noise proves an ineffective punishment, there is always silence. The American prison system offers both. Solitary confinement, which in 1890 came close to being declared unconstitutional by the United States Supreme Court, is currently used on at least 25,000 U.S. inmates, with an additional 50,000 to 80,000 kept in "restrictive segregation units." According to doctor and journalist Atul Gawande, "In all of England, there are now fewer prisoners in 'extreme custody' than there are in the state of Maine." Studies reveal that depriving people of social interaction for long periods of time can impair the brain as much as a traumatic head injury. The effects, not surprisingly, but paradoxically in view of Patrick Lincoln's experience, include heightened sensitivity to noise.

being callous, illustrates yet again how dualistic interpretations of noise complaints almost always fail to account for all of the parties affected. Is every person in prison a killer? Is every person in prison guilty? For that matter, is every person in prison a prisoner? Acoustical consultant Stephen Carter, who has done extensive work with prisons, puts the matter this way: "Even if we walked stooped over, dragging our knuckles on the ground believing that 'just desserts' equals noisy living environments, surely the staff who attempt to communicate instructions, advice, and encouragement in these spaces deserve a break." A survey conducted at prison facilities in four states showed that prison staff regard noise "as a significant contributor to tension or stress" and tend to worry more about their personal safety as prison noise increases. Inmates who participated in the same survey ranked noise second only to the quality of the food as a complaint. The current ACA (American Correctional Association) daytime standard limit for prison noise is 70 decibels, about twice the 35–40 dB recommended by the EPA for school classrooms and thus more than ten times as loud to the human ear. That's the standard; the reality is often worse. In the meantime, what Wallace Sabine did for the Fogg Lecture Hall in 1895 can be done with greater precision for cells and day rooms, given the will and the funding. Carter suggests that even so modest an improvement as a more judicious use of PA systems might yield the rehabilitative benefits of talking "in a normal voice to another human being."

Worthy as these efforts are, and much as they deserve public support, one marvels at the lateness of the intervention. How many of the men and women currently behind bars had schools or neighborhoods in which they were able to hear quiet conversation, birds, silence? I'm not saying noise causes crime. I'm saying that noise accompanies many of the inequities that do. *From noise you came and to noise you have returned*—we could write that over many prison gates, perhaps next to a shorter line from Dante's *Inferno*. I'd love to see an auditory profile for the childhoods of our most "hardened" criminals, along with complementary data for the attorneys, psychiatrists, and judges who prosecuted, diagnosed, and sentenced them. Perhaps along with all the other details recorded throughout our life span—birth weight, blood type, mother's maiden name—we ought to have a decibel reading for every child's crib, classroom, and front-door steps, with a fourth and final reading for the place where most of his or her adult time was served. What does "hardened" mean, after all, but a calcifying of some human faculty, and what hardens sooner than a child's ears?

Patrick Lincoln seems to have reached a similar conclusion. After the completion of his sentence for civil disobedience, he returned to his hometown of Harrisonburg, Virginia, to do work as a community organizer, work that has included monitoring police behavior in poor and minority neighborhoods and a campaign to raise awareness of conditions at the Richmond City Jail. "Of course I share responsibility for the School of the Americas, and the U.S. prison system at that, but when I did what I thought was the most radical thing possible, I didn't even know my neighbors' names or what issues were currently influencing the community" where he was attending college. As he was compelled to do in prison, he now finds himself "talking with people from different backgrounds than my own"—trying to jam, in other words, and to make something better than noise.

THE BEST IDEA WE EVER HAD

If prisons stand for the ultimate failure of conversation in America—or, if you prefer, the ultimate triumph of noise—then one of our best examples of successful engagement may be our national parks. Wallace Stegner is supposed to have called the park system "the best idea America ever had,"* though as with many other of our good ideas, it remains a work in progress.

Tradition credits the painter George Catlin with originating the park idea in 1832, but it was not until 1872 that Yellowstone became the first of our current 391 park units, and not until much later that the public would recognize its ecological values. As was true of other early western parks, the setting aside of Yellowstone had more to do with the preservation of visually stunning natural monuments than with any nascent environmentalism.

Not until 1934, in fact, with the establishment of Everglades, was a park instituted for the express purpose of protecting wildlife. And not until 1996 was Catlin's original vision of a prairie park of "monotonous" landscape, with

* It seems Stegner was quoting British Ambassador James Bryce, who made the comment in 1912.

"desolate fields of silence (yet of beauty)," partially realized in Tall Grass Prairie in the Flint Hills area of Kansas. The naturalist John Muir had foreseen the park idea's evolution when he wrote in 1875 of those tourists who "like children seek the emphasized mountains" but whose gawking after grandeur would eventually give way to a more mature sensibility in which "lowlands will be loved more than alps, and lakes and level rivers more than waterfalls." Whether Muir was also thinking with his ears I don't know, but his desire to exalt the "lesser" attractions of nature certainly implies its "weaker" sounds.

Seen against this background, the Park Service's establishment of a Natural Sounds Program in 2000 was one more step in a gradual evolution. It was not a step imposed from on high. In a 1998 study conducted by the University of Colorado, 72 percent of the Americans surveyed saw the opportunity to experience "natural peace and the sounds of nature" as a "very important" reason for preserving national parks. Only 1 percent saw it as having no importance at all. The survey did not inquire what else the 1 percent regarded as utterly unimportant, though I wish it had.

Like the Best Hundred Sounds program in Japan, the Natural Sounds Program was initiated partly in response to noise, which is increasing in parks, as in other sectors of the American soundscape, to the extent that peak-season decibel levels in the front areas of certain major parks rival those of New York City streets. Air tour and commercial overflights, auto traffic, park maintenance machinery, campground generators, snowmobiles, and other recreational vehicles are all contributing to the general commotion—often creating the impression that the parks are more "crowded" than they actually are. The more room we make for our machines, the less room—and quiet—we leave for ourselves.*

Some of the Park Service officials I spoke to referred to the headquarters for the Natural Sounds Program, located in Fort Collins, Colorado, as "Karen Trevino's shop." That was a fair description of what I found when I stepped

* The former park ranger and radical environmentalist Edward Abbey was making this point more than forty years ago, when he invited his readers to consider "the interesting fact that a motorized vehicle, when not at rest, requires a volume of space far out of proportion to its size." According to Abbey, the Park Service could "without expending a single dollar . . . multiply the area of our national parks tenfold or a hundredfold—simply by banning the private automobile."

through the door. Cases of sound equipment—cables, decibel meters, microphones—were laid out like a dorm room's worth of gear on the hallway carpet, not far from several bicycles that the mostly youngish staff rides to work. A few members of the team were preparing for several days of intensive work out in the field. Nearby and as animated as any of them was Karen Trevino.

Like other men and women I've met who work with noise, Trevino did not start there. She had originally planned for a career in international finance. "I took a leave of absence from law school and went traveling around the world for about six months. And it was on that trip that I saw some pretty atrocious things." One of those things occurred on a Yugoslavian tanker, where Trevino had scrupulously tried to make sure any loose trash she saw made its way into a garbage bag before blowing overboard, only to see one of the crew members heave the bag she'd loaded into the sea. Returning to the States, she switched her major to conservation, eventually finding her passion in the conservation of natural sounds.

"If the mayor of New York City is trying to make what people expect to be a noisy place quieter," she said to me, "what should we be doing in places that people *expect* to be quiet?"

As a step toward answering that question, Trevino and her crew calibrate sound level information and convert it into color-coded visual representations that allow a day's worth of sound levels, and even an entire park's sound profile, to be seen at a glance.* (The images reminded me of those holograms that show the hot and cold zones of a body, or the varying levels of activity in a brain.) The technicians also make digital sound recordings to develop a "dictionary" by which these visual depictions can be interpreted. Much of their research is focused on creating plans to manage the roughly 185,000 air tours that fly over U.S. parks each year—a major mandate of the National Parks Air Tour Management Act of 2000. Their first map and proposal, completed in 2009, was for Mount Rushmore, a 1,200-acre unit with 5,600 air tour overflights a year. Franklin Roosevelt once called this park "the shrine of democracy," an ironic (cynics would say "appropriate") designation for a

* Readers can see some of the completed profiles at http://content.lib.utah.edu/wss/.

place where some people pay homage on the ground while others look down on them from on high.

"When you think about it," Trevino says, "what's the highest tribute we pay in this country—really, in the world—of reverence and respect? A moment of silence. Now, that said, nature isn't always silent. It can be very noisy. And people in parks aren't quiet all the time." Neither are things like cannons in a historical park like Gettysburg—nor should they be, according to Trevino. "Our job from a public policy standpoint is asking what noises are appropriate, and if they're appropriate, are they at acceptable levels."

Trevino sees this as a learning process, not only for her young department but also for her. Some of what she's learned has passed to her private life. Recently she asked her babysitter to stop using the terms *indoor voice* and *outdoor voice* with her young children. "Sometimes it's perfectly appropriate to scream when you're indoors and to be very quiet when you're outdoors," she says.

Though the parks still have much to learn—and to do—in regard to noise, they have made some noteworthy progress. A propane-fueled shuttle system in Zion National Park has reduced traffic jams and carbon emissions and also made the canyon quieter. Muir Woods installed library-style "quiet" signs after social scientists discovered, somewhat to their surprise, that the ability to hear natural sounds—fifteen minutes away from San Francisco and in a park celebrated mostly for the visual magnificence of its trees—ranks high with visitors. In Sequoia and Kings Canyon, which has a major naval air station to its west and a large military air-training space to its east, military commanders join park officials on a five-day "Wilderness Orientation Overflight Pack Trip" in order to understand the effects of military jet noise on visitor experience and to foster some common sense of mission. "Both of us are protecting the values and resources of the United States," says Wilderness Coordinator Gregg Fauth, "and one of the values we protect in the wilderness is natural sounds. When I speak to the military, I try to relate to them that they go off to fight for the things we have here at the park."

The message appears to be getting through. Before the program started in the mid-'90s, rangers at the park were reporting as many as 100 prohibited "low flier" incidents involving military jets every year. Now the number of planes flying less than 3,000 feet above the ground surface is a fourth to a fifth of that. Complaints are taken seriously, especially when, as seems to have happened more than once, they're radioed in by irate military commanders from atop jet-spooked pack horses on narrow mountain trails. In that context, human cursing is generally regarded as a natural sound.

Sometimes the initiative to combat noise has come from outside the park system. Rocky Mountain, for example, has the distinction of being the only park in the nation with a federal ban on air tour overflights, thanks mostly to the unflagging efforts of the League of Women Voters chapter in neighboring Estes Park. Park Planner Larry Gamble took me to see the plaque they'd erected as a memorial to their efforts and to the importance of the natural soundscape. The League had chosen the perfect spot, with a small stream gurgling nearby and the wind blowing through the branches of two venerable aspens. From there Gamble and I walked up one of the glacial moraines to a place where we heard wood frogs singing in the lowlands below us and a hawk crying as it circled in front of snow-capped Long's Peak. But in the twenty minutes since we'd begun our walk, Gamble and I counted close to a dozen jets, all in audible descent toward the Denver airport. Their vapor trails striped the sky. Some have suggested that at our stage of history the drone of aircraft ought to be regarded as part of the natural ambient, much as the rumbling of lobster boats is considered part of the ambient at Acadia National Park. It is an idea I would have found hard to accept on our walk—and embarrassingly hard to refute. I'd flown into Denver the day before.

Probably the most intractable noise problem in our national parks comes from the sky. The reasons for this are both political and acoustic. The latter include the fact that noise in the air propagates better than ground-based noise, which interacts with and is attenuated by the ground and the objects on it. Also, during daytime hours, when the air near the ground is generally warmer than the air above it, ground-based noise tends to refract upward from its source. Not so with air-to-ground noise, which in certain conditions can actually be enhanced by temperature differences, usually after dusk and typically in the desert southwest, where the air on the ground gets much colder after sundown. Other factors can have uncanny effects on noise coming from the air. For example, the craters in Volcanoes Natural Park are free of noise-attenuating vegetation, while the porous volcanic rock tends to act as an acoustically absorbent material. In effect, a hiker there can find himself in a veritable sound chamber, with a helicopter for a ceiling fan.

The political factors in overflight issues can prove as problematic as the laws of nature. The skies above the parks are not managed by parks, any more than the skies above Indian reservations are managed by tribal councils. All commercial air space in the United States is governed by the Federal Aviation Administration, which has a reputation for safeguarding both its regulatory prerogatives and what is often referred to in aviation parlance as "the freedom of the skies." The number of U.S. passengers taking advantage of that freedom is projected to reach 1.1 billion, almost double what it is now, by the year 2025. But much of the controversy about aircraft noise in the parks has centered on lower-flying air tours.

A twenty-year dispute over air tours above the Grand Canyon has involved all three branches of the federal government and, for protraction and difficulty, makes the court case in *Bleak House* look like a session with Judge Judy. A breakthrough seemed likely when the Grand Canyon Working Group, which includes representatives of the Park Service, the FAA, the air tour industry, environmental organizations, tribal leaders, and other affected parties, managed to agree on two critical points: (1) The Park Service's proposal that "the substantial restoration of natural quiet" called for in the 1987 National Parks Overflights Act would be taken to mean that 50 percent or more of the park would be free of aircraft noise 75 percent or more of the time (with no limits established for the other 50 percent); and (2) a computer model of the park's acoustics would be used to determine if and when those requirements had been met. All that remained was to plug in the data.

The results were startling. Even when air tour overflights were factored out entirely, this model showed that only 2 percent of the park was quiet 75 percent of the time, due to noise from hundreds of daily commercial flights above 18,000 feet. Even if air tours were abolished, the park would still be awash in the noise of aviation. The Park Service has since redefined the standard to apply only to aircraft flying below 18,000 feet. The Working Group remains in virtual limbo.

Whatever the next step might be, the Grand Canyon controversy goes deep. It is not simply a conflict between the air tour industry and environmentalists, or between the Park Service and the FAA. It is also a characteristically American conflict between the desire to soar and the need to hear, between the "right" to go wherever we please as quickly as we please and the expectation of being highly pleased once we get there. In many ways it comes down to a question of speed, dear to the heart of a country with high aspirations and vast geographical distances, but also a bitter rival of intimacy, sensuous enjoyment, and contemplative awe. Dick Hingston, a member of the Grand Canyon Working Group and militant opponent of air tours, speaks of the Canyon as "a place of power, of extraordinary confrontation of the present-day mind with the incomprehensible. It is not accessible in twenty minutes of an airplane ride."

His sentiments were echoed by several members of the League of Women Voters whom I spoke to in a coffeehouse in Estes Park, Colorado. I'd come to learn more about the letter-writing campaign that led to the unprecedented and so far unique air tour ban over Rocky Mountain National Park and ended by asking why the park and its natural sounds were so important to them.

"Many people just drive through the park," said Helen Hondius, straining to be heard above the merciless grinding of a latte machine, "so for them it's just the visual beauty." For her and her friends, however, all of whom walk regularly over the trails, the place needs to be heard as well as seen. "It's like anything else," Lynn Young added. "When you take the time to enjoy it, the park becomes a part of what you are. It can shape you." No doubt John Muir would have agreed.

But, as Muir understood, not all of us begin at the same level of awareness. Robert Manning of the University of Vermont has worked with the park system for three decades on issues of "carrying capacity"—the sustainable level of population and activity for an environmental unit—and more recently on noise. He feels that the park system should "offer what individuals are prepared for at any given stage in their life cycle." In short, it should offer what he calls "an opportunity to evolve."

"I really admire those people"—and now that his children are grown, Manning is better able to be one of them—"who've developed their appreciation of nature to the extent that they're willing and anxious to put on their packs and go out and hike, maybe for a day, maybe for a two-week epic adventure, walking lightly on the land, with only the essentials. *But*—those people probably didn't start there. I bet a lot of them went on a family camping trip when they were kids. Mom and Dad packed them into the car in the classic American pilgrimage and went out for two weeks' vacation and visited fifteen national parks in two weeks and had a wonderful time."

Seen from Manning's perspective, the social task of the national parks is to provide an experience of nature that is both available to people as they are and suitable to people as they might become. Such a task is robustly democratic and aggressively inclusive, but it is not easily achieved. It obliges us to grow, to evolve as the parks themselves have evolved, and we may best be able to determine how far we've come by how many natural sounds we can hear, and by how much we desire to hear them.

NO FEAR, NO APOLOGY

Manning's credo, based on the evolution of the national park idea and suggestive of an ongoing conversation, becomes a little harder to hear when we move from the carefully regulated environment of the park system to the larger—and noisier—domain of all public lands, or simply to all open space. Increasingly the forests, wetlands, deserts, and seacoasts of the United States have been filled with the noise and fumes of "thrillcraft," variously known as OHVs (off-highway vehicles) or ORVs (off-road vehicles). These include dune buggies, swamp buggies, dirt bikes, ATVs (all-terrain vehicles), airboats, snowmobiles, and jet skis ("personal watercraft")—all of them loud.

The general public is probably most aware of the issue of snowmobile access to Yellowstone National Park, the latest phase of which was a 2008 federal court ruling that "the Bush Administration's decision authorizing snowmobile use in Yellowstone National Park violates the fundamental legal responsibility of the National Park Service to protect the clean air, wildlife, and natural quiet of national parks." But the Yellowstone dispute, like others involving the parks, is a microcosm of a much larger problem.

Thrillcraft illustrate several features about noise and at least one defining feature of Loud America. First, they illustrate how noise frequently accompanies grossly harmful environmental and social effects—in this case, the erosion of land, the degradation of air and water quality—while also being a harmful effect. Elk and bison frightened by snowmobile noise, for example, must expend precious winter calories in flight.

Thrillcraft also illustrate how noise gives disproportionate power to those with the ability to make it. It's estimated that ORV users account for only 5 to 7 percent of public land use, but their impacts on other users and on the land itself are far greater. As most of us know from experience, a speedboat on a small lake basically owns the lake, both spatially and acoustically.

Finally, thrillcraft provide yet another example of how the costs of noise are passed on to the public, or "socialized" as John Moyers puts it. In the 2000 study *Drowning in Noise*, environmental economist Charles Komanoff and mathematician Howard Shaw set themselves the task of developing a formula for computing the costs imposed by jet skis on other beachgoers. Jet skis were a good choice because of their peculiar noisemaking characteristics. They go closer to shore than other craft, they periodically "leave the water"—which both increases the engine noise by losing the water's muffling effect *and* produces the quantifiably greater annoyance of variable as opposed to steady-state noise when their hulls smack the water again—and, finally, their erratic maneuvers mean that there is no consistent load on the engine. "The result is a penetrating whining sound, rising and falling rapidly in pitch like a dentist's drill and demanding the attention of anyone within earshot." Factoring out other environmental costs,* Komanoff and Shaw estimated the annual financial loss to beachgoers at $900 million. There are at least 200,000 more jet skis now than the 1.3 million that were in operation ten years ago when the study was done. These cleaner-technology models are somewhat quieter

* In 1998 the California Air Resources Board stated that a jet ski operating for two hours produced the same exhaust emissions as a 1998 passenger car driven for 100,000 miles. With typical exactitude, Komanoff and Shaw determined that the figure needed to be adjusted downward, concluding that the emissions-per-hour rate of a jet ski at that time was only sixty times that of an automobile. Fair is fair.

than their predecessors, but an estimated 1.2 million of the noisier two-stroke jobs are still in use.

Perhaps the most interesting aspect of thrillcraft noise is the political "jamming" that comes from special-interest groups like the Personal Watercraft Industry Association, the American Recreation Coalition, and other organizations lobbying under the umbrella of the "Wise Use Movement." Their basic strategy can be summarized as defending the prerogatives of Genghis Khan while wearing the bib overalls of Tom Joad. The people riding the machines are "ordinary Americans" spending their hard-earned money on some harmless fun; the people opposing them are "elitists," a word that has to be the most commonly used cant term in American politics, and certainly the most commonly used with respect to noise.

Not surprisingly, the rhetorical noise masks several realities, the most obvious being that Tom Joad is riding a pretty expensive rig. A new jet ski, for example, costs between $6,000 and $12,000 and is purchased, in 42 percent of cases, by someone making in excess of $100,000 a year. Figures for snowmobile ownership are slightly more modest but comparable. As for Ma Joad's share in this democratic exuberance: A study of ORV users in Michigan showed that 94 percent were male. (And a quarter of these owned two homes.) The rhetoric also leaves out the fact that the ORV constituency enjoys financial contributions and political support from ORV manufacturers and the timber industry, which have a shared interest in "opening" as much land as possible.* If all this is too much information to take in at one time, you need only remember that the "elitist" is the woman in the canoe—or was, until she got knocked out of it.

It was after co-authoring his study of jet ski noise that Charles Komanoff heard the story of New York conceptual artist Alan Bridge, better known as Mr. Apology. Komanoff told it to me one summer morning when I visited him in his Lower Manhattan office, where I was immediately struck by the elitist absence of an air conditioner and the equally elitist presence of a bike.

* Under Ronald Reagan the road system of the National Forest Service grew from 225,000 miles of roads to 350,000, making the Forest Service "the largest road-building entity in the world." There are now 400,000 miles of roads in national forests, eight times the length of the U.S. highway system and enough to go around the world more than sixteen times.

From 1985 until 1995 Alan Bridge ran a toll-free telephone "apology line" where callers could dial and leave a taped anonymous confession. Many called to apologize for small mistakes or indiscretions, a few for terrible acts. "I want to apologize," said one caller with a thick accent, "but when I was in Israel for six months, I killed six Arabs at night with a gang of other Jewish settlers." By the time of his death in 1995, Bridge was receiving around 100 calls a day.

Bridge was killed in the water by a jet skier who was never identified. Reportedly he was seen circling back to Bridge's body, before taking off for good. Bridge's wife Marissa insisted that had her husband survived, he would have forgiven the person who hit him.

To claim that Bridge was a "victim of noise" or "a martyr for civility" might seem overblown, but that's how I tend to think of him. If his project seems dubious, that only serves to confirm what I feel. Not even Charles Komanoff could come up with a formula to compute the social benefits of a work of art so radically quiet it could do nothing but listen. Were I to write a novel called *Loud America* in which a character named Mr. Apology was struck and killed by a jet ski, reviewers would pan it for "hitting people over the head with heavy-handed symbolism." But this didn't happen in a novel; it happened off Long Island, and the only person who got hit in the head was Alan Bridge. He was 50 years old.

STURGIS

Perhaps because I'd been thinking about Mr. Apology, one of the most striking sounds I heard at the "World's Largest Motorcycle Rally," which draws upwards of half a million people to the South Dakota town of Sturgis (population 6,400) every August, was that of leather-clad men and women saying "I'm sorry" and "Excuse me" every time a tattoo brushed against my arm on the packed, baking sidewalks. I suppose I hadn't been expecting that. There was one tense moment when a Vlad the Impaler type with broad hairy shoulders brought his chopper to an abrupt stop and began mouthing words at me over the rumble, but I soon made them out as "Would you like to cross the street?" It's possible that several of the magnificent custom bikes I saw had been stolen (Sturgis police report that an average of a quarter of a million dollars' worth of motorcycles are stolen at the rally each year), but I saw very few being ridden like they were. Even on the open road, good manners seemed to prevail.*

Bad ones might have given me better rhetorical traction, like being able to say that Coltrane got surly as his music got loud, but loudness was mainly what I'd come for, and I wasn't disappointed. The streets of Sturgis were filled with the glare of chrome and skin, "bikini bike washes" and women with flames painted over their officially "covered" breasts, in-your-face T-shirts—

REASONS TO USE A TRAILER:
I MIGHT GET WET.
I HAVE TOO MUCH STUFF.
MY OLD LADY CAN'T TAKE IT.
IT LEAVES MORE TIME FOR VACATION.
I'M A SCARED PUSSY.

—that Halloween orange-and-black (also the colors for Harley-Davidson), skulls and chains, Visigoth heraldry on fields of red, white, and blue, and, laced with the smell of exhaust fumes, the greasy aroma of that carnival food so tempting to those (like me) whose idea of a suicide machine comes with cheese. I mention the other senses only to emphasize all that the noise was able to dominate. The din was inescapable, especially in the early evening when the riders who'd gone out on day trips to the Black Hills or Mount Rushmore joined the long roaring procession back to town for the nighttime rites. Most of the bikes—I've read estimates as high as 80 percent for the riding public overall, and would imagine an even higher number for an event

---

* My experience of the rally differs from that of Jeff LaRive of Hot Springs, South Dakota. LaRive wrote to noisestories.com after an incident in which he "politely" asked "four 50–60 year old bikers" at his local family diner if they could "cool it with the profanity" because "that's my 8 year old behind you." One of the men "stands up, bumps my chest while unsnapping his buck knife sheath and says, 'You got a problem?'" But, as more than one person has reminded me, get several hundred thousand people in one location, regardless of their lifestyle, and you're bound to have a few jerks. "Even at a conference for dentists"—although, given the rally demographics, it's possible these four men *were* dentists.

like Sturgis—were sporting modified exhaust systems, technically illegal but rarely challenged. In fact, the Oakland (California) Police Department recently earned Noise Free America's "Noisy Dozen Award" for outfitting forty-five of its Harley-Davidsons with modified pipes. Citing a fatal accident involving one of its officers, the police department claimed the change was for safety's sake; many motorcyclists make the same claim.* Nearly every vendor selling apparel at Sturgis seemed to feature a pin, T-shirt, or bumper sticker with the motto "Loud pipes save lives."

I talked to a couple of veteran riders who insisted that they did. "I'd probably be on the right front fender of a Cadillac" without them, said one. Others, more skeptical, explained to me that while most of a motorcycle's exhaust noise is "to the rear," most of a motorcyclist's danger lies with the car or truck directly in front of the bike. Riders of both persuasions also pointed out to me that a motorcycle's noise is as dependent on how it's being driven as on its equipment. Robert Kelly, who's worked as an editor for *Hot Bike* and *Street Chopper* and now represents the high-performance company S & S, put it this way: "I ride a Harley, very modified, very fast, very powerful. Has the potential to be very loud. The potential, I say. Because when I leave my home, when I ride down the street that I live on, I ride at a very low engine RPM. And I don't make a lot of noise. When I'm racing with my friends out on the back roads, I make unbelievable amounts of noise. The only person that has to deal with it is the guy that's losing the race behind me." And, perhaps, as I pointed out, someone who lives on the back road, though Kelly insisted that he was talking about roads *way* back. Anyway, if loud pipes do save lives, I was in the land of salvation. As if to underscore the point, one of the Christian motorcycle groups gave me a tract titled "He would have ridden a Harley." Allowing that quiet time "for prayer and meditation" is important, the club president insisted that when it comes to motorcycles, "the louder the better."

---

* No doubt the claim can be disingenuous, as revealed by some of the ads for modified pipes in a typical performance catalog, where you'll find such products as "Road Rage Exhaust," "Hooker Troublemaker," and the offer of "a truly 'Big Block Sound' that rumbles the earth." At the same time, you will find items like the "Performance 1st Mufflers from Billet Boys . . . designed to give top performance without waking the dead" and a product called "Peacemaker's Exhaust from National Cycle," which allows the rider to adjust his or her exhaust noise according to the situation.

Getting off the street to escape the noise of Sturgis was usually a case of jumping from the frying pan into the fire. The amplified music in the bars was louder than almost any noise I heard on the street. At a place called The Knuckle Bar I had a shouted conversation with a young waitress working a fourteen-hour shift. While she accepted the noise as part of the job, she said that it greatly increased her stress. Vendors and some riders, too, noted another effect that I hadn't thought much about before: the tendency of prolonged noise to induce exhaustion, an effect common to other stimulants. I noticed it myself, less often on the street than in the bars, where it sometimes brought me close to nausea. The Knuckle Bar waitress, by the way, was an ICU nurse down from Fargo, in town like a lot of other people were for the money to be made.

With a stock Harley-Davidson costing between $8,000 and $30,000, it is probably no surprise to learn that the average American biker is a college-educated white male in his 40s. In 2008, when the average income for all Americans was $50,233, the average income for a motorcycle rider was $61,190. Bikers from an older era are sometimes heard bemoaning the counterfeiting of their culture by the "RUBs" (Rich Urban Bikers) who fly or trailer in to the rally, some with rented escorts accessorizing their machines, and the arrogant, spoiled frat-boy types who buzz around on Japanese "crotch rockets." Probably, though, motorcycling is not a sport that lost its blue-collar soul so much as America is a country that lost its blue collars (and some of its soul into the bargain). More than once I found myself asking if what I was hearing in Sturgis was the apotheosis of the machine age or its last gasp. The best answer may be neither and both. As America continues to outsource its manufacturing jobs, it is likely to compensate by creating more and more forms of recreational noise,* including loud pipe noise, touting the sound as a matter of "saving lives" or freeing the wild man within, but in actuality mourning

* The two kinds of noise I'm contrasting here cannot be accounted for by decibels. The physical sound can, of course, but not the subjective "noise." The sound of manufacturing is often not "noisy" for people who hear it as generally purposeful (somebody's way of making a living) or specifically beneficial to them (*my* way of making a living). But the sound of a neighbor's "toy" may be a different matter. In other words, while the outsourcing of manufacturing jobs and the rise of a service- and information-based economy may result in a reduction of measurable sound pressure, it may *also* result in an increase in noise.

something more quintessential and *quiet*: the skilled hand that used to hold the wrench, or as I would put it (along with how many other of those 40-, 50-, and 60-year-old riders), my father's hand.

In addition to being the only person I saw in Sturgis wearing earplugs (though I was surprised to see so few people smoking and so many putting on sunscreen), I seemed to be the only person who'd dutifully read through the events schedule and thus stood patiently waiting, eyes aloft at the appointed hour, for the "B-1 Fly Over—In Salute to Our Troops and Veterans" on the afternoon of what the mayor of Sturgis had declared "Armed Forces Day." Ready as I was, I froze along with everybody else at hearing what was perhaps the single loudest sound I've ever heard in my life. Picture it: a street parked so thick with motorcycles that it reminded me of those diagrams of the lower decks of slave ships, and throngs of leather-vested, biker-booted riders mulling past the Loud American Roadhouse and the tattoo parlors and the young women in stiletto heels and black thong bikinis who for a price will let you have your picture taken snuggled up next to them, and suddenly the whole born-to-be-wild scene is stunned into silent stillness as that single, for want of a better phrase, "hell's angel"* comes screaming overhead, followed by a wake of half-embarrassed gasps, those glances that say, "I wasn't scared or anything, but holy *shit*."

There is an explicit tie-in between American motorcycle culture and the military, some of it historically based on class experience, some of it a marketing ploy by the motorcycle industry,† and some of it best left to the likes of Dr. Freud, but for just a moment every one of us there had a brief, albeit unintended, lesson on how it feels to be a civilian on the ground in Iraq or Afghanistan, on what Loud America and the "outlaw lifestyle" sound like on that level of engagement. I found it interesting that nobody cheered.

---

* According to club founder Sonny Barger, the name *Hell's Angels* originally belonged to a team of 1940s stunt pilots.

† Harley-Davidson's double-edged marketing strategy, which *simultaneously* identified its product with the outlaw image of the so-called one-percenter (precisely what it strove to disavow when the sport was still suspect to the American middle class) *and* with patriotic fervor was probably one of the most savvy maneuvers in the history of corporate marketing. One hundred dollars of Harley-Davidson stock purchased in 1986 was worth a little over $7,000 in 1998. In the ten-year period between 1996 and 2006, overall motorcycle sales in America increased by 250 percent. One Harley-Davidson ad refers to its product as "The Ultimate Freedom Machine."

Of course, some of the people on the street were war veterans who had been on that level or close to it, and I wondered how the noise of the jet and the overall noise of the rally might have affected them. According to people I queried at the Veterans Administration, loud noises are usually not as disturbing to veterans suffering from post-traumatic stress disorder as noises specifically tied to the incidents that caused it. I was told something like the same thing when I dropped by the Fort Meade Veterans Hospital just outside the center of Sturgis and asked about the rally noise. According to director Peter Henry, patients there can hear the rumble but do not complain. A sound that resembles the "Hueys" used in Vietnam could be a different matter for some of them, one reason Henry tries to have emergency helicopters land some distance from the main building. Many of the patients there were also riders, he said, as is he, and are "intrigued by motorcycles." He added that there were "some veteran and police officer motorcycle groups who actually make a point of dropping by to visit our patients," especially those in the nursing home, whom they sometimes will take for rides in sidecars. On my way in I had noticed a dozen or so bikes in the parking lot, and riders talking with residents on the grass. Although I tend to view the slogan "Support the Troops" with a jaundiced eye, wondering where the support leaves off and the self-congratulation kicks in, the oasis of quiet conversation that I saw on the Fort Meade Hospital lawn did nothing but move me.

To walk the streets of Sturgis by day and especially by night is to feel yourself in the very heart of Loud America, by which I mean, America and all its contradictions played loud. The nihilistic slogans of the FTW ("Fuck the World") genre side by side with the most mawkish expressions of patriotic sentiment, that weird Johnny Cash juxtaposition of the Good Book and the devil's playground, Elvis crying in the chapel and Elvis down in the Rumpus Room—it was all turned out on those streets, and turned *up*. Even the long predawn drive I made from Sioux Falls to Rapid City, over the moon-lit prairie landscape, Lakota chanting on the radio dial a station or two over from some fundamentalist preacher ranting about the "putrefaction" of the Roman Catholic Church, was a foretaste of what I would hear in Sturgis, which I suppose was essentially what James Baldwin had heard a generation ago in New York: "the nearly invincible and despairing noise, the sound of many tongues, all struggling for dominance." We tend to think of dominance as a masculine drive, and Sturgis is certainly the image of a man's world, but the main point here is that any struggle for dominance is going to produce both noise and despair.* For all of its characteristics of pilgrimage, the Sturgis rally is even more a pageant, the polar opposite of the Islamic *hajj,* where pilgrims of every class are required to don the same nondescript garment before entering Mecca as equals. That's not how you ride into Sturgis. The trouble with a pageant, though, is that there can be only one crown. However decked out you are, however well mounted or well hung, there must always be someone with a louder bike, a bigger muscle, a woman with longer legs cocked and spread behind him. According to Teresa Larue-Forbes, who runs the women's crisis line for the city, sexual assaults during the rally are nowhere near as common as instances of domestic abuse, of women abandoned at the end of the fest with no way to get home—and I wonder if these cases have as much to do with the telegraphing of misogynistic messages ("If you can read this, the bitch must have fallen off") as with some beta-male's despair at having lost the noisy struggle for dominance yet again. On a hill above town, at one of the campground sites, is a statue visible from the street. It's not Marlon Brando playing Johnny in *The Wild One,* or Peter Fonda playing Captain America in *Easy Rider,* not Sonny Barger or Doug the Thug of Hell's Angels fame. It's Rodney Dangerfield, the man who gets no respect.

But, "the Angels shall be Kings!" That's how Hell's Angel founder Sonny Barger ends his 2000 autobiography, and though he claims to be speaking "to those who long to ride—forever free," you know which angels he's talking about. He's certainly not talking about Rodney Dangerfield. Earlier in the book he recounts making his triumphal journey to Sturgis in 1982, when "I was riding at the front of the entire pack and felt as if no power could stop us. . . . People in the towns heard the roar of our bikes way before they even saw us. The local police just looked the other way; 'Closed' signs flipped over on the merchants' windows as they locked their doors; mothers grabbed their babies from their yards and ran into their houses." When he and his companions got to Sturgis "and walked into town strapped and tall," the crowd

* What may be most lacking in the cultural history of noise is a thoroughgoing and thoroughly intelligent feminist analysis.

"opened up like the Red Sea" because "[f]ifty thousand bike riders weren't about to mess with four hundred Hell's Angels." *And the Angels shall be Kings*—even as I write, a book reviewed in the current issue of the *New Yorker* speaks of "the kingly sport of motorcycling." Alexis de Tocqueville himself couldn't light up the big American question any better than old Sonny did, and this is it: *Is America better imagined as a country without kings or as a country in which everyone wants to be king?* Is Mount Rushmore really "the shrine of American democracy," or is Caesar's Palace?

The question is not unrelated to which kind of "jamming" informs our national discourse, or what kinds of sound are appropriate in our national forests and parks. It is not unrelated to the ways in which noise overpowers—and increasingly imprisons—the weak. I suppose we should be careful how we answer, because Tom Petty was only telling the truth when he said it was "good to be king," and because, let's face it, there is something innately and historically puritanical in the wish to kill one—and yes, in the wish to silence every noise. Still, at the end of the day, an America in which everyone secretly hopes to be king can never be far from despair. Or anything besides loud.

MATO PAHA

Had my only reason for going to Sturgis been a desire to experience a bike rally, I might have saved some time and money by visiting the smaller and for me much closer one in Laconia, New Hampshire, which takes place earlier in the year. But I also went to Sturgis because of a dispute surrounding a nearby mountain called Mato Paha in the Lakota language, or as most people know it, Bear Butte. Sacred to more than thirty Native American tribes, including the Dakota, Lakota, Nakota, Northern and Southern Cheyenne, and Northern and Southern Arapaho, the volcanic pile—similar in shape to Devil's Tower, but greener, like the wooly hump of a great green buffalo—has increasingly been squeezed by development related to the rally. The mountain itself is protected state land, with some of the adjoining acreage under the ownership of the Northern Cheyenne and Rosebud Sioux. But as the scale of the Sturgis rally has grown from a local racing event (started in 1938 by a club called The Jackpine Gypsies) to a world-class festival, campgrounds, bars, and outdoor music venues have moved ever closer to its base.

The resentment came to a head in 2006 when an Arizona-based entrepreneur and bike enthusiast named Jay Allen purchased a 600-acre track on the north side of the butte and declared his intention to build an outdoor rock amphitheatre and the "world's largest biker bar," which he intended to call Sacred Ground and mark with an 80-foot statue of an Indian. Allen nixed the statue idea and changed the name to The County Line in the face of opposition, but he got the permits and the liquor license he needed and, with the reported help of a Native American construction crew, got his place built. It was in its second year of operation in 2007, the year I came.

I had been told by a member of the Bear Butte International Alliance of a "prayer gathering" that would be taking place during the rally, though my attempts to nail down details proved difficult. I arrived with no clear idea of where I was going or when I was expected. But when I took a ride out of town for a closer look at the mountain, just before I entered the gate to the park I spied in the distance what seemed like two good omens: a herd of grazing buffalo and, even more promising, a single white tepee.

I parked my car and walked up a gravel drive toward what revealed itself to be a small encampment centered on the tepee, a few standard camping tents, some cars, and perhaps half a dozen men, women, and children, most of whom looked unmistakably Indian. The tall young man with long black hair who approached me could have played the role of the warrior son of a prominent chief; his wife, whom I would meet half an hour later, was no less striking, or serious.

Philip Gullikson instantly made me welcome though he hadn't been expecting me. The idea behind the camp, which he credited to his wife, Marcella Gilbert, was a quieter witness to the sacredness of Bear Butte than some of the earlier opposition. In setting up the camp, and in continuing to call for a four-mile, bar-free buffer zone around Bear Butte (as proposed in a bill killed by the South Dakota legislature a year before), the Indians were certainly protesting noise, though for Gullikson the word denoted more than sonic pollution.

"Prayer is the only kind of noise that's positive," he told me. He named some other kinds that weren't. The biggest and most negative for him was "the intergenerational trauma" created by the dominant white culture's repeated attempts "to kill the Indian in order to save the man," a noise he compared to the traumas of war and to the traumas of those women abused and abandoned at the rally. Against these combined noises, but not altogether impervious to them, were the numinous energies of the mountain, which he claimed had been sacred to Native Americans for 30,000 years. As Gilbert would explain: "There are certain places on this planet where the energy is strong and pure. It's not everywhere, it's more concentrated in certain areas . . . and this is one of those places. So, when we mess with the vibration of that place it affects the universe, it affects everything, everybody. It could affect the city of Sturgis, it could affect the animals that live around here, it could affect the water sources, the weather."

I was invited to come back to the camp for dinner later that evening. This would allow me to meet more of the people and to hear a talk by Arvol Looking Horse, "the nineteenth-generation keeper of the sacred pipe." I arrived at the hour I'd been told but soon realized that it was going to be several hours before anything would begin. Settling down with two other men on the tinder-dry grass in the stingy shade between two rusty cars, I listened as a man called Jay Red Hawk explained that our clocks were now on "Indian time," which I took to mean that three hours, give or take, was no big deal. I wasn't particularly amused, but as the day wore on the phrase would take on a deeper significance.

After a while others began to arrive, some whom I recognized from earlier in the day, some new. A family of ranchers came with an American flag and several of the men, one shirtless and long braided, raised it on the end of a tall grey trunk of a dead pine, like a western-style Iwo Jima, with Bear Butte in the background and the buffalo grazing placidly nearby. Apparently Red Hawk's nephew had walked into an Iraqi crossfire several days earlier and been seriously wounded. One of the young women from the ranch told Red Hawk that she would stitch the man's name into the flag if he wanted. I was told that it was traditional for Native American servicemen and women to come to Bear Butte prior to reporting for duty and when they returned home.

Gradually the shadows fell and the stars came out over and around the dark shadow of the butte. Sterno cans were lit and a hearty potluck supper was laid out and served. I saw Marcella Gilbert telling several of the children that they were not to run. Someone offered a joke, but its innocence seemed to reinforce the mood rather than disturb it.* I'm at a loss to describe the unaffected reverence of the people there—beyond saying that they were *quiet*. Repeatedly throughout our conversations before, during, and after that evening, I heard the case made that Indians were merely asking for the same show of respect for their sacred places as whites wished for theirs: "Would you have strippers and heavy metal music near a church?" was a typical remark. What struck me almost as much as the justice of the argument was the naiveté of its comparison—for it seemed these people might have had a more "churchly" sense of religious decorum than many Christians do. I saw no glad-handing that night, no passive-aggressive hugs. No one suggested that White Buffalo Calf Woman, the giver of the Sacred Pipe, would have made an awesome Miss Buffalo Chip.

Everybody was silent when Arvol Looking Horse rose to speak. He had no microphone clip on his shirt. His words came out like exhalations of thick smoke, without theme or thread that I could follow, only a rope of related strands unraveling in the air, the sacredness of the Black Hills, the dangers of global warming, the need for inter-tribal respect, the importance of avoiding violence and listening to young people, the difficulty of praying over all the noise. From off in the distance, toward the town of Sturgis, and also from Route 79, which passes closer to Bear Butte and which the Indians were asking bikers to avoid, came the constant drone and roar of motorcycles, interrupted now and again by sirens and marked by endless corpuscular arteries of white and red light. It was as if we were the remnant of some sacked city and had escaped into the mountain while barbarian hordes rampaged on the burning plains below. For my hosts, of course, there was no "as if," only history. They were on Indian time. Over my shoulder I glimpsed the dark silhouette of one of the Cheyenne security people, looking out toward the noisy conflagration, while in the firelight in front of me, children slept clutching the folds of their mothers' long skirts.

Had a rock concert been going on in one of the venues near the mountain, I'd have seen it strafed with search lights and pulsing with strobes; I'd have heard hoarse words screamed above the engine drone. I was happy not to. At the same time, I couldn't help but reflect on how the noise from the highways

* How do you tell a "rez dog" from a "city dog"? Throw them both in an oven. The city dog will jump out saying "Man, that was hot!" The rez dog will say, "Good sweat!"

seemed to heighten the experience. Without the motorcycles, this would have felt like a Sunday school picnic; tonight it felt like a last stand against the world. That's not to say the Indians ought to rethink their opposition to the rally's encroachment, or that the rally is somehow doing them a big favor. I only mean to acknowledge that along with the lines we feel compelled to draw between quiet and noise, wrongs and rights—the sides we have to take, according to Graham Greene, if we are to be human—there are also mysteries, cold currents in the warm air, and noise is one of them.

Jay Allen may be another. The next day I drove to his bar, the Broken Spoke County Line, which looms like a pinewood cathedral on the plateau facing the back side of Bear Butte. Vendors of every kind filled the enormous paved lot in front of it, not unlike a cathedral fair. Over one of the hangar-sized entrances stretched a banner reading "God bless our Troops." Another equally wide entrance on an adjacent wall allowed patrons to ride their motorcycles through the building and between the two main bar counters inside. On top of one of these, a woman dressed in black and red with little horns on her head pursued a white-suited cowgirl, striking the latter's exposed buttocks with a riding crop. The patrons cheered and snapped pictures, as motorcycles rumbled in and out behind them. Loud as it was, I still jumped to hear a pistol shot fired behind me; a skit about the Wild West had just begun, ending almost as soon with a barrage of gunfire and a litter of bodies. A pierced and formidable-looking bartender, with "White Girl" written across the front of her belly shirt and a few tattoos below that, brought me my beer and promised to give my card to Jay Allen.

After further help from Nicole the bartender (herself the rider of "a stripped-down, hard-tail chopper," a bike with no suspension), I was directed by a bouncer to Allen, who was just then getting out of one of the restored 1930s pickups he had parked around the lot. He smiled affably and shook my hand. There was nothing of the biker wannabe about his appearance, which I would describe as dude-ranch Redford when his wrinkles were in their prime. The more I talked to Allen, though, the more I believed that his parents must have been reading *The Great Gatsby* when they named him Jay. He described himself as a lover of sunsets, an admirer of Native American culture, someone who was glad to have played "the villain" because "I helped bring

Native Americans together." He owned no fewer than five Broken Spoke Saloons around the country and held the same number of land-speed records for motorcycle racing. (One night I caught him on his cell phone coming home from a race and he exulted through the faulty connection, "My partner and I are on a mission from God!") If I took the most literal view of Marcella Gilbert's description of the powers emanating from Bear Butte, then I might be tempted to ask how else those energies might seize someone like Jay Allen except with the dream of building "The World's Biggest Biker Bar" on 600 acres of sacred ground, a place where he could exercise his "wonderful power" to make "insecure people" feel welcome—"and I know it's not earthshaking and there are people with far deeper meaning in their lives, but I'm 53 years old and it seems like that's where I've evolved"—and who still hoped that after a time of "healing" the "Native Americans would come there and have powwows someday. That would be the ultimate." William Carlos Williams said that "the pure products of America go crazy," and I do believe it, but there's another way to think about the line, which is that sometimes there is more purity than meets the ear in Loud America's craziness—more potential music, in other words, than a person like me might be willing to admit.

Out on the lot, where I seemed to be under some kind of surveillance by a man with a pager and a clipboard, I accosted various patrons and asked if they'd heard anything about the controversy over Bear Butte. Only one couple had, vaguely, and were sympathetic to the Indians, though that hadn't stopped them from coming to the bar. Since that time the word seems to have gotten out, and more bikers have pledged to boycott the Broken Spoke County Line and not to ride Route 79.*

---

* Three years after my visit to Sturgis, the fight over Bear Butte continues. According to Tamra Brennan, who helped to organize the prayer camp in 2007, most of the people I met there have disappeared without a trace. She also reports that more and more riders are responding to the campaign to avoid driving on Route 79 and to boycott The County Line. In addition to using the Internet to spread the word, Brennan has done street canvassing during the rally and reports that not a single rider, including a few with Hell's Angels patches, "has gotten in my face." Notwithstanding recurrent disputes over the renewal of his liquor license, Jay Allen remains at his bar. Campground-sponsored air tours over Bear Butte seem to be the next phase of the fight, though Brennan says most of these were staved off in 2009 thanks to sympathetic intervention from—get ready for this—a representative of the Federal Aviation Administration.

Looking at the dispute from a distance, I want to believe that there can be a workable compromise, that local developers like Allen and their Native American opponents need not be at odds. Allen, who is not without a sense of humor, told me a story of introducing himself to a towering Native American seatmate on a passenger flight by saying, "I guess you ought to know that I'm your 'enemy.'" Taken aback, the man replied, "Are you Crow?" They began to talk about Bear Butte; Allen says they parted friends. With a jolt I recognize how my hopefulness would sound to someone on "Indian time." She would think she was back at the first Thanksgiving, listening to an after-dinner speech. "In a country as vast as this New World"—on a mountain as grand as Bear Butte, in a place as big as Yellowstone National Park—"my friends, there ought to be enough room for us all." Red Man and White Man, snowmobiler and snowshoer, rez dog and city dog—we can all find our place.

Except that history said otherwise. A single windblown tepee hugging the base of Bear Butte says the rest. The cant that goes with noise, and with every imperial project that behaves like noise, is always about "multiple use," mutual respect, live and let live. But loud noise does not live and let live. Hitler's loudspeaker does not work like a radio station that plays all your requests. "Live and let live" in any dispute involving loud noise or big money has the same meaning as it did in the historic confrontation between Indians and whites: "I'll live, and if you do what you're told, I might let you live too."

Which brings me to the AK-47. I made one more stop before I left town. I wanted to see the famous (or infamous, depending on who you talk to) Buffalo Chip Campground, one of those establishments within earshot of Bear Butte. Some of the rally's biggest rock concerts are held there. John McCain made it a campaign stop in 2008 and suggested that his wife might compete in the Miss Buffalo Chip pageant. Just outside the campground I came upon a concession run by Chris Berg, who offers customers the opportunity to fire a machine gun or ride in a historic Russian tank.

Berg, who runs the business with his wife, is a good-natured, unassuming young man, a long way from anyone's stereotype of a "gun nut." He takes a historic interest in military weaponry and claims that many of his customers do as well. While I was deciding whether $30 was too much to spend on roughly that many seconds of automatic weapons fire, a middle-aged man who'd chosen to shoot without ear protection came back from the range and asked when the ringing in his ears would stop. Berg said in about an hour, and I did not volunteer any additional information about noise-induced hearing loss. Instead I filled out the obligatory questionnaire regarding past arrests, incarcerations, drug use, etc., and told Berg I was opting for the AK-47.

I approached the range, where attendants were waiting to assist me, and saw the targets, all of them faces on the current American enemy's list. If I had the chance to eliminate someone like Osama bin Laden, wouldn't I take it? Hadn't I applied that question, at least theoretically, to certain leaders much closer to home? And must every spontaneous impulse be subjected to scruple? And so I stepped into the firing hut, capped the headphones over the earplugs I was already wearing, uncapped them again so I could hear the instructions, shouldered my weapon, and with several jerking spasms spent the quickest thirty bucks of my life.

It was only after I'd handed my empty weapon back to the attendant, and wished Berg luck for when his National Guard unit shipped out to Iraq, that I looked up beyond the shooting range and over the thundering road and saw Mato Paha against the blue sky. Nestled in the grass of its foothills was a small encampment of people who had welcomed me, fed me, and quite possibly trusted me and whose peace I had possibly helped to disturb. I'd love to say that I did so by way of admitting my complicity with the larger noise of the rally, but the truth is, I hadn't given them a thought.

What I'd done instead is make noise for the same reasons everybody else does. Keen for a novel experience of The Loud, I'd merely embraced its most hackneyed justifications. I "needed it for work." I wanted it for fun. Plus, after all my traipsing around in the hot sun with a tape recorder and a notebook, I'd *earned* it. I felt like calling Mr. Apology right about then, but he was long dead, and besides that, I had a plane to catch.

SPRINGFIELD, USA

The first motorcycles manufactured in the United States were made in 1901 at the Indian Company in Springfield, Massachusetts, now the location of the corporate headquarters for handgun-makers Smith & Wesson. (I know about the Indian bike because my grandfather drove one in his youth, when he ran with a pack that called itself "The Knockout Club.") The state's third-largest city after Worcester and Boston, Springfield was once known as "The City of Homes," many of which still stand out today with their elaborately painted Victorian porches, though for several decades the City of Homes has been losing its manufacturing base and showing the effects. The median income in Springfield is well below the national average and considerably below that of Massachusetts. Several years ago the city made national news when a baby shower ended in a gunfight. These days I think of it mostly in connection with Herb Singleton, principal of Cross-Spectrum Labs, "sound and vibration consulting." When I first began working on this book, Singleton took me for a drive through the neighborhoods of Springfield, which he knows well. He was born there and, after graduating from MIT and working at an acoustical engineering firm in Boston, he did what Thomas Wolfe said you couldn't do and came home again.

I met Singleton for the first time at an international conference on noise in Honolulu, where he delivered one of the most interesting papers I heard during several days of interesting papers. Subtitled "A Tale of One City," it had to do with the unsuccessful crackdown on loud car-audio systems that Springfield officials had undertaken in 2006. More broadly it had to do with the need for acoustical engineers (most of his audience) to grasp "the reality . . . that perceptions are just as important as empirical data in shaping citizens' attitudes toward noise control." Most of the perceptions Singleton addressed had to do with race and class, which figure heavily in a place like Springfield, where African American, Hispanic, and Asian citizens make up more than 40 percent of the population.

Singleton himself was a minority at the noise conference. He was one of only a few presenters who based his research on the place in which he lived and on the politics of that place. He was one of only two black people I saw in several days of sessions. And he was a minority of one, a singleton indeed, in being the only presenter who dared to appear without a "PowerPoint presentation." Instead, as befitting someone talking about his hometown, he showed us slides of the city as he spoke.

Singleton's "A Tale of One City" was straightforward, insightful, and, for anyone who cares about the future of Loud America, a bit discouraging. Complaints about "boom car" noise, mostly from racially mixed but also from predominately black communities, led to several public hearings around the city. But quite soon the city leadership realized it had gotten off on the wrong foot. Some residents were understandably concerned that the crackdown would target minority drivers and be used as a pretext for shaking them down. Why, for example, was the city focusing on boom cars, typically driven by young, often minority males, and not on motorcycles, typically driven by middle-aged white males? Police in Springfield were well aware of these perceptions—so much so that they were reported to issue traffic citations only to drivers of their own color—and thus were reluctant to enforce existing noise codes. Their sense that noise was a low-priority issue, their fear of applying the "plainly audible" standard for fear of its being rejected in court as "too subjective," a lack of training in the use of sound meters, and a lingering contract dispute between the city employees' union and the city administration also contributed to police reluctance. Finally, not all of the city's seventeen neighborhoods had the same noise problems. In certain districts loud bars were considered more of a noise annoyance than stereo systems. In the most affluent suburbs, far away from the busy intersections in the city center and possessing no bars but those in finished basements, noise was scarcely perceived as a problem at all. I recalled the aristocratic MP who spoke in defense of organ-grinder noise in nineteenth-century London, confident that no such thing would ever be heard on his manor lawn.

Eventually, the city tried to broaden its noise focus, but misgivings remained, as did much of the noise. Singleton had participated in the discussions all along, but had not worked as a paid consultant for the city or any community groups. His recommendations, reiterated in his paper, included educating the public on the health effects of noise, "simpler objective methods" for measuring it, and better training for police. Above all, he stressed that while "acousticians can help communities cope with the technical aspects of noise control . . . we all need to be aware of underlying societal issues that slow or block progress."

As Singleton drove me through the city's deteriorating downtown, he remarked, "All beeper stores and nail salons." He pointed out the fried-chicken joint where, a month back, an argument over a spilled soda had led to the fatal shooting of several people. He called off the names of different districts as we crossed borders invisible to me, the Bay Area, "predominately black," where I saw my first Nation of Islam Temple; the North End, which in Singleton's childhood had been ruled by a gang called "The Latin Kings"; Pine Point, "more working-class white," where I saw the noisy bars and clubs mentioned in his paper; and Forest Hills, once a working-class neighborhood of privately owned homes, now changing to a neighborhood of rental properties and Section 8 housing. We were headed there for an interview with a longtime resident named Susan Poole.

She met us in a coffee shop, a petite, intense middle-aged woman dressed in slacks and a tweed blazer. She had lived in her current home, two houses from a busy corner store, since 1977. Contemptuous of what she called "back-porch complainers," Poole was active early on in the boom car debate. Singleton had approached her after one of the meetings, and she had invited him to her house to hear the commotion for himself. He had sat in her living room and heard the sound—and felt the vibration—of cars parked up the street at the corner store with their engines on and their sound systems throbbing. It was obvious that Poole remained grateful for his interest.

Depending on your perspective, Poole is best characterized as a community activist, a busybody, a model citizen, or a crank. She shovels the snow off her sidewalk all the way up to the corner. She also drives away kids who are hanging out on the corner. She makes a point of welcoming new people to the neighborhood but also makes it clear to them that hers is a street where residents prize quiet. She decries the absentee landlords who take Section 8 money but don't care about the behavior of their tenants, who evict the ones that cause too many police complaints, then sell out and find hassle-free properties on noisier streets. She attributes most noise problems to what she sees as a widespread "lack of respect." She tries to be respectful herself by prefacing requests to turn down car stereos by telling drivers that she likes their music if that happens to be true. She also takes photographs of drivers who refuse to turn their music down. "'You're just doing this because I'm a Puerto Rican,' they tell me." She insists this is not so. After all, she took her former paperboy to court for blasting the street with his car speakers, and he was white. "You feel like you're being held hostage," she says, "like you're in your own home or you're in your own yard and you want to get away from it, but it's your house, your home, your yard."

My take? Basically what I'd call her is an American who is financially and perhaps spiritually disinclined to adopt the classic American solution to just about any problem, the solution that "made this great country what it is today." She won't move.

Listening to Singleton as he listened to her was an exercise in watching someone practice his profession as an "acoustical consultant" in the highest sense of the term. He knew the noise was real, he knew the excuses for it were bogus, and he knew the politics were complicated, just as Poole did, so she trusted him. At the same time, he knew the nuances and was not afraid to point them out. When Poole praised the black police chief of a neighboring city for his tough stand on crime, he gently noted that the crime rate in that city remained higher than Springfield's. When her complaint that "the race card gets pulled too often"—a complaint with which Singleton concurred—led her to remark that she wished certain minority youth were more careful not to give "their culture a bad name," he remarked that "nobody expects Britney Spears to be a credit to her race."

Back in the car, I asked Singleton how it felt to be a black man in a field that remains overwhelmingly white. Before he became better known, he said, people at professional conferences would ask, "Do you work here?" It bothered him some, but "it's gotten to the point that as long as you don't accuse me of stealing something, I can deal with it." We checked out a few more of the neighborhoods, and a few markers of special meaning to Singleton, the ravine where he'd lost his sneaker as a kid, the hobby shop where he and his friends used to go for comic books ("I bet we put that man's children through college"), the apartment where he'd lived upstairs from two noisy tenants when he first moved back to town. "He drank and she ran around, and when she came home he'd beat her." He showed me his mother's house in a suburb of modest, well-kept houses called Sixteen Acres. "When I was a kid, I knew that my parents and my grandmother lived in different neighborhoods, but I didn't understand what made them different."

Now he is trying to foster that understanding.

Without taking anything away from New York City, I'd like to revise my earlier statement that if noise abatement can make it there it can make it anywhere. If Loud America finds its way around noise and everything noise announces, it will do so in places like Springfield. And it will do so through insights and personal qualities embodied in someone like Herb Singleton.

I mean, first of all, that a problem like noise will never be successfully addressed in America apart from the larger issues of race, class, and economic justice. Singleton would hasten to add that we'll also need to have a better understanding of the health effects of noise, and he's right, though that understanding will take us back to race and class and to the way that "health" gets distributed in our society.

I'm also suggesting that someone like Herb Singleton can hear the soundscape with ears that surpass the capabilities of the most sophisticated sound meter. That's because he actually *lives* somewhere. There are A-weighted decibels and then there are I-weighted decibels, which are all the more meaningful when somebody can point to the place where his sneaker disappeared twenty years ago. It's a scientific maxim that "movement creates noise": Start your car, drive down the road, you start vibrations, you create noise. But perpetual displacement also "creates noise" by unraveling the human connections that make it worth anyone's while to address noise, to be careful not to make noise, or to hear a neighbor's stirrings as anything much besides noise. Yes, it's kind of noisy here in Hither, but not to worry, because we'll soon be zipping over to Yon. Not only in Singleton's knowledge of a specific community, but in his obvious affection for it, there is something worthy of esteem. Tocqueville's despot, who doesn't mind if his subjects don't love him so long as they don't love one another, is well served when his subjects can barely get to know one another.

Finally, I saw in Herb Singleton the capacity for diversified democratic *jamming* that allows us not only to engage with people different from ourselves—as Singleton is doing with Poole, and she with him—but also to acknowledge the different parts of ourselves, as Singleton does every time he turns on the stereo system in his car. Before I got back into my own, I reminded him of his promise to let me hear his subwoofers. It so happens he likes a big bass sound, and in fact traces his career as an acoustical consultant to a childhood fascination with speakers. "So this is actually what you'd call a boom car?" I asked. It was.

We were parked away from any houses and had our windows rolled up. As I would have expected, Singleton did not blast me through the roof, and as I also would have expected, he played an eclectic mix: a passage from Stravinsky's *Firebird* followed by a taste of The Youngbloodz rapping "Damn," which didn't strike me as a very Singletonian anthem (he has since said as much) but might have stood for the bottom line in Springfield and in countless forgotten communities all around Loud and languishing America. If we don't give a damn about justice, why should they give a fuck about noise?

After a track by Lil' Jon and the Eastside Boyz, I asked Singleton if he'd be willing to let me hear one of my favorites on his system. He was happy to oblige. I think he liked what I played. "Who is this again?" he asked after a few minutes. "John Coltrane," I said, though thanks to the subwoofers my ears were mostly fixed on a musician I'd not paid enough attention to before. It was Jimmy Garrison, Coltrane's bass player, the one who'd stayed.

CHAPTER 8

Sustainability and Celebration

You must change your life.

—RAINER MARIA RILKE

Happy are the people who know the festal shout.

—PSALM 89:15

Assuming we can agree on two things—that a just and sustainable future for all humankind is worth searching for, and that finding it will require our making certain changes in the way we live—we might be able to agree on a few more.

First, we might agree that we will need to live closer together than many of us do now. We will need to become more neighborly, and in more than one sense. There will be more of us, for one thing—there are well over 6 billion of us now, about double the population of forty years ago—and there will be strong environmental incentives for reducing our need to travel long distances, for another. When we do travel over long distances, we will be traveling in company more than we travel alone, again as neighbors in closer proximity.

We will also want to preserve the arable land required to feed us—especially once we have abandoned, by choice or by necessity, the petroleum-based industrial farming model America has exported to the world*—which will

* At present it takes half a gallon of oil to produce one bushel of Midwestern hybrid corn. That doesn't count the oil it takes to move the corn around. On average, an item on the American dinner plate has traveled 1,500 miles and changed hands six times.

probably mean that even the houses of farmers will be sited closer together. In this "denser" world, "P & Q" (an abbreviation for "peace and quiet" that I first heard from an Irish acoustical consultant) will be almost as hard to take for granted as potable water.

Second, we might agree that we will need to live with people who are different from us. Living closer together in a pluralistic world of porous borders, and possibly of fewer habitable zones, will not mean living in homogenous groups. Density will include diversity, which is the best thing going for density. But density combined with diversity rarely exists without tensions.

Third, we might agree that we will need to find alternative sources of energy. We will need to diversify the ways that we *power* our civilization or else we will need to watch as its cities are flooded by the sea and blasted by natural disasters. We have seen the future, or one plausible version of it anyway, and its name is Katrina.

Fourth, we will need to learn how to live with less—not less love or art or adventure, but certainly less in the way of consumption and reproductive opportunity. Otherwise even impressive environmental accomplishments will amount to a zero-sum game. The history of noise abatement provides plenty of analogies: Cars, planes, and trucks are all quieter than their predecessors of thirty years ago, but there are many more of them now, so noise levels continue to rise. Ditto for carbon emissions. If the people of China and India drove nothing but hybrid cars, at the current U.S. rate of 0.8 cars and trucks per person there would be a billion hybrids in each country and more than 400 cars per square mile. The same scenario with cars powered entirely by gasoline is unthinkable.

For the whole world to achieve an American standard of living the whole world must become hell. For the whole world to achieve even a modest standard of living, the people of the developed world must restrict their wants enough to create an environmental margin sufficient to accommodate that added development. As many have noted, "we" can hardly tell "them" to put a freeze on their development because we're quite satisfied with things as they stand—not when the average Chinese produces 2.7 tons of carbon a year to the average American's 20, not when a family in Tanzania takes a year to consume the same amount of fossil fuel as an American family uses between midnight on New Year's Eve and sundown on the second day of January.* Bill McKibben puts the matter nicely when he suggests that while everyone in a sustainable (he prefers to say "durable") world might not be able to own a refrigerator, every village in the world ought to have a refrigerator in which to store medicines.

Fifth, we will need to take stock of the basic assumptions, the big ideas, the loud and grandiose ways of thinking that have brought us to our present impasse. To do otherwise is to risk curing our sickness with our disease—calling it by another name, of course, prefacing it with the word *green,* no doubt—but dying from it just the same.†

I can think of several ways to summarize what I've said so far. But I can think of only one that doesn't involve some abstract word (*responsible, sustainable, intentional*), only one that gives the slightest nod of acknowledgment to our sensuous natures, only one that comes close to giving a specific answer to the question most of us tend to ask in the face of any awesome challenge: "What are we to do?"

We must learn to live more quietly.

I would be curious to know your first reaction to such a statement. I would be curious to know what your imagination sees and hears. My hunch is that what many people and certainly most Americans would see and hear is a murmuring dystopia of repressed, shabby-looking people, bored to tears and boring each other to death—a dismal soundscape interrupted now and then by a squabble in a ration line and the ominous tolling of a rusty tower clock.

What I hear instead is a world bustling with vitality. I hear tools, animals, conversations, song—festive outbursts of the "collective joy" that Barbara Ehrenreich chronicles in her recent book *Dancing in the Streets,* "our distinctively human heritage as creatures who can generate their own ecstatic pleasures out

---

* More of the same: The average American uses 6 times as much energy as the average Mexican, 38 times as much as the average Indian, and 531 times as much as the average Ethiopian. I am hardly the first to note the irony that if global temperatures continue to rise the Ethiopian will die first, either of drought or of drowning.

† Wendell Berry summarizes the effect of bringing "Big Ideas, Big Money, and Big Technology into small rural communities" as follows: "The result is that problems correctable on a small scale are replaced by large-scale problems for which there are no large-scale corrections. Meanwhile, the large-scale enterprise has reduced or destroyed the possibility of small-scale corrections."

of music, color, feasting, and dance." I hear silence, too, but not a deadly one: the silence that precedes an exchange of wedding vows as opposed to the silence that precedes a volley from a firing squad.

NOISE AND CARBON

The history of noise recounted in this book is in many ways an implicit history of fossil fuels. In 1585, when Shakespeare would have been 21 years old, the city of London burned 24,000 tons of coal. The energy it generated was equivalent to what the United Kingdom currently consumes in about half an hour. In and around these mounting heaps of coal is an increasing quantity of noise. If the results of one National Institute of Occupational Safety and Health study can be taken as normative, 90 percent of coal miners will suffer hearing loss by their 50th birthday. Coal plant operators, along with their counterparts in coal-fueled power plants, are also exposed to unhealthy noise levels.

Needless to say, coal is not the only source of energy, carbon emissions, or noise. Petroleum-based transportation, manufacture, and agriculture have all grown dirtier and louder in the aggregate, even in the face of amazing technical strides in the design of individual machines. That is because we have continued multiplying the machines and their applications. The majority of our industrialized farms now work—and sound—like factories. U.S. corn yields grew by 346 percent between 1910 and 1983; agricultural energy consumption during the same period grew by 810 percent. Much of that energy came from petroleum, the steady supply of which seems to depend more and more on the deafening—and hardly carbon-neutral—machinery of war. Noise and carbon are married for life, for better and for worse, which means not merely for worse. Planting trees along highways and building more energy-efficient homes reduces CO_2 emissions while yielding quieter highways and homes.

The connection between harmful amounts of carbon and noise is more complex than we might at first think. When, for example, we start losing our heads to the notion of a "quieter, cleaner electronic age," someone living next to a datacom substation says "I can't sleep."* There are good reasons he can't. Denser packing of the electronics in these facilities leads to increased heat and a more aggressive use of HVAC equipment—and fossil fuel. To be wired is to be plugged in, and those little black holes lead somewhere, not infrequently to a seam of coal.

But noise does more than herald these connections. Noise not only accompanies carbon emissions; noise is an indirect cause of carbon emissions. Les Blomberg often speaks of noise as the hidden cause (and not so hidden result) of sprawl. The suburbs and the exurbs, along with the vast transportation networks that make them possible, are at least partly the result of a yearning for quiet. As much as it is about a bigger house or a more fulfilling occupation, the so-called American dream is about the hope of leaving Hell's Kitchen for Heaven's Gate, of "getting out" and "getting over" to some promised land where having to drive everywhere is the price you pay for being able to enjoy P & Q somewhere. And many of us are glad to pay it. About 45 million Americans have a strong preference for a quieter environment; the U.S. Census Bureau reveals 4.4 million households wishing, if not always able, to move because of noise. A recent survey of national attitudes in Great Britain revealed that more prospective home-buyers were concerned about having noisy neighbors than about living on a flood plain—a fact worth the attention of any climate-change prophet who hopes to effect a change of heart with talk of rising seas.

With this in mind, we might ask if a more effective way of communicating the idea behind a "smaller carbon footprint" is with the auditory metaphor of a lighter footstep. This is not merely "a matter of semantics"; it is potentially a matter of action. Tell someone to leave a smaller footprint, and you're figuratively telling her to lose a whole lot of body mass or a few toes. You're pointing to an anatomical measurement that none of us can help. "I'm a size eleven. You have a problem with that?" But ask someone to tread more softly and

---

* From a resident of St. Paul, Minnesota: "I've been trying to get Xcel Energy to stop a noise from a substation that's been going strong for 9 months. I've had county officials tell me 'you're screwed.' . . . Xcel tell[s] me these things take a long time . . . a neighbor tell[s] me it doesn't matter because the world is going to end in 2012, substation workers stare me down. I've become well-versed in white noise machine technology in these months. I'm tired and starting to lose it."

you're saying that with care and forethought there is something she can do. Not only that, but you've gone beyond a mere figure of speech, because some of what she can do literally comes down to *making less noise.*

At this stage it hardly needs saying that the consequences of doing nothing, or simply too little, about climate change promise to be disastrous.* To avoid the worst of those consequences George Monbiot, the author of *Heat: How to Stop the World from Burning,* estimates that the rich nations of the world must reduce their total carbon emissions by 90 percent by 2030.

With an agenda of that magnitude, noise might seem scarcely worth the bother. Then again, it wasn't that long ago that governments were saying the same thing about climate change. In practical terms, they're still saying it. But if the relationship between noise and carbon can be firmly established in the public imagination, we will have a quantity of sound bytes that no procrastinating political leadership can ever hope to match. Every noise amounts to a wake-up call.

Those already awake seem to fall roughly into one of two groups, and noise has some relevance here as well. The first group views climate change primarily as a problem of chemistry. We have all this carbon and it's gotten out of hand. The other group sees climate change as symptomatic of something much more profound: It's the civilization that's out of hand.

You can think of the difference as two assessments of a failing circus. One view holds that the circus is in trouble because of the elephants. They take up a lot of room, these elephants; they eat a lot of food, and they leave a lot of unpleasant waste behind them. This is too bad, because people are fond of elephants and elephants also do a lot of useful work around the camp, but it's pretty clear that we need to find a way around the elephants. In the "entrepreneurial" version of this view, we put a few rubes in elephant suits and have *them* put up the tents. We call it "alternative energy." We give each of them a candy apple and a free pass to see the two-headed calf. We profess the confi-

---

* A conference of UK scientists concurred that an increase of global temperature by 1.5 degrees over preindustrial levels would expose 400 million people to water shortages, 5 million to hunger, and would kill 18 percent of the world's species. Apparently the consensus is that 2 degrees is the critical threshold; beyond that we're in big trouble. We've already gone .6 degrees toward the tipping point.

dence that with a little luck, a little razzle-dazzle, and not a little haste getting out of town, we'll be flush by the time we reach Memphis.

The other view asks what an elephant is doing in Memphis. It asks whether the whole concept of "freaks" and "rubes," chains and cattle prods, a few high-fliers making their connections over our heads, and a ringmaster forever cracking his whip isn't a little nuts to begin with.

It would be foolish to suggest that one of these views overrules the other, just as it would be foolish to suggest that the problems of noise can be solved entirely by technical or political means. Nevertheless, I think that a thoughtful consideration of noise will tend to steer us toward the more radical appraisal. Whenever I have tried in this book to trace noise to ultimate causes, I have usually wound up at one of two destinations: *a denial of the body,* which manifests itself as a desire to abolish the physical limitations of time and space through speed; or, a denial of our equality with other people, *a contempt for "the weak."* In essence, they come down to the same thing. If you wish to exceed the limitations of time and space, to go faster, to be closer to invincible, to *make some truly awesome earth-shaking noise,* what's the most time-honored way of achieving the goal? The people who made the first great noise on the plains of Mesopotamia could have told you the answer in a New York minute. You get slaves. You go to war, and you bring back captives who will power your civilization. Animals help too, but human beings, though slightly deficient in horsepower, are much higher tech.

To escape the limitations of their bodies, our civilized ancestors used other people's bodies. And it was a noisy proposition from the get-go, with all those swords, whips, chains. Eventually they made machines, and the machines made noise too, though the people who heard most of it ran the machines. As the machines became more efficient, more "intelligent" even, the status of the people who ran them became one of increasingly greater subservience, to the point that we have virtually replaced the old feudal pyramid of kings, knights, and serfs with one of kings, machines, and machine-minders in that descending order. (The yeomen and their troublesome unions we have taken care of by outsourcing.) If this sounds a little overblown to you, it's been a while since you've been to a fast-food restaurant, or since you've stopped for gas at a typical convenience store and watched as a young woman or man was beeped, buzzed, and blipped at like some Information Age Step-n-Fetch-It by the register, the gas pump, the deposit safe, the door ringer, the lottery-ticket machine, the manager's voice on the speaker phone. Perhaps you've never asked yourself which comes closer to feeling like a daughter or a son to you, that harried clerk or your laptop? Which "voice" is dearer to your ears: his, hers, or its?

Those who fail to see much relevance here to the problem of global climate change probably tend to see the latter as a matter of chemistry. I tend to see it as a matter of noise, of that wish for "disembodiment" and that contempt for the weak that are the roots of noise. That said, it's possible that the chemical view, in addition to being the more modest, will prove the more humane. Referring to Leo Tolstoy's utopian-agrarian asceticism, Anton Chekhov wrote to a friend in 1894: "Prudence and justice tell me that there is more love for mankind in electricity and steam than in chastity and abstention from meat." I love that line, though I do wonder if Chekhov would have held the same view in our noisier, more imperiled world, and if he did, whether he would have added jet fuel to the electricity and the steam.

NOISE, CARBON, AND AIR TRAVEL

In 1946 the U.S. Supreme Court settled what is sometimes referred to as the "chicken suicide case" but is more formally known as *United States v. Causby et ux.* Thomas Causby was a chicken farmer whose 2.8-acre farm was close to a small airport in Greensboro, North Carolina. This does not seem to have bothered Causby (i.e., "He never complained"), but in 1942, large four-engine bombers began using the airstrip, some of them flying in as low as seventy feet above Causby's farm. The terrified chickens flew into the walls of their coops and within two weeks 150 of them were dead. Though the jets never came onto Causby's property, and though Causby could not claim to own the airspace over it, the Supreme Court ruled seven to one that the government had effectively appropriated Causby's land in violation of his Fifth and Fourteenth Amendment rights and awarded him $2,000.

Two updates seem in order. In the years since 1946, chickens, at least those raised on factory farms, have stopped flying; human beings, at least those who can afford to eat free-range chicken, have been flying a lot. The average American family in 1991 flew twenty-five times farther than its counterpart in 1951. Though I came of age during that era, and have flown close to twenty times in my life, I have never gotten over the idea that I live in an era when upright featherless bipeds believe it is their God-given right, or should I say Darwinian prerogative, *to fly*. Wrap your head around that, and something as unthinkable as a melting polar ice cap seems pedestrian. Then try wrapping your head around the pejorative meaning we've given to the word *pedestrian*.

The relationship between noise, carbon, and air travel is probably here to stay, notwithstanding some remarkable improvements. Two generations of U.S. aircraft have been retired, some with miles left to go on their clocks, due mainly to problems of noise. A new generation of "quieter, greener" planes is already in development.* In the meantime, aircraft noise continues to affect millions of people† even as flying remains the most polluting form of transportation per passenger mile and the fastest-growing source of greenhouse gas emissions. Its contribution could grow as much as 200 percent in the next twenty years.

Given what's at stake for our planet and our posterity, some have suggested a radical rethinking—at the very least, a rationing—of the "right" to fly. (The rethinking might begin with something so basic as the fact that as many as half of all air flights are for trips of less than 500 miles.) Some of the most insistent voices seem to be in Great Britain, where anti-aviation groups like HACAN‡ Clear Skies and Plane Stupid have held demonstrations at major

* The "Silent Aircraft Initiative," a study undertaken by the Cambridge-MIT Institute and funded by the UK Treasury and Rolls-Royce, Boeing, and Marshall Aerospace, maintains a website at www.silentaircraft.org.

† Figures for the number of people in the United States affected by aircraft noise vary depending on definitions and metrics—even within agencies of the federal government. The FAA (which works with a day-night average of 65 decibels) claims that over the last thirty-five years it has reduced the number of those affected by aircraft noise by 90 percent to around half a million people. Yet, a recent report co-authored by the Department of Transportation (of which the FAA is a part) says that those affected by "serious aircraft noise" was 24 million in 2000 and will be over 30 million in 2025.

‡ Heathrow Association for the Control of Aircraft Noise. HACAN merged with Clear Skies in 1999.

airports (their members were legally barred from public trains prior to one scheduled action in 2007), and the Anglican Bishop of London has called flying "a sin." George Monbiot, who offers some compelling reasons why planting trees might not be the "offset" we think it is,* has gone so far as to say, "If you fly, you destroy other people's lives."

For me to say amen to this would be adding the sin of hypocrisy to the crime of murder. What I will suggest, and firmly believe, is that our exercise of "the freedom of the skies," especially in the United States, will be our single most accurate and audible indicator of how seriously we regard the threat of climate change. No less important, it will be the single most accurate predictor of how seriously the nonflying public regards it.

As long as air travel is viewed as sacrosanct, many people "on the ground" will see the issue of climate change as what, in fact, it has largely been: a bonanza for companies like Enron and copy for writers like me. In contrast, imagine the effect of a single conference of "leading climate-change scientists" adopting so "dotty" a resolution as that of the Bishop of London. The Duke of Windsor gave up a crown for the woman he loved and people are still moved by his renunciation. What might happen were our dukes and duchesses to give up flying for the planet they loved?

In regard to noise, I think it is impossible to overestimate the psychological effect that a quieter sky would have on the public imagination. The only thing more compelling is the disaster that says, "It's too late." The crisis on most of our minds most of the time is not climate change; it's mortal change. Yes, the globe is warming, but I'm not, and I'd like to see San Francisco once more before I'm on ice. But imagine a day when no planes flew, not because of a terrorist attack the day before but out of respect for the days to come. Aside from any dividends it might pay in terms of carbon reduction, the cessation of noise in a place where we have grown used to hearing it—and where human beings have looked for answers since the beginning of time—might go a long way toward leveling our heads.

* In addition to noting that "a tonne of carbon saved today is far more valuable, in terms of preventing climate change, than a tonne of carbon saved in twenty years' time" (i.e., when a planted tree will mature), Monbiot points out that trees can be felled or destroyed by fire, whereas the effects of the carbon emissions they were supposed to offset remain.

NOISE AND WIND POWER

Another ancient place for seeking answers is in the wind, which in addition to providing a viable source of alternative energy is likely to spark some of our most contentious noise disputes in the years to come. Future debates over air travel are likely to revolve around carbon, not noise, but in debates over wind energy, noise will be front and center.

An analysis of the merits of wind energy is beyond the scope of this book or the expertise of this author. My discussion here rests on the assumption that wind power is a valuable source of renewable energy that we cannot afford to do without. As Charles Komanoff wrote several years ago in an essay for *Orion* magazine: "In the new, ineluctable struggle to rescue the climate from fossil fuels, efficiency and 'renewables' (solar and biomass as well as wind) must all be pushed to the max. Those thirty megawatts that lie untapped in the kitchens and TV rooms of Adirondack houses are no longer an alternative to the Barton [New York] windfarm—they're another necessity."

That said, it has become increasingly clear that wind turbines produce a devilishly complex form of noise that, combined with the imprudent siting of certain wind installations, is making some people sick. Of course, we can say that these people and their ailments are trifles compared to the bigger issues that face us. In other words, we can take our cue from those who say that three-legged frogs are a trifle compared to the needs of a chemical-plant worker with three kids to support. On the face of it, both statements feel right to me, but it's largely due to such feelings that we're in the fix we're in.

It has also become increasingly clear that the noise effects of wind turbines have routinely been denied by ignorant or unscrupulous developers. In league with them are any number of disingenuous politicians hoping to avoid the hard work and political fallout of a sustainable energy policy by certifying their pale-green credentials with a few visually imposing monuments. The irony is that the tone, methods, and motivation of their denials are uncannily similar to the denial of global warming itself. If only we could find a way to generate renewable energy by lying, we could electrify our ice caps and keep them too.

Nevertheless, it is probably too easy to attribute all of the denials of wind tower noise to duplicity. As Dutch wind-researcher Frits van den Berg puts it: "What had surprised me from early on was that people in the wind power business seemed to know so little about their raw material, the wind." And if they knew so little, those who trusted them often knew less when they endorsed their projects. The testimony of a Mr. Bellamy in Caithness, Scotland, is not unique. "I'm not the moaning type and I have no problem with the look of the windmills," he says. "I'm not anti–wind farm." But Mr. Bellamy is pro-sleep. "The problem is particularly bad at night when I try to get to sleep and there's a strong wind coming from the direction of the turbines. They just keep droning on. It's a wooh wooh type of sound, a ghostly sort of noise. It's like torture and would drive anyone mad." Other descriptions of the noise from other sites include "a cement mixer in the sky," "sneakers in a dryer," "a giant baby's heartbeat during ultrasound," "a train that's coming but never arrives." Some who have visited wind farms and stood close under the towers have said that they don't hear much of anything beyond the gentlest rhythmic swish. On several occasions, and in several countries, I've said the same thing.

That points to the first of many factors that make wind tower noise so perplexing: Some of it may not be heard. That is to say, some of it may be of such a low frequency as to be inaudible to the human ear, but not undetectable—or "unwanted"—by the human body. The effects of "infrasound" and "whole-body vibration" are still under investigation, but researchers in Portugal, Germany, and elsewhere have done studies of "vibroacoustic disease," which includes such symptoms as abdominal pain and chest pain, lump in the throat, and a general feeling of discomfort that can include "eyeball resonance," the "urge to urinate," and changes in breathing pattern. Anyone who finds all this a little dubious should be aware that military interrogators have known for some time that "a combination of persistent low frequency noise, infrasound, and visual strobing [also an effect of turbine blade shadows on the interior surfaces of homes] can destabilize the human body," whose various organs emit their own sound frequencies.

The "cocktail effect"* of audible sound, infrasound, and visual strobing from wind towers has been noted by rural physicians on both sides of the At-

* Not to be confused with the "cocktail party effect," the ability of a listener to direct his or her attention to a particular sound—the voice of one's spouse in a crowded room, for example—while not regarding the other voices or what they are saying.

lantic. Not surprisingly their anecdotal observations are often dismissed. As Dr. Nina Pierpont, who practices medicine in upper New York State, says: "A typical approach to wind turbine disturbance complaints, world-wide, is *noise first, symptoms second*: if an acoustician can demonstrate with noise measurements that there is no noise considered significant in a setting, then the symptoms experienced by people in that setting can be, and frequently are, dismissed."

That is not to say that acousticians and psycho-acousticians can't help us to understand why wind tower noise (objective and subjective factors combined) can sometimes seem so out of proportion to measurable wind tower sound. Among these factors, Les Blomberg notes the following:

- Wind turbines are overhead sources. Overhead sources are difficult or impossible to block with barriers, and they enter houses both from above and the sides. . . .
- Wind turbine noise often is more prominent in the evening and nighttime. Typical noises tend to correlate better with times of the day when most people are working. Wind turbine noise often is not masked by wind due to wind gradients.
- Wind turbine noise is unpredictable. People cannot know ahead of time when the noise will be present, so that they can plan around the noise.
- Wind turbine noise is not reciprocal. Typical rural noises have no impact on wind turbines, but wind turbines impact rural life.
- Wind turbine noise is unique and unusual in a rural environment. There is nothing equivalent to it.
- Wind turbine noise is not constant. It has a time-varying component that various people have described as beating, swishing, or thumping. . . .
- In rural areas, wind turbines are audible at a greater distance than almost every other rural noise source.

Additional subjective factors have been noted by Swedish researchers Eja Pedersen and Kerstin Persson Waye. Not surprisingly they've observed that those "who think of wind turbines as ugly are more likely to appraise them as not belonging to the landscape and therefore feel annoyed, also by the noise." More disturbing is Pedersen and Waye's suggestion that the noise can become a "hindrance to psycho-physiological restoration"— in other words, an obstacle to recouping from daily stress in those domestic spaces where most of us go to recoup. Among those she surveyed, Pedersen noted: "For some informants, the intrusion went further into the most private domain, into themselves, creating a feeling of violation that was expressed as anger, uneasiness, and tiredness." Those feelings would almost surely be enhanced in the absence of the "fair process effect"—that is, when neighbors of wind farms feel they have, to quote one noise consultant's assessment of people living near a wind farm in Mars Hill, Maine, "been screwed."

MARS HILL

I went to visit a handful of the Mars Hill people on a blustery day in December, driving due north through miles and miles of moose country on a desolate stretch of Maine interstate. Mars Hill is the name of both a town (population around 1,600) and a hill, the latter just 21 feet shy of counting as a mountain and supposedly the first point of land in the United States to receive the sun at certain times of the year. "The mountain belongs to everybody. It's our namesake," explained Wendy Todd, who grew up on a farm here and returned with her husband to build a house on a section of her parents' land. "It's privately owned land," she added, "but it's our landmark. When you move to Mars Hill, it's part of the package."

The package now includes twenty-eight industrial-sized wind turbines, each roughly 370 feet high from base to blade tip, one of which looms over the Todds' home. The other towers are sited 800 to 3,200 feet from the most heavily impacted households. This is a much smaller distance than the mile minimums recommended by the UK Noise Association and the French Academy of Medicine.* When I arrived, the turbines were moving briskly in the gusty wind but were not audible above it.

---

* Many, including George Monbiot, feel that siting wind turbines offshore remains the best long-range plan. There are challenges involved in offshore construction, but with noise impacts out of the way, offshore turbines can run faster and generate more electricity.

The half-dozen people I met with included a farmer, two teachers, and a retired carpenter. All reported being affected by the noise, which they claimed exceeded the limits stipulated in the permit application and certainly exceeded what they'd been told by town officials and representatives of the wind company. "We weren't supposed to hear anything," said Ella Boyd, whose family has farmed on Mars Hill for close to a hundred years and who no longer enjoys working in her garden on windy days. "I believed it because I didn't think our town government would lie to us." After examining the permit application that was submitted by the town and the developer to the Maine Department of Environmental Protection, the affected neighbors discovered acoustical data *predicting the noise they'd been told they wouldn't hear.*

Of course, the noise issue is complicated by the process that brought the noise. I asked the group directly if they thought their irritation with the process might be contributing to their subjective experience of the sound. Wendy Todd, who had been excited about the proposed wind project after she and her husband visited a wind farm on Prince Edward Island, said yes. "I think anger and rage were added when we found out the truth. They knew all along that people on the north end of the mountain were going to have problems, and they chose to ignore the reports."

Her neighbor Kevin Jackins, a high school teacher, was less inclined to credit the negative side of the fair process effect. "Noise is noise to me. When I have to run a fan four feet from my head in order to sleep at night because I'm being bombarded by the pulsating noise—these things are supposed to save electricity, and they're costing me more. Kids are pretty objective, and when kids get woken up in the middle of the night because they can't sleep, and they come and cry, they don't understand the politics and all this stuff. All they know is that they can't sleep at night."

His neighbor and fellow teacher Dorothy Miles, who reported the most acute symptoms among those at the table—headaches, rashes, depression—insists that the noise is the cause even though she does not consider herself particularly noise sensitive. One of her children used to have a garage band that played "heavy-metal, head-banger music" and practiced at her house, but "the boys were respectful." Her husband Peter added, "It was from seven to nine, and then 'everybody outta here, because I'm going to bed.'" Now, he says, even when it's quiet and the turbines don't turn, "you're always waiting for that other shoe to drop."

As we know, uncertainty can be a major subjective factor in the experience of noise. And plenty of objective factors were here along with it; indeed they'd been assigned estimated measurements in the permit. Still, the more I listened, the more I found myself coming back to the "fair process effect" and wondering how much it might be exacerbating some of Dorothy's reported symptoms. After retiring, she and her husband had moved to Mars Hill to build the house they'd saved for all their lives; now they weren't sure they could sell it if they wanted to. "The wind towers were visually repulsive to people down by Sugar Loaf Mountain," she said. "That's where the tourists live. They wouldn't put them down on the Cape because they didn't want Walter Cronkite and Ted Kennedy to have to look at these things." Where Dorothy wanted justice, her neighbor Wendy wanted truth. "If it's acceptable to do this to people, tell them. Tell them, I'm sorry, but this is the way it's going to be. We're doing this for the greater good, and it's going to screw you over. Wouldn't that be better?"

WINTER BEANS

I heard almost the same comment in the Netherlands, when my wife and I took a journey with Frits van den Berg, formerly the director of the "science shop" at Groningen University and now a consultant with the municipal health service in Amsterdam and director of the Wind Farm Perception Project. "If the government would say, okay, we know all the facts, but still we think it's very important to go on with wind energy, then okay, that's a political decision and I'm not going to fight it. But as long as they are deceiving people, I'll fight it."

Van den Berg's 2006 doctoral dissertation, "The Sound of High Winds," addressed a puzzle experienced by people in Mars Hill and seemingly "unknown" to wind developers: why turbines sound different at night, especially on "calm" summer nights. By taking exhaustive measurements near the Rhede Wind Farm, located in Germany near its border with the Netherlands, and by applying the principles of atmospheric physics to what had been contemptuously dismissed as "a new Dutch disease," van den Berg was able to offer what he seems to regard as a fairly modest explanation. (He takes more pride in his related findings on how atmospheric conditions affect microphone noise.) His analysis is far more detailed than can be summarized here, but basically it comes down to the simple fact that "due to strong winds at greater heights coupled with very light winds at ground level, the turbines can be a lot noisier in a night time atmosphere than they are in daytime." If you're asking, "How could anyone not know this?" you're asking one of van den Berg's questions.

He offers several answers. One is that newer turbines, while quieter than their prototypes, "now reach higher into less familiar parts of the atmosphere." Another is that as remote locations become more scarce, wind farms are being sited closer to human habitations. But his study also takes account of political factors, including the "positive ring of the term 'sustainability'" and how it is used as a jamming device to drown out reasonable objections. Not least of all he notes the tendency of acoustical consultants to avoid "biting the hand that feeds"—in this case, the hand of an energy developer.

For about the first ten minutes of a meeting he arranged for me with several wind-farm opponents in the polder region of Holland, where iconic Dutch windmills of previous centuries still dot the landscape, I believed that I too might have made a new discovery. At least it was new to the United States as far as I knew. It had to do with winter beans. As nearly as I could tell, either the noise of wind turbines was killing the beans or else the bean plants were an ingenious Dutch strategy for mitigating the noise. It seemed the beans were being planted everywhere across the landscape. After scribbling a number of notes, but before giving too much thought to how I might smuggle some winter bean seeds past Dutch customs officials, I realized what my hosts were talking about. A winter bean is a wind turbine, pronounced with a heavy Dutch accent.

One of the people at the table was an Amsterdam newspaper journalist named Ina Vonk. The farmer next door to her had recently planted a 200-meter "winter bean" in his back field and she was not happy with it. The smaller tower it replaced had apparently not disturbed her. She showed me a photograph of her little house with the wind tower looming monstrously over it, like a frame from a science fiction movie or the work of a Dadaist photographer. "I cannot be a grandmother in my own house," she said, explaining that she does not let the children come because of the noise. "I can't put my garden in, because it goes in your body, the vibration. The health department tells me to put rubbers in my mattress"—a suggestion she compares to imposing curfews on women in order to prevent sexual assault. "I don't want to live in a prison in my own house."

Vonk invited us to stop by her house later on, but she was not at home when we arrived. The tower was every bit as large as it looked in the photograph, but seemed no closer to her house than to that of the *windboer* (her term, which literally means "wind farmer," but seemed to be pejorative). Knocking at the farmhouse door, we met Bernard Schuijt, the *windboer*, who in the United States might have been sized up as a ski instructor, or, after a few minutes of conversation, as the owner of a ski resort. Van den Berg would later say that in the Netherlands "he is what we call 'the healthy farmer,'" an entrepreneurial type who "doesn't see that others can't live the way he does."

Indeed Schuijt had little sympathy for his neighbor. According to him, she had refused both his fair-market offer to buy her house and other offers of financial partnership in the turbine. He also saw her as exemplifying "the core of the problem: we have people who've made money in the city and go to the country to be quiet. 'Shit, what's happening?' they say. 'Shit, they're working! It should be quiet! *I'm* here now!' Go back to Amsterdam and take out your decibel meter." Van den Berg gave what some would regard as a typically Dutch response: "She has her view of what should go on. You have yours." Before Schuijt could say the words, van den Berg added wryly, "But you belong here." Schuijt did not hesitate to reply, "I belong here."

Even if his great-grandfather hadn't built his farm, I'd have been inclined to agree with him. Although his characterization of wind tower noise as proper to a working agricultural landscape struck me as disingenuous—farmers are not generally known for operating threshing machines at 3 in the morning from 400 feet in the air—I could not put the *windboer* in the same category as his counterpart in the United States. He was at least living in the shadow of his own wind tower and in the range of his own noise. His neighbor was, too. She was perfectly justified in asking whether he had a right to get rich "on my back, " an essential question in any discussion of noise and property rights; however, the question needs to be asked not only of this one farmer but also of those who buy his energy and the other crops (corn and tulips) that grow on his land. Like many noise disputes, the problem with this one is not that it's a mountain made out of a molehill but rather that it's a molehill obscuring a mountain: the ways we "make use" of one another and of those resources that constitute our common birthright.

Schuijt would not let me use my tape recorder during our interview but he cheerfully invited us to climb the steps at the tower's base and enter its amazingly hot control room. "They are pretty impressive machines really," van den Berg would say as we traveled on, "and I think they are part of the solution." But without significant changes in patterns of energy consumption, he tended to agree with the English Queen's Councilor who'd called them "window dressing." Van den Berg drove the car we'd rented but does not own one himself, relying on bicycle and train to travel between his job in Amsterdam and his home in Friesland. He finds it "astounding" that in 2004, two years before his dissertation was published, all the wind-generated energy in the Netherlands for the entire year was equal to two months' *growth* in Dutch energy consumption. He finds the same patterns worldwide, which in his view means that the contribution of wind power will soon become "negligible" in terms of reducing greenhouse gases.

Whether or not van den Berg is correct in his assessment, one thing seems certain: The growth in noise will not be negligible, especially if the towers continue to be sited close to people's homes. That is likely to happen so long as acoustical expertise, corporate money, and government coercion are arrayed against scattered and internally divided rural communities, with the ultimate effect of making what ought to have been the preeminent symbol of a cleaner, quieter civilization into a symbol of noise and its potential to overpower. "I want people to fight on an equal basis," van den Berg said. I doubt that a sustainable world will be possible until they do.

NOISE AND NIMBY

In order to explain what I mean by that last statement I will need to address that epithet of scorn used both by and against environmentalists: NIMBY, referring to any person who says, or who can be caricatured as saying, "not in my backyard." Some of the people mentioned in this chapter might be construed as NIMBYs, if for no other reason than the fact that the noise sources they hate are almost literally in their backyards. And some of them would readily admit, as I do, that in many ways a NIMBY deserves to be scorned.

Wendy Todd of Mars Hill said as much, in a tone I could only describe as penitential. When she'd first heard that wind towers were coming to her mountain, she liked the idea. She had heard rumors of noise, yes, but someone told her it would only affect people six miles from the towers. "So I thought, I'm not six miles. I don't care. It doesn't affect me. How many times have I thought, you know, there I was, one of the ones who said as long as it doesn't affect me, let her go. And now it affects me."

It's easy to think of the *N* in NIMBY as also standing for "noise." Perhaps more than anything else, noise is what the NIMBY does *not* want in his or her backyard. Essentially I'm a NIMBY every time I turn on a light switch in my quiet country home. Let someone else live under the humming power lines, the thumping wind turbine, or the decapitated West Virginia mountaintop; just don't make that sort of racket around here.

Even when noise is not the primary concern, it often emerges as the NIMBY's best weapon. My hunch is that there are at least as many visual objections to wind turbines as auditory ones, for example, but since no link has been established between even the ugliest art and sleep deprivation or elevated blood pressure, the decibel meter is usually the first gun pulled.

Nevertheless, and while not denying any of the hard truths I've stated, I would make the counterintuitive argument that our contemptuous use of the term *NIMBY* amounts to the Achilles' heel of the environmental movement—a failure of imagination that may actually be at the root of our failure to address climate change in any serious way.

I base the argument on three premises. The first is that everyone needs to have a backyard or its closest equivalent. As a matter of basic distributive justice, everyone needs a part of earth that they care for and that gives them pleasure. How this looks in a village, a suburb, or a city is going to vary, but everybody on earth needs to be able to point to an accessible patch of the natural world and say "This is mine." When Jane Jacobs determined that Robert Moses wasn't going to run a highway through Washington Square Park in Greenwich Village, what was she saying if not "Not in my backyard"? Is there anyone who wishes she hadn't?

Second, people need to love their backyards. They need to love their backyards with the same particularity as they love their own children—not to the total exclusion of other children, which would ultimately hurt their own children, but with the passion and partiality that are of the nature of love. My quibble with the most common form of "tree-hugging" is that it rarely involves hugging the same tree. In some cases it doesn't involve hugging any tree. Knowing a tree, watering, pruning, and *loving* a tree is not the same thing as having the image of one on your tote bag or the wallpaper of your computer desktop.

My third premise is that people must spend more time in their backyards—that is to say, out of their cars and away from their wall sockets. For this to happen, there must be sufficient loveliness in those spaces, and a certain level of maturity in the persons who claim them. But these conditions are not sufficient in themselves. It requires a community to foster and protect them. It takes a whole village to create a habitable backyard. And here we come back to noise. Why should I stay in my backyard if it makes me sick to be there?

This came home to me not long ago when I received an invitation to spend some time in an artists' colony. I was honored, and as the invitation carried a stipend of money, I was tempted. I went so far as to make discreet inquiries as to the possibility of my getting the money in exchange for not putting anybody out by showing up, but that didn't fly. In the end, neither did I, and I'd love to tell you it was because of the carbon. The truth is, I simply couldn't bear the thought of leaving my backyard. Does this mean that I am more grounded, mature, or reliably inspired than the people who accepted the invitation?

Hardly. It simply means I've got it good where I am. Let that change, and I'll be booking the flight tomorrow.

Americans have always prided themselves on their "freedom of movement." This is a real and precious right, one that did not exist in serfdom and does not exist in totalitarian states. But it ceases to be a right as soon as movement becomes compulsory. Refugees, exiles, and homeless people are also "free to move." If I may borrow for just a moment the persona of a Yiddish comic, with rights like these, I should need wrongs? In the end, the right of free movement is meaningless without the corresponding right of staying put.* The right of not being driven out of your home by violence, oppression, and intolerable noise is at least as fundamental as the right to move. Politically they exist on a par; environmentally the right of staying put may be the more valuable.

With profound respect for and great interest in the work of those scientists who travel to the Arctic Circle and the Amazon River basin to take the temperature of the ailing patient Earth, let me suggest adding one more critical ecosystem to the chart and one more measure to the diagnosis. Take your ears into the parks, backyards, and village greens of America and *listen*. You can learn as much about the health of the planet from a backyard in Hoboken as from any place on Earth.

But what would happen to the world if everybody turned into a NIMBY? One thing that might happen is that we would stop putting window dressing around the issue of global climate change. What might happen is that communities and individuals within communities would have to come to the same table and negotiate what they wanted and what they were willing to give up to get it. Tranquility in one place could no longer be purchased at the cost of noise in another, at least not on the sly. Because *everybody's a NIMBY now*. Everybody has a backyard, loves their backyard, and is committed to keeping it habitable. Instead of an energy developer and an absentee landowner bellying up to the bar with a few good old boys in the local gov'ment, every stake-

* The United Nations Declaration of Human Rights makes explicit mention of the right of movement in Article 13: "Everyone has the right to freedom of movement and residence within the borders of each State." The right of staying put is implicit in Articles 12 ("No one shall be subjected to arbitrary interference with his privacy, family, home or correspondence"), 17 ("No one shall be arbitrarily deprived of his property"), and 24 ("Everyone has the right to rest and leisure") but is never stated with the same clarity as the right of movement.

holder—and who on this shrinking, warming earth is not a stakeholder?—has a voice in determining where the energy comes from, where the garbage goes, and how we divvy up the silence and the noise. Think of it as sacrifice with no altars, only alternatives.

Sacrifices come in two kinds, loud and quiet. In the loud version we run around screaming about the latest dire omens, capture the victims, sound the trumpets, invoke the gods, convene the planning commission, and throw the chosen few into the volcano. In the quiet version of sacrifice, a neighbor says to another neighbor, perhaps not even aloud because the neighbor might not have been born yet, "This is something I can do without."

SACRIFICE AND FESTIVAL

If words like *sacrifice, sustainability,* and *quietness* sound deadly to our ears, that is partly because we live in a society that needs them to. You can't maintain an economy of shopping malls and garbage dumpsters, of obsessive consumption and contrived obsolescence, without first setting up a vocabulary in which the opposite of "a real blast" is "a total bore."

I don't believe this. I believe that just as lowering one end of a seesaw raises the other, reducing the levels of waste and noise in the world will raise the level of festivity. There are several good reasons for thinking so. First, in a society requiring sacrifices, we will need the diversion. If I can't fly to Cuba, then at the very least I need someone who can play conga drums. Second, we will need each other to make these diversions; the neighbor with the drums is going to want someone with a trumpet, or someone whose garden grew a few more potatoes than his did. Third, in a world of scarce fossil fuels it's likely we will be more physically vigorous, stronger in the legs, which means our celebrations are likely to be more physical too. I might even learn how to dance. Finally, if we have managed to avert the disasters now threatening us and get to know more of our neighbors into the bargain, we will have plenty to celebrate.

At present the society we inhabit might be described, and sometimes is described, as one in which "every day's a party"—for somebody—which means that it's also a society in which every day has the potential of making a lot of noise for somebody else. The society I'm imagining is one in which some days are celebrations—for everybody—which means that such a thing as noise rarely exists. *If a tree falls in a forest with no one to hear, does it make a sound?*—we can ask a more interesting question. If a Maypole goes up in a village, with everyone gathered around it, does it make a noise?

"OH I HAVE BEEN TO LUDLOW FAIR"

"Just follow the noise," the desk clerk at my motel told me—propitious directions, given my agenda, though she might also have told me to keep an eye out for Portuguese flags. It was a balmy day in early September, and I'd driven down to Ludlow, Massachusetts, to check out the Our Lady of Fatima *festa*. Apparently in years past the amplified music at the event had led to citizen complaints in nearby Wilbraham, such that town officials from both municipalities had gotten together for a chat. I wasn't expecting anything too momentous, probably no more than the three children in the legend were on the day they walked out to watch the sheep and supposedly met the Virgin Mary, but I thought "it might be different," as they say, especially after having been at the Sturgis bike rally only a few weeks before.

There wasn't much noise to follow as I walked toward the festival grounds. None in the factory I passed, which had recently closed along with others in blue-collar Ludlow, and none along the residential street that would be filled with cars by sundown. Eventually I came to a long recreation field, at the top of which I saw a carnival midway and, above the tents and beside the dormant Ferris wheel, the mission-style parish hall with its stucco walls and red tile roof. As I walked closer I saw the square parish church and, against its back wall, the covered stage on which Mexe Mexe and Portugal's Sol Brilhante, "in its first U.S. appearance," would rock the crowds at night. That's when I would go to work, beginning with a call to Herb Singleton, who'd kindly offered to come over with his meter and measure the sound. For now it was pleasant just to stroll the quiet grounds, noting the uncanny way that flapping flags seem to claim a deserted public space, like candles left by worshippers in an empty church. At the end of the midway, across the parking lot from the stage, were the tents and eating pavilions, reminiscent of the firemen's picnics I knew as a kid in New Jersey. Past these was a sloping grove of oak trees, marked by life-sized cement statues, including one of the three Portuguese children raptly attending the Virgin Mary. A small shrine stood at the top of the hill. A group of women were gathered outside, chatting as they added flowers to the platform on which the Virgin's statue would be carried in procession the next night. They smiled and greeted me, but, when I lingered, said that if I liked I could go into the shrine.

The grove was soon full of families with picnic blankets; the midways, with roving groups of kids. I might not have been the only person there who was neither Roman Catholic nor Portuguese, but I seemed to be the only person who was there alone. That was one big difference between the Ludlow *festa* and the rally in Sturgis, where there were plenty of lone riders, but not the most crucial. The near absence of motorcycles in Ludlow wasn't the most crucial difference either. Nor was the simple fact that being Portuguese or being Catholic was just that, a simple fact, as opposed to an identity certified by a machine.

The crucial difference was the presence of all the generations. I had seen hardly any children in Sturgis, fewer women than men, no babies, nobody very old. The people I met there, as in Ludlow, were for the most part courteous, but it occurred to me how the nastiest forms of noise almost always accompany some kind of human segregation, usually but not always adult male segregation: Armies, mobs, gangs, vandals, vigilantes, cliques, cabals, clacks, juntas, almost never examples of "mixed company," are almost always loud. One of the best reasons to "celebrate diversity" is that the best celebrations are diverse. That's not to deny that people in diverse gatherings won't naturally gravitate to birds of their own feather, as Malcolm X observed on his life-changing pilgrimage to Mecca, and as I observed in Ludlow: the boys playing soccer on the grass, the old women caressing the Virgin's feet inside the church, the firemen joking in the beer tent, the young girls strutting their stuff in front of the cotton candy stand, the older, often shorter married couples strolling arm in arm with formidable dignity among their peers. But every so often there would be that "breakout," some pigtailed girl leaping up into her uncle's bricklayer's arms, some young blood halting mid-swagger to kiss the air on either side of his aunt's proffered cheeks. And the aunt smiling down at some child in such a way that if you wanted a naturalistic explanation for the Fatima miracle, it was right there in her face. Again I thought of Les Blomberg's observation that behind every noise dispute there's a breakdown in community. There could be no noise here.

But working-class Ludlow is not the same community as upper-middle-class Wilbraham, and the band had yet to play. The sound tech laughingly told me that his motto was "Annoying you is our business" but then explained what sound strategies he'd adopted to prevent "a repeat of last year." The two cops on duty told me that their noise standard was simple: If they could make conversation with each other standing at the edge of the parking lot "dance floor," the band wasn't too loud. When I told that to Herb Singleton, he said it was a very sensible standard, "if it's enforced." After Mexe Mexe began to play, he discreetly took some measurements: 85 dB by the side of the church, 90 in the area near the sound control booth, 100 near the stage. Not exactly Metallica, though I wouldn't have wanted my kids as close to the stage as some were. And of course no one could hear the voices of the kids.

Singleton and I drove across the river to Wilbraham and were surprised to find the music more audible than he estimated it would be. We were even able to make out the words and some of the instruments from three-quarters of a mile away, as estimated by his map—an effect Herb attributed to variations in air temperature over the river. He repeated his favorite maxim that there is simply no substitute for on-site sound measurements. Like the people who make it, noises seldom behave as they should.

To be honest, I had a hard time seeing much of a noise issue. I tended to agree with the manager of my motel, whom I overheard telling two guests, "It's only two nights of a weekend when most people have off work." The next day I talked to the captain of the Wilbraham Fire Department, Tom Laware, who proudly let me know his background: not Portuguese factory workers exactly, but quarry workers and railroad men. He felt the few complaints about last year's *festa* had come from upper-middle-class people who stayed in town "for three years or so" and moved on to complain somewhere else.

He said there used to be a Six Flags race track that was audible in Wilbraham and a train line that still went through town, and as far as he knew, no one had complained about those. Several times he stressed how "one person gets the others going."

The only fault I could find with the amplified music, aside from any damage it might have done to vulnerable ears, was its tameness compared to the wildness of an older sound. I would hear that sound the next night during the brief interval between the conclusion of the Sunday-evening outdoor mass and the start of the electric music. The mass is the culmination of the festival, when hundreds, perhaps thousands of candles are lit in the darkness at the close of the service, and the statue of Mary, as if riding a slow-moving flow of molten lava, is borne down the hill by a procession that continues out of the grove and through the streets of Ludlow. After I had watched the tail of the procession merge into the city lights down the street, I turned back toward the midway, dormant during the mass but now gearing up for the last round of amusements. From the darkened field behind the tents and concessions, I heard strains of music and voices singing, raucous singing.

Had this been going on during the mass? They were huddled in a circle, men and women of all ages, though the men in the center with accordions and castanets were young, perhaps in their 30s. I had no idea what the men were singing, but from the looks on the faces of those ringed around the singers, I would not have been surprised to learn it was a song traditionally sung to bridal couples on their wedding night. In the light leaking from the midway I saw the approving faces of a handsome old couple, his face like an old mariner's, and hers turned to him with a look that seemed to mean "Do you remember this?" Anyone who saw those faces could never be afraid of growing old.

And anyone who heard that music could never go to a rock concert with all its well-rehearsed high jinks and engineered special effects and find it anything but tame. This is not to knock the bands I heard, which were good. It's only to say that knocking them would have felt like mocking someone's well-made tuna salad sandwich after eating a fresh-caught brook trout. I could almost understand why the folkies booed at Newport '65, because once the amps were turned on, this show was done. It could not have gone on much longer anyway, not at that intensity, even though it had probably been going on in one form or another for 10,000 years.

I didn't just happen upon the place where the singers were. I'd come back to it as instinctively as a nesting swallow to a mission belfry, like a third-generation Portuguese-American anesthesiologist who's seen all his favorite bands live from orchestra seats in Montreal and L.A. and flown to Lisbon and San Paolo and heard a Papal mass at St. Peter's to boot but whose heart yearns every September to eat salty breaded codfish, fried, baked, and smothered with tomatoes and onions in Ludlow, Mass. In the heat of that afternoon, in that same "thin place," I had witnessed something even more powerful, a tarantella-type dance of old and young, all the dancers with castanets and someone playing an accordion. This was not one of the scheduled cultural events, this was something off the lot and "behind the tents," where the real celebrations often take place. Among the dancers was a compactly built man in his 60s, dancing with the vigor of someone a third his age, and drenched with sweat. After every several rotations he would cry out a word that sounded like "Ab-i-TOO!" in what can only be described as an ecstasy. A woman I took for his wife clung to him with her eyes tightly shut, clearly determined not to let go, but utterly spent, hoping he was strong enough to swim for them both. He cried out again—what did he have to say that demanded such passion and why was I close to tears every time he said it? All I could think of was a line from Homer: "Let the trial come, my tough heart can take it."

As if this wasn't enough, in the midst of all the whirling a young man's wrap-around sunglasses fell off his head and with one of the grandest public gestures I've ever seen, he kicked them contemptuously to the sidelines, like a matador too proud to let himself be gored by anybody's bull. This is what I say to "cool." With my foot I say it! Of what use is "cool" when you're dancing with a vital young woman while a man old enough to be your grandfather is shouting out his heart?

Finally, I turned to a woman next to me. "What is that man saying?" I asked her. "What is that word he keeps saying?"

"He says 'turn,'" she told me, without taking her eyes from the dance. "He's calling to them. 'Turn and turn again.'" And so I did, after my fashion, which meant turning the word over and over in my mind. In Hebrew it is *šûb,* often rendered "repent." Fleshed out in Rilke's German it's *Du mußt dein Leben ädern*: "You must change your life." According to the little booklets and holy cards for sale on the church lawn, that was the substance of the message the three shepherd children had heard when their Celestial Mother had appeared to them in her avant-garde skirt. Of their visions I can't be sure, but I was fairly sure of my own. I'd heard a snatch of what the world might sound like if we turned down the noise, and it was nothing to fear.

CHAPTER 9

The Most Beautiful Sound in the World

I frequently hear music in the very heart of noise.
—GEORGE GERSHWIN

We are all conflicted, compromised, and confused.

If the intellectual revolutions of the last hundred years have meant anything, they have meant that. Nothing is quite so clear as we'd like it to be. The nature of time, the meanings of texts, the reasons we do the crazy things we do—and the lovely things as well—no one is sure.

If each of us could hear the sounds inside his or her own mind, would they sound more like music or like noise? If your mental processes could be rendered as an audio track, would it sound more like a piano sonata or a demolition derby? "Why not ask yourself," you say. Fair enough. I believe my track would sound more like noise. Signals going every which way, discordant notes playing at the same time, screech, feedback, reverb. Stations all mashed together. Blocks of white noise fleetingly broken by snatches of rock and roll. I'm so shattered. Conflicted, compromised, and confused.

This might sound like a final capitulation to noise. "Noise wins." Noise is the fullest expression of what we are, the authentic voice of our age. Russolo and his Futurist friends were right. Hooray for the howler, the croaker, the roarer! Hooray for the hiss and the boo, the jet engine's roar, the soccer thug's curse, the prisoner's scream. Let's put a biker bar *on top* of Bear Butte, a heliport on top of every kindergarten, a loudspeaker at the corner of every street. Let's "feed off the aural glut" and belch as loudly as we can once we're full.

But there is another way to think about this.

With so much noise to contend with, both within ourselves and without, we might do well to meet each other in greater quietness. We might do well to tone things down. We might want to pull a few plugs from the wall. We might want to make sure, now and then, that our political rasp—and long live political rasp—is neither rudeness nor the repudiation of those humanistic values that make rasp possible, purposeful, and enduring. We might do well to take the motto of that Dutch anti-noise campaign of the 1970s—*Laten we zacht zijn voor elkaar*, "Let's be gentle with each other"—to heart.

Some of us may say that our hearts are too fierce for that. We may feel that the Dutch motto is too precious, too tame. But maybe only tame people think so. Sometimes in my imagination I see two samurai meet. I would not call either of them tame. Each carries a blade made from hundreds of layers of tempered steel. A man can cut a gun barrel like a celery stick with one of those blades. He can amputate a human limb with a single stroke. Two samurai meet and only one of them will leave the meeting alive. Perhaps neither will leave undamaged. The stakes are high, the outcome is unknown. They meet in silence, and in that silence they hear a cicada sing. In that silence they bow.

I want to believe that in time we will be able to do without the sword and all of its descendents, that our descendents will be able to abolish those distinctions that set the samurai above the peasant, the business-class passenger above the child learning to read. But I doubt we can do without the bow. And I am certain we cannot do without the silence.

WHAT WE TALKED ABOUT WHEN WE TALKED ABOUT NOISE

I am also certain that we cannot do without certain noises, but before taking a farewell listen to some of the least dispensable, we should take a few moments to recall where we have been.

This book started out with several premises: that noise is a relatively "weak" issue, not as important as water pollution, for example; that it has a disproportionate effect on the "weak" (marginalized people, small creatures,* and simple pursuits); and finally, that this tendency to affect the weak is another reason for its status as a weak issue. Conventional wisdom (along with the World Health Organization) says that people begin to care about noise when they become more prosperous and better educated. No doubt that is true. But it is at least as true to say that governments begin to care about noise when it affects more prosperous and better-educated people.

Based on these starting points, I suggested that noise can help us understand some of our bigger issues, the reason being that many of those issues—war, hunger, racism, sexism, climate change—have come about through a contempt for whatever we regard as weak, small, quiet, or useless. The extent to which we regard noise issues as "precious," in the pejorative sense of the word, is the extent to which we will squander those things we ought to hold precious in the positive sense of the word: fragile ecosystems, manual skills, local cultures, neighborhoods, children. Even if noise is not killing us—though some compelling evidence suggests that it can—noise offers a way to understand the attitudes that are.

Noise also compels us to seek our understanding through different filters. I can think of few subjects that lend themselves so readily to a multidisciplinary approach. Physicists, musicians, historians, psychologists, artists, engineers, and philosophers have all lent their ears and their expertise to its challenges. Noise is a complex phenomenon that reveals our complexity as human beings. It is both easy and hard to define, objective and subjective, new as the latest gizmo and old as the most ancient myth. Its subjective aspects put us in touch with our prejudices, our fear of difference, our deep-seated need to be acknowledged. Its objective aspects put us in touch with

---

* Several times throughout the book I've quoted Fred Woudenberg, head of the Municipal Health Service in Amsterdam, on noise. His most memorable quotation for me, and perhaps the most essential to this book, had to do with why he had given up his career as a psychologist doing research in psychopharmacology. "I was doing animal experimentation in a laboratory with rats, and I thought, well, I [will] investigate for myself it it's worthwhile what I'm doing, if it has some benefits for society. And, if it's worth the trouble to torture these small animals. And in the end I came to the conclusion that it wasn't."

our nature as physical beings. Our noisiest inventions, after all, have the common aim of reducing the restrictions of time and space, which are also the conditions of living in a body. This is not to suggest that we can liberate our bodies by abandoning every device that makes more noise than a darning needle or a hoe. It is merely to suggest that in some ways our mechanized civilization is at war with our bodies. Noise is often the sound of that battle.

As noise affects our bodies, it also affects the body politic. Noise is political. It makes its first grand appearance in the *polis*, the city, just as politics do. Noise is political, first of all, because loudness is powerful. Noise is political because peace and quiet are forms of wealth, subject to the laws of supply and demand, and because how a society divvies up its wealth is a basic political question. Inherent in every "unwanted sound" produced by a human source lies the question: What kind of a society do we want?

A society I don't want is one in which cruelty and gross injustice are superficially "subverted" by petty noise. By itself noise is seldom revolutionary. You can start a revolution in a café, a union hall, or a barracks, but the only thing you can start in a stadium is a riot. In the context of a loud society, quiet resistance may be the most radical option of all.

Not only do I consider the notion that noise is revolutionary a piece of silliness; I also consider silliness an essential characteristic of noise. Celebrity culture, militarism, technological fundamentalism, arrested adolescence, academic posturing, single-issue politics, conspicuous consumption, senseless destruction—all noisy in various ways—are also silly in much the same way.

By silly I don't mean light-hearted or even frivolous. I mean what a mother means when she says to her child, "Now you're being silly." She means: "You know what you're doing, but you're pretending that you don't." Or she may mean: "You know what you're doing is harmful, but you're pretending that it isn't." After a while one begins to notice how often noise accompanies that sort of pretense. George Orwell once said that the goose-stepping parades of totalitarian countries would never make it in England because the common people would laugh at them. In other words, the common people would see such displays as silly. The same can be said for many forms of noise (but not for all populations of common people).

Apparently there was a yacht club in New York that liked to fire a cannon at the close of the day. Neighborhood complaints led to a court hearing, but the judge did not seem predisposed to side with the complainants. She did not want the city to overreach its authority, and she respected that some of the yacht club members were war veterans for whom the cannon ritual had sentimental value. (I'm not sure I get this, but I'm assuming none of the yachtsmen had returned from active duty with PTSD or tinnitus.) The judge also seemed dubious about the logarithmic nature of decibels, the notion that 100 decibels is not twice as loud as 50 decibels, but thirty-two times as loud. She said she wanted to hear this cannon for herself.

So she went to the harbor and the yachtsmen fired their cannon. As the reverberation died and the smoke cleared, the judge said: "If I were a mother with a child doing homework, I would not like this."

She was not a silly judge. In her framing of the issue in terms of the way ordinary people live with ordinary predicaments, in her consideration for the welfare of children, the judge offered her court a stunning glimpse of that most endangered of North American species: the human adult.

By this I don't mean to imply that grown men firing a cannon from the docks of a yacht club are necessarily being silly, only that wrapping noise in a mantle of patriotic sentiment or altruistic privilege is. This happens all the time. Referring to a recent Denver ordinance forbidding modified motorcycle exhaust pipes, a spokesperson for a company that makes custom motorcycles told me that now the "Toys for Tots" program was in jeopardy because bikers would no longer feel free to organize big runs for the purpose of delivering toys to the sick little children of Denver. Perhaps you should wait a moment before saying that you never heard of such a thing. One of the landmark court cases involving noise in the United States was *Ward v. Rock Against Racism* (1989), in which the U.S. Supreme Court ruled by a vote of six to three that the police commissioner of New York City had not violated the First Amendment rights of musicians performing at a Rock Against Racism concert when he insisted that they use sound equipment and an audio engineer provided by the city.* I'm not about to suggest that fighting racism is silly, any more than I'd say it's silly to take toys to ailing tots. But can someone help me understand the relevance of a sound system to the fight—or the relevance of rock music, for that matter, which may be the most apartheid cultural institution left in the Western world? If pro tennis looked like rock music, Serena and Venus Williams would be fetching stray balls.

Before this book is out, or some time shortly thereafter, I predict you are going to see a pink breast cancer ribbon on the fuselage of a supersonic jet, either that, or a Free Tibet sticker pasted along the snout of a leaf-blower. I'd say you're going to see a tractor-trailer-sized banner that reads "God Bless Our Troops" over the entrance of what claims to be the world's largest biker bar, but I've already seen that one. Again, I'm not calling anybody's cause, cannon, or hobby silly. I'm saying that whenever loudness gets caught on rectitude, whenever the border of the American flag or the hem of Jesus' garment just happens to have gotten stuck to somebody's un-muffled tailpipe, then somebody knows exactly what he's doing and is trying to pretend that he doesn't. Somebody's mother needs to tell him that he's being silly.

To have a good ear for noise, you need a good eye for silliness. You also need a heightened awareness of pain, because noise is often painful, and because much of politics revolves around pain's uses, abuses, and abolition. Writers and philosophers like Elaine Scarry, whose *Body in Pain* I have cited; Judith Shklar, who begins her book of political essays *Ordinary Vices* with a chapter called "Putting Cruelty First"; and Richard Rorty, who locates the liberal ethos in a refusal to engage in humiliation, provide what may be our best touchstones for understanding the politics of noise, assuming that older im-

* Apparently the city had imposed the conditions on all performers in the Naumberg Acoustic Band Shell in Central Park after repeated warnings to lower the volume at a 1984 RAR concert met with no results. (The tactic of cutting off electricity nearly produced an ugly result.) The Court ruled that since the city had made no attempt to interfere with the content of the concert and had imposed the same restrictions on other acts, it was not interfering with the right of free speech. Justice Thurgood Marshall, who wrote the dissenting opinion and whose judgment I'm not inclined to take lightly, said that by applying restrictions ahead of the concert the city was placing an "impermissible prior restraint" on free speech. "Government no longer need balance effectiveness of regulation with the burdens on free speech. After today, government need only assert that it is most effective to control speech in advance of its expression."

peratives, such as doing unto your sleepy neighbors as you would have them do unto your sleepy self, no longer carry weight.

FROM BIPED TO IPOD

If noise is as central to human experience as I've argued, then it ought to have been with us from the beginning. It has been, as testified by our oldest book (*The Epic of Gilgamesh*) and by oral traditions of indeterminate antiquity. Like a dramatic drum roll, it has often announced important junctures in human history: the first cities, the inventions of the wheel and of gunpowder, the industrial and electronic revolutions, the conquest of the sky, and the warming of the globe. Often expressive of the might of empire and the prerogatives of religion, noise has always been there to tell us who's boss and what counts. Hitler's use of the loudspeaker is a prime example. The Buddha's noble silence is a prime exception.

Though noise has been with us for as long as human beings have been awakened by unwanted sound, noise took a quantum leap with industrialization. To say that noise became louder is generally true, though certain noises have disappeared or been reduced due to new technologies and to a greater awareness of the scientific properties of sound. It is probably more accurate to say that noise became more pervasive, no longer confined to work, to the working day, to the temporal auditory experience. Electric lights, recreational machinery, personal listening devices, earworms—all have contributed to a collective condition of virtual tinnitus.

As industrial societies became more aware of their impacts on the natural environment, noise came to be seen as a pollutant with adverse effects on human health and the well-being of nonhuman creatures. Like other pollutants noise has followed the path of globalization. In the most literal sense, development has raised noise levels around the world through increased mechanical production, transportation, consumer products, and the dislocation of rural populations to city centers. In a figurative and perhaps larger sense, these changes have tended to act *like* noise by "drowning out" the distinguishing notes of local cultures.

At the same time, the greater exchange of information and products between countries can lead to a sharing of noise solutions. The attempt to build a "green" highway system in Ireland, the "Best Hundred Sounds" program in Japan, the ongoing research on noise and health in the European Union, the attempt to craft workable noise policies in places like Delhi, Accra, and New York City, the controversies over industrial wind production in rural areas of the United States and in more populace regions of the Netherlands, the emerging research on the effects of noise on fish, birds, and mammals will be worth following in the years ahead.

Not least of all, research on the effects of noise on childhood health and development will need to be brought to the fore. Studies like those conducted by Gary Evans in the United States and the RANCH project in Europe need greater funding, attention, and follow-up. This is a conceptual as well as a moral imperative. A failure to take account of the most vulnerable victims of noise is a failure to understand what noise essentially is.

That is because people who escape notice are the least likely to escape noise. The grinding noise of the city, the explosive noise of war, the bustling noise of lifestyles beyond their reach—these fall on the ears of the weak with disproportionate force. The Roman citizens awakened in the night, the poor people set upon by aristocratic hooligans in the streets of eighteenth-century London, the slaves wondering when the revelry at the Big House would end in some malicious mischief, the housemaid in the basement several floors below Carlyle's soundproof study, the mill girl who sang "inside her head" because she could not hear herself singing aloud above the din of her machines are the historical antecedents of people who live with excessive noise today.

In the consideration of past and present noise disputes, especially when those disputes appear divided along class lines, we often fail to account for anonymous, isolated parties who were also affected. My favorite illustration is the dispute over carnival barkers recounted in Emily Thompson's *Soundscape of Modernity*. In 1907 the New York commissioner of police banned the use of megaphones by carnival barkers on Coney Island. Like others in his trade, Pop Hooligan, reportedly the oldest barker on the island, was outraged.

"What would Coney Island be without megaphones? How are you going to get a crowd to come in and see the boy with the tomato head and the rest of the wonders if you don't talk to them? I will see this Czar and make him abrogate his order." Pop Hooligan was undoubtedly correct in seeing the ordinance as a privileged attack on his livelihood and way of life. And so we have a classic, class-based noise dispute of the kind that historians quickly parse as a conflict between fastidious bourgeois elitists and their social inferiors, between the straitlaced folks in "Pleasantville" and the earthy proletarians trying to hold their corner of the street. That's a fair assessment as far as it goes. But the question I have tried to raise in this book, and the question I want to leave with my readers, is this: What about the tomato-headed boy? What did *he* hear? What was *his* soundscape like? What were *his* prospects of being heard over the noise of the protesting barkers and the police whistles blown against them?

Look closely at almost any noise dispute and you will find a tomato-headed boy.

One reason that I chose to write about the admittedly more negative and narrow topic of noise instead of the richer and more inclusive topic of auditory culture is that I wanted to remain at the gritty, bodhisattva level where the tomato-headed boy clasps his hands over his ears. I wanted to keep faith with the people who cannot move away, hire a lawyer, or "contextualize" their noise annoyance in some abstract theory of the soundscape or in some trope of acoustic transgression. I wanted to keep my mind's ear open to the people who can't afford the plane ticket, the concert ticket, the car speakers, or the car.

I will leave you with a riddle. *What else besides mortality and a shared interest in noise do acoustical engineers, musicologists, sociologists, neuroscientists, sound artists, alternative energy developers, cultural historians, legal experts, wildlife biologists, and authors like me have in common?*

What they have in common is the knowledge that success in their respective fields will enable them to live in quieter places than other people. On the day when the full weight of human expertise is brought to bear rigorously on *that* problem, the problem of noise will all but fade away.

LOUD AMERICA, VULNERABLE EARTH

Americans are supposed to believe in equality, and at their best I think they do, but what Americans believe in most is infinity. Maybe the distinction is not so clear-cut as that; maybe it takes a sense of infinity to believe in equality, a sense of infinite promise to believe we can be neighbors in the fullest sense. Nevertheless, the innate American suspicion of limits, especially limits on personal freedom, has not made us a quiet people. Neither have the goals of world leadership and unlimited economic growth.

Noise is well-suited to identifying those aspirations that make Americans who they are. It is especially well-suited to identifying those places of tension where aspirations conflict, where equality and infinity begin to grate. One of those places of tension occurs between the potential of the individual and the good of the community, between the need of a genius like John Coltrane to hitch his wagon to a star and the need of his fellow musicians not to fall out of the wagon. It is the place where "free" and "brave" run the danger of becoming little more than loud.

A related place of tension occurs in our politics. When does the rasp of political discourse degenerate into a noisy repudiation of the discourse itself? When does the "jamming" that Wynton Marsalis sees as characteristic of both jazz and American democracy become "jamming" in the sense of alienating noise? What do we lose when members of a band can't hear what other members are playing?

One of the areas in which America has done some of its more positive jamming is in its National Park system—"the best idea America ever had"—where issues of access, appropriate use, and the stewardship of national resources continue to be debated. A concern for the preservation of natural sounds is now part of that debate. At issue is how well a population of diverse tastes and abilities can own and occupy the same public space. It is a debate worth having, and it is in large part a debate about noise.

Finally, in a nation that began by a declaration of independence from a king, noise raises the question of whether we are to be a country without kings or a country in which everyone wants to be king. Noise disputes are often particular expressions of that fundamental question. Kings rarely walk quietly through the world. In fact, they rarely walk. They ride. When they can, they fly. They are great believers in infinity.

It is foolish to suggest that a truly democratic America would be utterly quiet. To wrestle with questions of justice and freedom will always make some noise. Even to keep those questions alive will require making noise in the form of protest. But to identify every noise as an expression of protest or as an instance of Americans struggling to be true to themselves is mere silliness—especially when it serves as a commercial ploy. Struggling to be a good neighbor, to make a place for a new neighbor, to respect other creatures, to preserve conversation—these are no less authentically American than the desire to open the throttle and pull out all the stops. When Native Americans can pray in quiet on Bear Butte, toasted with a round of cold beers at the Loud American Roadhouse in downtown Sturgis, we will be that much closer to the Promised Land.

Not so fast, says Nature. There is no promised land without a sustainable future. As we chart a course toward that goal, noise will be as instructive as it is in identifying the tensions in American democracy. The carbon-based economy, the growth-obsessed economy, the corporate-dominated, transportation-intensive economy must always be a noisy economy. Put quotations marks around *always* because such an economy cannot last. One way to imagine an alternative is to think in terms of a quieter world: quieter habits of consumption, quieter interactions between powerful and less powerful nations, quieter—because more locally based—production of energy and food.

The acid test for such a project is whether the "alternatives" are truly alternatives or merely clever disguises of the same attitudes and methods that have brought us to the brink of environmental catastrophe. Here, too, noise will serve as a useful gauge. The extent to which noise concerns are shunted aside in the name of "more urgent priorities" will be a reenactment of the ways in which air, water, climate, flora, fauna, and posterity have repeatedly been shunted aside in the name of The Looming Crisis and The Next Big Step. The most dangerous noise in the world today comes from a superlative.

One benefit of a quieter world would be the pleasure of hearing more sounds, including those of celebration. The equation of sustainability with dullness is our civilization's tragic lie. The more we quiet the noise of competition, consumption, and conquest, and the more deliberately we do so, the more we can share in celebration. Sometimes I imagine all the people I've met and corresponded with while writing this book sleeping peacefully through a whole night, then rising with their strength renewed in order to join with all their neighbors in dancing the next night away. In such a world there is abundant silence and abundant sound. There is precious little noise.

SOUND AND SERENDIPITY

One of the pleasures of this project has come from hearing unexpected sounds. I went to Ludlow, Massachusetts, for the purpose of observing a reportedly noisy festival and wound up hearing vestiges of celebration more ancient and visceral than anything coming out of the amplifiers. I went to Carlyle's house to see a soundproof study but was most impressed by the anything-but-soundproof basement chamber of his housemaid.

One of these surprises came during what I can only describe as a fool's errand. While planning a trip to the United Kingdom, I read of a study that had determined the noisiest and the quietest cities in England, respectively, Newcastle-on-Tyne in the north and Torquay on the Cornish coast. Almost as soon as my wife and I were in London, however, I was told by a knowledgeable source that the study had "serious flaws." Since I planned to drive east anyway in order to visit some wind farms in the region, I decided we might as well keep our plan to visit Torquay. The setting for a number of Monty Python sketches and for the TV comedy series *Fawlty Towers,* as well as the site of what may be the oldest remains of a modern human ever found in Europe, the seaside resort seemed an attractive destination, something I could write about. And the idea of the quietest city in England, however faulty its supports, intrigued me.

Unbeknownst to us, we happened to be traveling during a national school holiday toward one of the most popular vacation destinations in Great Britain. By the time we reached the center of Torquay, I remarked to my wife that probably the reason this city was so much quieter than Newcastle is that in Newcastle the traffic actually moves.

I'll spare you the details of our escape. Suffice it to say that by the time we managed to get out we were half-starved. After some fruitless searching, we stopped in the nearby suburb of Preston at a small fish-and-chips place called "The London Fryer," with the punning image of a fat tonsured monk on its overhead sign. The place was a study in "loud," beginning with its abundant Halloween decorations, which seemed to have been chosen with an eye toward promoting the appetites of cannibals. The radio speakers were going full tilt, with Sheryl Crow singing what I assumed was the company theme song, "All I Wanna Do Is Have Some Fun," followed, with a nod to the decorations, by Stevie Wonder's "Very Superstitious." The server was a stout mercurial woman who called us "lovies" on the way in and took our orders with additional terms of endearment. "That's for darling, now what about for sir?" she asked above the radio. The portions she brought us were obscenely huge, a saucy kick in the bum to "heart-healthy," and altogether delicious. Later, when my wife said apologetically that she was not able to finish her plate, the server remarked, with something like a wink in her voice, "I'd have thought you'd be used to large portions." As my wife is quite petite and not especially ravenous in her appearance, we assumed the subtext to be "You being American and all." I can't say the other customers were loud, though they clashed with any sense of a "clientele." Seated near us was a taciturn Anglo-Saxon couple, dressed as for a high tea with the Queen. Two leather-jacketed young men with shaved heads stood by the take-out counter, and just before we finished, two salty codgers took the table nearest us and commenced chatting up my missus on the superiority of the Fryer's steak-and-kidney pie. I couldn't think of the last time I'd so enjoyed a lunch out. The only thing that could have made it better is if Thomas Carlyle himself had walked through the door, adding his horrified expression to that of the severed rubber head hanging close by.

Such experiences, common to all travelers, were at once "incidental" to my research and vital in helping me to understand it. They helped me to consider the relationship between serendipity, surprise, and signal, to ask what would happen in a world where there was no such thing as "unwanted sound." The result could only be a sad and ironic constriction on our wants, for it is sometimes in the midst of unwanted stimuli that we discover a new pleasure and acquire new tastes. As Gershwin put it, sometimes we hear music in the heart of noise.

That makes for a worthwhile qualification in a book as anti-noise as this one, and a good thing to keep in mind in a culture of focus-group marketing and niche broadcasting, gated communities and identity politics, one in which we are increasingly able to hear, see, taste, and smell what we want—or are led to believe we want—to the exclusion of everything else. Although I want to be able to sleep at night, and to hear myself think in a library, and to hear the conversation of my companions at table, I also want—and need—some random signals, some unbidden information, in a word, some *noise*. Yes, I want a few "Thou shalt nots" that maintain the P & Q of private spaces, and the inclusiveness and diversity of public ones, but I am made uneasy by any "Thou shalts" that seem overly prescriptive.

This is yet another reason—in addition to my solidarity with noise sufferers—that I've chosen to take a narrower, anti-noise approach over the more affirmative soundscape approach, much as I respect the latter. Admittedly with the advantage of thirty years' hindsight, I shudder to reread the rhetorical question near the close of R. Murray Schafer's *The Tuning of the World*: "Why could not everyone choose his or her telephone signal?"—a matchless illustration of the need to be careful about what you wish for. The quotation is taken from a chapter called "The Acoustical Designer," which ends with a visionary passage on "The Utopian Soundscape." With respect, no thank you. Give me the comparatively flat-footed city ordinance keyed to decibel levels and certain hours of the night. The notion of an "acoustical designer" or a "utopian soundscape" makes me nervous.

SOUNDING THE RETREAT

We are all conflicted, compromised, confused. One would think that something called a "soundscape retreat" would have made me nervous too. It did. The word *retreat* especially. More even than I dislike noise, I hate all orchestrated "experiences," all petri dish encounters. I believe I've gotten along well with every supervisor I've had for the past forty years, from foremen in glue factories to bishops in tweed skirts, but mention the term *session leader* and I'm ready for a fight.

Still, just as I was beginning this book, I learned of a soundscape retreat sponsored by the Canadian Association for Sound Ecology deep in the wilderness of Ontario. I would get to meet R. Murray Schafer. I would get to meet Bernie Krause, the wildlife biologist who advised the National Park Service on the early phases of its natural sound programs and whose justly celebrated recordings of natural soundscapes have brought the remotest regions of the world into the ears of thousands of people. I might even get to hear some wolves.

After a flurry of e-mails and phone calls in which I made it clear that I was not into any kind of touchy-feely business, co-ed bunking, compulsory nakedness, saying nice things (or not nice things) to the person standing to my right or to my left either, and after I was assured that the only thing even remotely in that line would be a few "ear-cleaning" exercises, I paid my fee, bought the most expensive article of clothing I've ever owned (a pair of high sturdy boots), and headed north. In Ottawa I rendezvoused with the sponsors and several other participants, accepted a ride north that took "Drive em like you stole em" to a whole new level, and finally set grateful feet on muddy ground in a wilderness compound named Haliburton. (Supposedly it bore no relation to the corporation, though the name made me nervous just the same.)

The people I met were all courteous, intelligent, and I should probably add tolerant, for I was scribbling every minute and it was probably clear from the get-go that I was not recording happy thoughts. In fact, I began to wonder if I wanted to write about noise after all. With a few exceptions it seemed that what most of these people were interested in was electronic recording equipment. They were talking "shop," of course, as all enthusiasts do, but after a while I felt that the only thing in their shop was wires. Why belabor how isolated I felt? Only to underscore my previous point about surprise and serendipity. This was to prove one of the most important trips I took.

For one thing I did get to hear wolves. Well before dawn several of us walked to the very end of a logging road, fanned out, and waited in silence. Up on a ridge looming in front of us, behind a fence that supposedly enclosed acres of forest, lived a protected pack of wolves. Chilled by the cold March breeze blowing through the pines, we watched as the sky turned cobalt and lost all but its brightest stars.

Before any bird stirred it sounded: one unmistakable note in the stillness. Asked to translate, I would render it as "I'm up." Then another, longer note, more wide awake, insistent. "Very up." A second voice answered: "Yeah, me too." And then the whole chorus, full throated, as if alive for the first time since time began. It was not the most beautiful sound I would hear that weekend, but it was beautiful. Strange to think that relative to human noise it was not the sound of strength. The wolves of history were weaker than the wolves of legend. Socialized in packs where gestures of submission are met with mercy, the wolves offered little resistance as they were pulled from homesteaders' traps and subjected to unspeakable cruelties. What our predecessors had heard as the primal unwanted sound, a noise so frightening that it turned them into images of the very fiends they imagined, was now a sound we longed to preserve. I took out the little pocket recorder I use for interviews in the hopes of capturing a sample, suddenly and humbly envious of those standing around me with better gear. The experience reminded me that all discussions of sound, music, and noise begin and end where we were standing, ears bent to the natural world.

If hearing wolves for the first time was a high point of the weekend, I suppose that one of it low points for me was when R. Murray Schafer commanded us to stand up and receive a lesson in "how to howl like wolves." Given that most of the people in attendance seemed to consider themselves Schafer's disciples, this was completely appropriate. As I did not, I stepped quietly out of the room. (This was not the only time I did, though I tried to be a good sport.) For the sake of the person I was partnered with—a gentle, utterly unpretentious man who builds instruments out of found objects and hosts a Canadian radio program about sound art—I got down on my hands and knees (with someone else calling the orders this time) and crawled with my eyes closed into an open space where I stood up and wandered around with nothing but my partner's voice and my ears to guide me. With considerable effort I refrained from raising my hand to ask—when a professor gave a talk about something called a *flaneur*, which from what I gathered is a person who helps bring about a socialist transformation of society by pilfering his employer's office supplies—what possible good could come from a socialist society peopled by screw-offs and petty thieves. But like much that happened at the retreat, this was all to my benefit, for the professor was preparing me to understand those cultural theorists who talk about the subversive qualities of noise. They mean that noise is another way of making off with the paper clips.

After the session on the *flaneur*, I stepped outside for a bracing gulp of Ontario air and heard another beautiful sound, accompanied, as beautiful sounds sometimes are, by a beautiful sight. Two boys, I'd say 10 or 11 years old, were perched atop idling all-terrain vehicles, ready to take off into the wilderness that spread out before them. Such a curious sensation came over me just then. I was supposed to hate these contraptions. On some level I did. They ruin hiking trails, erode topsoil, and fill the woods with exhaust fumes and noise. They are the same machines that enable lazy trout fishermen to clean out the remotest brooks, close cousins to the machine that killed poor Mr. Apology. Damnable things.

But after all the supervised howling and crawling, the sight of those two boys waiting to take off on their brash machines filled me with a crazy kind of exultant gratitude. I wanted them and their noise to exist forever. I wanted them to ride full throttle into the sunset. No fear. No surrender. And no—under any circumstances—retreats.

It's possible my sentiments were not so regressive as they seem. It's possible that the temples of a quieter, carbon-free civilization will contain shrines with sacred Harley-Davidsons, straight pipes included, and every person will in his or her youth be granted a brief, enchanting taste of rolling thunder. During a tour of Lancaster County in Pennsylvania, I was told that Old Order Amish teenagers are notorious for cracking up cars since they are not required to take up buggy reins until they've become full-fledged, adult members of the church. It seems they make the most of the interlude. There's a parable there, I suspect: one that hints at the capacities of a quieter, less silly, and more sustainable society to restore the prerogatives of youth, if only so that young people might understand, by means of a sharper contrast than they often see now, the dignity of adulthood.

But the ATVs are probably superfluous to that vision. It may even be that such machines interfere, not only with the woodsy peregrinations of old hermits like me, but with carefree abandon itself. Several times in Europe we were awakened by the noise of young people in the streets—shouting to one another in Amsterdam's Vondelpark in the middle of the night, disembarking from the trains in the Dutch city of Baflo (our host had warned us, "You may hear the kids coming home from their dances on the train"), and most memorably in Dublin at Halloween. You don't know Halloween until you've seen it in Dublin, in the country where it began. It's the busiest night of the year for the Dublin Fire Department and surely one of the loudest since the Easter Rebellion. The sky is full of fireworks—official and otherwise—and the sidewalks packed with the sexiest witches you've ever seen. In all these instances what struck me was the absence of *devices* (except for the pyrotechnics in Dublin): I heard no bass boom, no exhaust rumble, no squealing tires, no machines impotently raging against The Machine. I was awakened just the same, of course, but the sound that woke me was a human voice, and in the majority of cases a joyful one. It put me in mind of a passage in Robertson Davies' novel *Fifth Business* in which a Spanish lodger at a hotel complains of what he assumes are the noises of a riotous honeymoon (the man and woman are actually having a row). Through the door he calls: "Señor . . . 'zis honeymoon—oh, very well, very well for you, señor, but please to remember there are zose below who are not so young, if you please, señor!" Next morning a bouquet of flowers and a note arrive at the door: "Forgive my ill manners of last night. Love conquers all and youth must be served. May you know a hundred years of happy nights. Your Neighbor in the Chamber Below."

No, I haven't gone soft on noise, nor have I forgotten my irritation at being wakened night after night in graduate school, when I too was young and still on my honeymoon. My point here is that "youth must be served," and no healthy society can forget it. The boys on their ATVs reminded me of that, though by the time my thoughts had gotten that far, they were long gone, leaving nothing but a faint reverberation behind them. And still I had not heard the most beautiful sound in the world.

AT LAST

That came, ironically enough, in the midst of one of those group "exercises" that sometimes irked me. This one was not quite so controlling as some of the others, but it was worth going through all the others just to hear what I heard this time around. I remain thankful for it still.

Our leader for the session had instructed us to take a walk along the gravel road by the compound, to leave the building one at a time with intervals between us, to stop at designated places along the road, and to listen. A modest request, and one I was happy to honor since the road went through the woods. At each stop I heard the songs of several birds, foliage ruffled by the wind, branches and tree trunks rubbing together. After completing the "sound walk," we were to come back to the room without talking, one at a time in the same order as we had left, and attempt to sit in our metal folding chairs without making a sound.

That was impossible, as most of us would have realized at the start, but we all made an honest effort. Our ears were no doubt hypersensitized from listening as intently as possible to every sound we had heard in the preceding twenty minutes. As each person took his or her seat, we heard the creaks and rustlings of matter and flesh, snaps and buttons, gravity and friction, making their presences known with audible insistence. Within the context of the exercise, these sounds could only be called "noise," for they were "unwanted" and we were expending considerable effort not to make them. Yet, weren't we glad for them too?

It was then, at almost the beginning of my explorations, at the lowest ebb of my confidence and on what I confess was the least favorite of the journeys I would make, that I heard what I scarcely knew I'd been listening for: the most beautiful sound in the world. More beautiful than the cry of wolves, more beautiful than an ATV leaving some *flaneur* in the dust—it could never be separated from noise, never be "pure" of commotion any more than my seatmates and I could achieve perfect silence in the act of sitting down. John Coltrane made this sound in his middle period—and in his early period, and in his final period. McCoy Tyner continues to make it past his 70th birthday.

The Lakota and Cheyenne make it in their vigils at Bear Butte, and their biker sympathizers make it when they choose a less traveled road. Roy and Louise Abbott have made it at The London Fryer for the past twenty-eight years, with "mushy peas" and pickles on the side. Every time the phone rings in his cluttered Montpelier office, Les Blomberg makes it, and every time he rides his bike to work. A severely handicapped boy in a wheelchair frequently made it in the library where I went to do my research, continuing to make it even when his mother gently told him, "Shhhhh." I suppose that in her way she was making it too. Most certainly Rosa Parks made it when she refused to give up her seat on the bus, and Bob Marley made it when he told people to "get up, stand up," and make it for themselves—the most beautiful sound in the world: the sound of a human being trying to do his or her best.

Sitting Quietly at the Back: A Set of Resources

A TIME LINE OF NOISE HISTORY

c. 3.5 million years ago Hominid ancestors leave two or three sets of footprints in volcanic ash, possibly after having heard the volcano erupt.

c. 3000 B.C.E. *The Epic of Gilgamesh,* in which gods flood the earth to quiet the noise of humankind.

c. 1550 B.C.E. Egyptian awareness of tinnitus.

c. 1250 B.C.E. Israelites march around Jericho making noise and toppling the walls.

c. 850 B.C.E. Homer describes the noise of battle.

c. 500 B.C.E. Buddha asks his monks to turn down the chit-chat.

c. 250 B.C.E. Chinese text recommends that criminals be tortured with bells. "Ring, ring the bells without interruption until the criminals first turn insane then die."

44 C.E. Julius Caesar outlaws wheeled vehicles from operating "within the precincts of the city, from sunrise until an hour before dusk."

c. 50 Roman philosopher Seneca writes about the noise of the gymnasium downstairs from his apartment.

117 Roman poet Juvenal writes that the "perpetual traffic of wagons" on the nighttime streets of Rome "is sufficient to wake the dead."

384 Augustine of Hippo observes with some fascination that Ambrose of Milan reads silently to himself.

900s Silent reading has become normative in the West.

1200s Genghis Khan uses "Greek fire" (perhaps gunpowder) to incite terror on the battlefield.

1252 Church officially permits the use of torture. Screaming increases.

1253 Royal statute commands all English Jews to say their services in a low voice (*submissa voce*) so as not to be audible to Christians.

1300s Anonymous poet complains that "[s]wart smutted smiths, smattered with smoke, drive me to death with the din of their dints."

1377 The priory of St. Mary of Bethlehem (Bedlam) takes on the care of the mentally ill.

1400s Physicians claim that noise can damage the ear.

1407 English Archbishop Arundel defends the songs and music of Canterbury pilgrims against complaints that they are too noisy.

1500s Wife-beating prohibited after the hour of 10:00 P.M. in Elizabethan England.

1530 Erasmus recommends masking the "horrible" noise of farting with a judicious cough.

1598 German jurist Paul Hentzner travels to England and notes that the people there are "vastly fond of great noises."

1608 Elizabethan playwright Thomas Dekker writes about the sound of a bear-bating ring: "[T]he very noyse of the place put me in mind of *Hel*."

1617 Fynes Moryson notes that the Irish "are by nature very Clamorous."

1627 Francis Bacon records the experience of temporary hearing loss due to a falconer's whistle.

1661 Bern, Switzerland, passes a bylaw against "shouting, crying, or creating nuisances on the Sabbath."

1712 First practical steam engine.

1713 Bernadino Ramazzini notes noise-induced hearing loss among coppersmiths in Venice.

1738 First cast-iron rail tramway.

1741 William Hogarth engraves *The Enraged Musician*, which depicts sources of street noise that cause the hapless musician to clasp his ears.

1791 Invention of gasoline engine.

— Revolutionary French begin to take down church bells to make cannons.

1798 A woman in Newark, New Jersey, complains that she is unable to sleep for several nights due to "noise in the streets."

1816 René Laënnec invents the stethoscope, which reveals our "noisy" insides.

1833 German philosopher Arthur Schopenhauer writes to civil authorities urging strong measures against those who disturb theatrical performances. "I have for a long time been of the opinion that the quantity of noise anyone can comfortably endure is in inverse proportion to his mental powers."

1847 Daniel Webster pays tribute to the Northern Railroad, dismissing complaints about its annoyance, including his own. A good sport for "progress."

1853 Henry David Thoreau writes of railroad noise at Walden Pond.

1854 English writer Thomas Carlyle moves into his new "soundproof study" (though he finds it less than soundproof, merely "supportable").

1862 Combat-induced hearing loss recognized by the U.S. government as a service-related disability.

1863 Bloodiest and perhaps noisiest military engagement on U.S. soil fought at Gettysburg.

1864 British Parliament debates the noise of street musicians.

1865 Arguably the most noisome of all American noises—the sound of a whip striking the flesh of slaves—is officially silenced with the end of the Civil War.

1872 Yellowstone becomes the first of the United States' national parks—that is, one of the first recognized arenas of recent history for contesting the meaning of public space.

1874 English writer and critic John Ruskin leaves Italy to escape "entirely monstrous and inhuman noises in perpetual recurrence . . . wild bellowing and howling of obscene wretches far into the night."

1877 Hermann von Helmholtz distinguishes music from noise.

1878 Doctors testify before a grand jury about the noise of elevated train tracks in New York.

1883 Eruption of Krakatoa, perhaps the loudest sound ever heard on earth, audible for 3,000 miles.

— Oscar Wilde writes that "America is the noisiest country that ever existed."

1888 Noise recognized as an indicator of mechanical inefficiency.

1890 Superintendent of U.S. Census declares that the "frontier" no longer exists: the beginning of a growing recognition that noise problems cannot always be solved by "lighting out for the territories."

— Pedestrians begin complaining about the "noiselessness" of bicycles.

1894 Premiere of Richard Strauss's opera *Guntram*, a flop "mainly because the orchestration drowned out the singers."

1895 Physicist Wallace Clement Sabine is given the "impossible" task of improving the acoustics at the Fogg Lecture Hall in Boston. Birth of modern architectural acoustics.

1896 Women of Aurillac, France, sign a petition protesting the tenor bell of the town church for "battering us about the head and instilling sadness and grief in our hearts, banishing the sweet thoughts and tender feelings that we harbor toward sex."

1903 Wright Brothers fly the Spirit of St. Louis at Kitty Hawk. Aviation age is born.

1906 Julia Barnett Rice founds the Society for the Suppression of Unnecessary Noise in New York. Mark Twain accepts post of honorary chair of "Children's Branch."

— American composer Charles Ives composes *Central Park in the Dark*, "incorporating representations of city noises." See 1912 below.

— Victor Talking Machine Company introduces new-model Victrola phonograph.

1907 New York commissioner of police places a ban on barkers using megaphones on Coney Island.

— Federal legislation ("The Bennet Act") forbids the unnecessary blowing of whistles in harbors and ports.

1908 German philosophy professor Theodor Lessing publishes his essay "Noise: A Lampoon Against the Din of Our Lives" and helps to found *Deutscher Larmschutzverband* (German Association for Protection from Noise).

1910 League for the Hard of Hearing founded in New York.

1911 Chicago peddlers riot to protest silencing ban.

1912 Charles Ives retreats to Connecticut, declaring New York a noisy "Hell Hole."

1913 Italian Futurist painter Luigi Russolo publishes *The Art of Noises.*

1914 Acoustic scientist Wallace Sabine identifies noise as a "modern acoustical difficulty" with "urgent need" for "abatement." Among other contributions to the problem of office noise, he advises the Remington Company on reducing the noise of its typewriters.

1916 British surgeon Dan McKenzie publishes *The City of Din: A Tirade Against Noise.*

1918 First World War ends, having produced hundreds of shell-shocked veterans.

1920 Japanese governor says, "My first impression of New York was its noise."

1922 George Owen Squier patents a way to play a phonograph over electric power lines: "the technological foundation for Muzak."

— Russian composer Arseny Avraamov presents his *Symphony for Factory Whistles.*

1925 Bell Labs invents the audiometer.

1926 *Forum* magazine conducts investigation of noise in New York; one of first uses of audiometer.

1927 French George Antheil presents his *Ballet pour Instruments Mecanique et Percussion* in New York.

— Industrial researcher A. W. Kornhauser determines that typists produce 1.5 percent more lines in a noisy office than in a quiet one.

— First of the Nazis' Nuremberg rallies, with "unending shouts of 'Heil' joined with the music and the sounds of the fanfares and the beating of the drums."

1928 Decibel comes into popular use.

— Founding of the Acoustical Society of America.

— Almost two out of three American families own a car; one out of three owns a radio.

— Treaty of Paris outlaws war, a declaration interesting to consider in the context of similarly well-intentioned but ultimately futile initiatives to "outlaw" noise during the same era.

1929 Appointment of New York Noise Abatement Commission.

— Stock market crash inaugurates Great Depression, and deals a blow to the noise-abatement movement. People have more pressing worries.

1931 British physicist G. W. C. Kaye defines noise as "sound out of place."

1932 New York Noise Abatement Commission dissolved.

1933 Theodor Lessing (see 1908) assassinated in exile by two Czechoslovakian Nazis.

1934 Muzak Corporation born.

1935 German government sponsors a "noiseless week" to bolster "the strong nerves" of the nation.

1938 First Sturgis, South Dakota, motorcycle rally.

1945 Atom bomb dropped on Hiroshima, which survivors close to ground-zero remember as a "silent flash."

— At least 40,000 American GIs return with hearing loss.

1946 *United States v. Causby*. U.S. Supreme Court rules that the noise of military jets, which killed a farmer's chickens, violated his Fifth Amendment rights.

1948 French radio engineer Pierre Schaeffer presents *Concert of Noises*, part of which contains "the huffing, chugging, and whistling of six locomotives he had recorded."

— Bell Labs scientist Claude Shannon publishes "A Mathematical Theory of Communication" and launches the "digital age" of ceaseless "communication."

1955 With no noise beyond that of a quiet refusal, Rosa Parks sparks a social revolution.

1958 For the first time, more people cross the Atlantic Ocean by airplane than by boat.

1965 Bob Dylan "goes electric" and is booed at Newport Folk Festival.

— Beatles drowned out by their fans in Shea Stadium.

1966 McCoy Tyner leaves John Coltrane's band, claiming that its new style makes it impossible for him to hear and respond to the other musicians.

1967 Japan sets noise standards through "Basic Law for Environmental Pollution."

1969 U.S. National Environmental Policy Act (NEPA) mandates a review of environmental effects for any major development receiving federal funds.

1970 British acoustician Douglas Robinson develops a precise formula to analyze the phenomenon in which varying noise is generally more disturbing than steady noise.

— New York anti-noise activist Robert Baron publishes *The Tyranny of Noise*.

1971 Detroit promotes "muscle cars."

1972 U.S. Noise Control Act (NCA): "The Congress declares that it is the policy of the United States to promote an environment for all Americans free from noise that jeopardizes their health or welfare."

1974 EPA publishes Levels Document, disputing the adequacy of a DNL (day-night average sound level) of 65 decibels as an acceptable maximum level for exposure to aircraft noise.

1975 New York noise researcher Arline Bronzaft finds marked differences in student reading levels between children on the train-track side of a public school and their peers on the other side.

1976 English rock group The Who sets a concert loudness record of 126 dB at 32 meters. (The record was thereafter broken by other groups; e.g., Manowar's 129.5 dB in 1994.)

— EPA reverses its 1975 noise rules for supersonic transports to allow landing of Anglo-French Concorde at Kennedy and Dulles Airports.

1977 R. Murray Schafer publishes *The Tuning of the World.*

— Jacques Attali publishes *Noise.*

— U.S. National Research Council (NRC) publishes *Guidelines for Preparing Environmental Impact Statements on Noise.*

1978 Congress passes the Quiet Communities Act as an extension of the Noise Control Act of 1972.

1980 World Health Organization begins to address the health effects of community noise.

1981 Boston acoustics firm Bolt, Beranek & Newman issues its comprehensive report, *Noise in America: The Extent of the Problem.*

1982 Reagan administration closes the U.S. Office of Noise Abatement and Control.

1987 National Parks Overflights Act, the interpretation and enforcement of which are still under discussion and in dispute.

1989 U.S. troops use loud music to force the surrender of Panamanian president Manuel Norriega.

1993 Loud music and the recorded sounds of dying rabbits played nonstop for fifty-one-day siege of Branch Davidians at Waco, Texas.

1994 President Clinton issues Executive Order 12898 mandating "environmental justice."

1995 World Health Organization publishes a preliminary set of guidelines for community noise.

— In one of the most symbolic noise events of the century, "Mr. Apology" (New York conceptual artist Alan Bridges) is killed by a jet skier.

1996 European *Green Paper Report* addresses the environmental impacts of transportation.

1999 President Clinton estimates that over 2.3 million American are affected by aviation noise.

— Louisville International Airport in Kentucky dissolves and relocates the entire suburb of Minor Lane Heights as a way of solving its airport noise problem.

2000 Congress passes National Parks Air Tour Management Act.

— U.S. Park Service begins its Natural Sounds Program.

2001 Terrorists fly planes into World Trade Center, arguably changing the way many Americans hear the sound of aviation noise.

— RANCH project—"the first cross-national epidemiological study . . . to examine exposure-effect relations between aircraft and road traffic noise exposure and reading comprehension"—with data from the Netherlands, Spain, and the UK.

2002 European Union directs members to devise noise maps for areas near major railways, roads, and airports and for areas of dense population.

2003 BBC reports U.S. Army use of Metallica's "Enter Sandman" and Barney the Purple Dinosaur's "I Love You" in the interrogation of Iraqi prisoners.

2004 Acoustic weapons used in the second battle of Fallujah, Iraq.

— World Health Organization issues final draft of LARES (Large Analysis and Review of European Housing and Health Status) report.

2006 International Noise Awareness Day observed for first time in country of Ghana.

2007 Revised New York City Noise Code goes into effect.

— Protests against climate change and noise at English airports.

— Feature film *Noise*, in which the protagonist violently deactivates car alarms.

2012 Next round of European Union noise maps to be completed.

2020 European Union target date "to avoid harmful effects of noise exposure from all sources and preserve quiet areas."

2050 Date by which the United States is to have reduced its carbon emissions by 80 percent (and presumably a good portion of noise into the bargain), according to the Obama energy plan.

COMMON TERMS USED IN DISCUSSIONS OF NOISE

absorption the changing of sound energy into some other form, usually heat, after striking or passing through certain materials, e.g., acoustical tiles.

abutter applied to persons whose property lines are adjacent to those of a proposed development or existing installation. Abutting and hearing are not synonymous. The noise of a given installation may be audible to someone who is not an abutter.

acoustical consultant a person who gives informed advice on the actual or predicted behavior of designated sound sources within a given domain.

acoustical engineer someone qualified to design or modify structures or landscapes in order to alter the way sound waves act in a given place. Membership in the Institute of Noise Control Engineering (INCE) and certification by its board are strong indications that an engineer is fully qualified.

acoustics the science that studies the behavior of sound.

ambient the "normal" or "everyday" sound levels of a given environment; sometimes used exclusively of "natural" sounds (wind, water, etc.), sometimes inclusive of human-made sounds characteristic of the area under measurement.

annoyance one's subjective irritation, which may vary. Lack of annoyance does not necessarily mean lack of adverse effect, and vice versa.

architectural acoustics the application of acoustics to architecture in order to determine the behavior of sound sources within a building (e.g., music in an auditorium) or to mitigate the sound of exterior sources.

attenuation the decreasing level of sound as it travels between source and receiver. Conditions of terrain and atmosphere and the presence of structures will influence the degree of attenuation.

audible sound you can hear (or someone with normal hearing can hear). Not all sounds that affect an organism are audible to it.

A-weighting used of decibel readings measured or computed in such a way as to deemphasize very low and very high frequencies and to emphasize the middle and higher frequencies in a way similar to the de-emphasis and emphasis of human hearing. See also "PNL" and "weighted decibels."

broadband used of a sound that contains high, low, and middle frequencies, without specific tones or peaks in any one frequency. Most of the sounds we hear everyday have broadband characteristics.

community noise usually used for all noises except those within the industrial workplace and in natural settings; sometimes called "environmental noise."

decibels a logarithmic measure of sound pressure, usually written as dB.

DNL day-night average sound level, with weighted values for the nighttime hours, when noise is generally most disruptive; used throughout the United States, except in California.

dose-response a ratio of an amount of noise to the measured or reported effects on a human subject or subjects; also called "exposure-response." Researchers sometimes plot noise effects along a "dose-response curve."

emitter synonym for noise source.

environmental noise See "community noise."

FAA Federal Aviation Administration, the agency with sole jurisdiction of U.S. airspace and aviation noise.

frequency the number of oscillations per second of a sound wave or of a vibrating object.

habituation a term used to express the (often dubious) idea that noise annoyance decreases as one grows "used to it."

Hertz the basic unit of measure for frequency, usually written as Hz. The range of normal human hearing falls between 20 and 20,000 Hz.

impulsive used of noise that comes in "bursts" of short duration. Gun shots, car horns, breaking glass all produce impulsive sounds.

intermittent used of noise that is characterized by variable and often unpredictable occurrence and intensity.

ISO International Standardization Organization; often appears as a prefix to numerical standards set by this organization.

Lden "day-evening-night sound average," a more refined version of DNL, with weighted values for the evening (often 7:00 P.M. to 10:00 P.M.) and nighttime (often 10:00 P.M. to 7:00 A.M.) hours, when noise is generally most disruptive; used in California and the European Union.

Leq "equivalent sound level," the average noise level during a designated period, usually less than twenty-four hours. Leq5 would be the average noise level during a five-hour period.

Lmax the instantaneous maximum sound pressure level during a specific period of time.

masking the obscuring of one sound by another sound. Sometimes an undesirable sound is intentionally masked by a desirable one, desirability being determined by those in control of the masking.

neighbor noise used of that subset of community noise that comes exclusively from neighbors in nearby dwellings; often the most annoying and intractable of all noise sources.

NIHL noise-induced hearing loss.

noise event the occurrence within space and time of a given noise; also called "sound event." The resulting effects on humans or animals may or may not be confined to the duration of the noise event.

noise footprint the area acoustically affected by a given noise source.

noise map a geographical representation of the acoustic "terrain" of an area, often expressed with different colors, and often plotted for areas with high levels of transportation noise.

nosocusis hearing loss attributable to disease. See also "presbycusis" and "sociocusis."

NRC noise reduction coefficient, a number used to rate the effectiveness of sound-absorbent materials. The NRC was designed to rate the effectiveness of a material in reducing the frequencies most likely to interfere with speech. It gives no information relevant for high and low frequencies.

OSHA Occupational Safety and Health Administration.

PNL perceived noisiness level; rating of the "noisiness" of an aircraft. The rating is calculated using the frequencies of the sound, adjusted by the sensitivity of normal human hearing. Decibel measurements are often weighted (see also "weighted decibels") in order to achieve readings that approximate PNL.

presbycusis hearing loss attributable to aging. See also "nosocusis" and "sociocusis."

private nuisance used of a noise (or other annoyance) that interferes with a landowner's use or enjoyment of his or her property or with a right associated with it. In law, a private nuisance is a tort or civil wrong.

propagation the way sound travels over a given space or through a given medium.

public nuisance used of a noise that is judged to interfere with the public welfare and is thus prosecutable as a crime. An individual cannot initiate action in a public nuisance case.

pure tone a sound with most of its energy in a narrow frequency range.

receiver the person, creature, or instrument that registers a sound event.

reflection the bouncing of sound waves off hard surfaces.

reverberation multiple reflections (see "reflection") such that sound repeatedly reaches the receiver for a while after the sound source has stopped emitting energy. Particularly and intentionally noticeable in concert halls, reverberation may produce an unsatisfactory environment for listening to a speaker or to a small group of musical instruments.

sociocusis noise-induced hearing loss, as distinguished from hearing loss attributable to aging or disease. See also "nosocusis" and "presbycusis."

sound level meter an electronic device that measures decibel levels. Sound meters vary widely in sophistication and price, with high-end models costing thousands of dollars.

soundscape the acoustical parallel to a landscape, with the difference that it commonly includes both the sounds of a given environment and the way those sounds are subjectively perceived, valued, and interpreted by the people who hear them. The term is not always used in its broadest sense in this book.

standard a figure (e.g., >30 dB) or principle ("will not interfere with a reasonable person's enjoyment of his or her property") used as a guide or stated as a goal in sound ordinances, agency rubrics, and permitting processes. Standards are not the same as "statutes" unless incorporated into the language of the statute; statutes are only as good as their enforcement.

steady state used of a sound that does not vary in intensity.

stochastic resonance the tendency of small amounts of noise or other random signals to restore order to an operating system.

stressor anything that causes stress. Noise is a recognized stressor.

tinnitus a chronic ringing or buzzing in the ears that does not originate with an external sound source but may have been caused by a traumatic noise event, by a physiological disease, by certain drugs, or by psychological factors.

turbulence disordered movement of air, as with wind that is highly variable in speed and direction, which can have unpredictable effects on heard or measured sound levels.

vibration the rapid linear motion of a particle through space or an elastic medium about a point of equilibrium. All sound is vibration. The sounds of some frequencies may induce vibrations in buildings. Vibrations can have physiological effects even when they do not generate noise.

weighted decibels decibel readings that assign greater value (i.e., "give more weight") to some frequencies than to others.

white noise a sound that contains equal energy within all frequency bands of equal width, just as white light contains all colors within the visible spectrum. White noise is sometimes used to mask other sounds. See also "masking."

WHO World Health Organization. The WHO has been instrumental in establishing noise pollution as a global health hazard.

DECIBELS IN EVERYDAY LIFE AND EXTRAORDINARY SITUATIONS

Certain military weapons 185
Song of humpback whale in water 170
Twelve-gauge shotgun 160
Firecracker 140
Sound reaches level of "pain" 140
Loudest rock concerts 135
Civil Defense siren at 100 feet 130
Titanium golf club striking ball 128
Japanese Taiko drumming performance 120
Sound reaches level of "discomfort" 120
Chain saw 110
Concorde landing (3,280 feet from runway) 110
Jet flyover at 1,000 feet 110
Video arcade 110
MP3 player at maximum volume 105
Power lawnmower at 3 feet 95
Loud factory noise 90
Ice cream truck with amplified music 70–90
Apartment next to freeway 88
Low range for human hearing damage 85
Central hospital nursing station 82
Food blender at 3 feet 80
Urban high-density apartment 78
Vacuum cleaner at 3 feet 70
Handheld electronic games 68–76
FAA acceptable day-night average sound level 65
Normal conversation 60

Air conditioning unit at 100 feet 55
Light traffic 50
Distant bird calls 45
WHO guideline for average daytime hospital noise 45
Refrigerator 40
Rural residential 39
WHO guideline for average classroom noise 35
Wilderness area 35
Soft whisper at 5 feet 30
WHO guideline for average nighttime hospital noise 30
Ticking of watch near one's ear 20
Barely audible sound 10
Threshold of hearing 0

ORGANIZATIONS THAT DEAL WITH NOISE

This is but a small sampling of the many organizations that deal with noise. An Internet search will identify a number of others, including, perhaps, one in your area.

AMERICAN SPEECH AND HEARING ASSOCIATION (ASHA)

This is the professional organization for speech pathologists and audiologists. Its journal, *ASHA Leader*, often contains up-to-date articles on noise and its effects on hearing.

220 Research Boulevard
Rockville, MD 20850-3289
301-296-5700
www.asha.org

ACOUSTICAL SOCIETY OF AMERICA

Founded in 1928, this society is virtually as old as the decibel. It publishes the influential *Journal of the Acoustical Society of America.*

Suite 1NO1
2 Huntington Quadrangle
Melville, NY 11747-4502
516-576-2360
http://asa.aip.org

CENTER FOR HEARING AND COMMUNICATION (FORMERLY LEAGUE FOR THE HARD OF HEARING)

Since its founding in 1910, the League for the Hard of Hearing has been a tireless advocate for the hearing-impaired and for better education regarding noise and hearing loss. The new name went into effect in April 2009.

50 Broadway, 6th Floor
New York, NY 10004
917-305-7700
2900 W. Cypress Creek Road, Suite 3
Ft. Lauderdale, FL 33309
954-601-1930
www.chchearing.org

EPA (ENVIRONMENTAL PROTECTION AGENCY)

Though the EPA's Office of Noise Abatement and Control (ONAC) closed in 1982, the agency remains responsible for oversight of the Noise Control Act of 1972 and the Quiet Communities Act of

1978. It also provides useful information on its website, including some of the history of ONAC.

Ariel Rios Building
1200 Pennsylvania Avenue, NW
Washington, DC 20460
202-272-0167
www.epa.gov

EUROPEAN ACOUSTICS ASSOCIATION (EAA)

The merged journal *Acta Acustica United with Acustica* is a highly regarded publication that includes articles by world-renowned authorities on noise and sound. See www.Acta-Acustica-United-with-Acustica.com.

FRIENDS OF SILENCE

Neither a noise organization nor a very noisy one, Nan Merrill's Friends of Silence publishes a free newsletter dedicated to the appreciation of physical and spiritual silence, an appealing feature of which is pertinent quotations from writers, artists, and philosophers. You have to walk to your mailbox to get it.

11 Cardiff Lane
Hannibal, MO 63401

HALT OUTRAGEOUS RAILROAD NOISE (HORN)

Robert B. Simpson's organization serves those who find either side of the tracks the loud side.

P.O. Box 494
Mt. Tabor, NJ 07878-0494
simpson@carroll.com

HEARING EDUCATION AND AWARENESS FOR ROCKERS (HEAR)

Dedicated to promoting noise awareness among musicians, the group offers a video, *Can't Hear You Knocking.*

1405 Lyons Street
San Francisco, CA 94115
415-409-3277
www.hearnet.com

INTERNATIONAL COMMISSION ON BIOLOGICAL EFFECTS OF NOISE (ICBEN)

The commission meets every five years to review the negative health effects of noise. Conferences feature papers and displays on a wide array of health-related topics. The papers and proceedings are available online at http://icben.org.

INTERNATIONAL INSTITUTE OF NOISE CONTROL ENGINEERING (INCE)

This organization's annual Inter-Noise conferences bring together scholars and engineers from around the world to discuss issues of engineering, public policy, and auditory culture. The emphasis is on the technical side, but the approach is interdisciplinary and international. The INCE-USA co-sponsors Inter-Noise and holds its own conference, NoiseCon, once a year in the United States.

Business Office, INCE-USA
9100 Purdue Road, Suite 200
Indianapolis, IN 46268
317-735-4063
ibo@inceusa.org

NIOSH (NATIONAL INSTITUTE FOR OCCUPATIONAL SAFETY AND HEALTH)

Part of the Centers for Disease Control and Prevention, "NIOSH is responsible for conducting research and making recommendations for the prevention of work-related illnesses and injuries."

Centers for Disease Control and Prevention
1600 Clifton Road
Atlanta, GA 30333
800-232-4636
www.cdc.gov/niosh/topics/noise

N.O.I.S.E. (NATIONAL ORGANIZATION TO ENSURE A SOUND-CONTROLLED ENVIRONMENT)

Affiliated with the National League of Cities, this "coalition of locally elected officials and industry stakeholders" aims to be "America's community voice on aviation issues."

415 Second Street NE 210
Washington, DC 20002
202-544-9844
contact@aviation-noise.org.

NOISE FREE AMERICA

Founded and directed by political science professor Ted Rueter, Noise Free America is an aggressively activist organization that claims chapters across the United States. NFA publishes an online newsletter and a manual, *How to Fight Noise*. It also issues a "Noisy Dozen" list each month.

1971 Western Avenue (# 1111)
Albany, NY 12203
877-664-7366
www.noisefree.org

NOISEOFF

Founded in 2004 by Richard Tur and based in Queens, NoiseOff maintains a website focused on specific noise sources (boom cars, motorcycles, etc.). See www.noiseoff.org.

NOISE POLLUTION CLEARINGHOUSE

Readers of this book will recognize the name of NPC's director Les Blomberg, who, in addition to maintaining one of the world's largest databases on noise, frequently advises individuals and community groups involved in noise disputes. NPC's projects have included an initiative for Quiet Lawns and the review and rating of appliances for their noise emissions. Blomberg's emphasis is on promoting awareness and a culture of respect over direct activism. NPC's website lists many noise-related organizations.

P.O. Box 1137
Montpelier, VT 05601-1137
888-200-8332
www.nonoise.org

NOISE WATCH

This Canadian-based organization began in 1994 when Cindy Davidson wrote to the *Toronto Star* asking people concerned about noise to contact her.

59 Manresa Court
Guelph, Ontario N1H 6J2
519-826-5833

OCCUPATIONAL SAFETY AND HEALTH ADMINISTRATION (OSHA)

The Office distributes publications relating to industrial noise and hearing conservation. See www.osha-sic.gov/Publications.

OREGON HEARING RESEARCH CENTER

This group operates an educational program for elementary, middle, and high school students called "Dangerous Decibels."

Mail Code NRC04
3181 SW Sam Jackson Park Road
Portland, OR 97201
503-494-0670
www.dangerousdecibels.org

THE RIGHT TO QUIET SOCIETY

The full name continues "for Soundscape Awareness and Protection." Based in British Columbia, where the soundscape movement began, the society aims for "a world where quiet is a normal part of life and where it is possible to listen to the sounds of nature without the constant intrusion of machine noise and artificial stimuli."

#359, 1985 Wallace Street
Vancouver, BC V6R 4H4
604-222-0207
www.quiet.org

UNITED KINGDOM NOISE ASSOCIATION

Founded by John Stewart and Val Weedon, the UKNA has published accessible documents on a number of noise issues, ranging from public housing to the siting of industrial wind farms. The organization also publishes a newsletter, *Noise News*.

Broken Wharf House
2 Broken Wharf
London EC4V 3DT
0207 329 0774
info@unka.org.uk

WORKERS' COMPENSATION BOARD OF BRITISH COLUMBIA

Research on occupationally induced hearing loss among sawmill workers in British Columbia has been some of the most conclusive to date. The Board offers "The Hearing Video" for education purposes.

Box 5350 Stn Terminal
Vancouver, BC V6B 5L5
604-232-9704
www.wcb.bc.ca

PRACTICAL CONSIDERATIONS FOR NOISE DISPUTES

ANSWERS TO ARGUABLE STATEMENTS

Obviously the answers that follow are only as good as the reasonableness of the party making the noise. In general, and especially in cases of neighbor noise, it is always best to avoid an argument if possible. Assume the best ("You may not be aware of this, but . . . ") and give a person credit for good intentions. Remember that not all people have the same sensitivity to noise. Some people are simply being honest when they react to your complaint with surprise. It doesn't necessarily mean they won't respond positively.

After a while you get used to a noise.

Studies show that most people do not get used to noise annoyance. In fact, over time the annoyance is likely to increase. People tend to stop complaining once they realize that no one is going to take corrective action.

Noise annoyance is all subjective anyway.

As a measurable acoustical phenomenon, a loud sound is anything but subjective. Audible vibrations are not subjective. Noise-induced hearing loss is not subjective either. Nor are sleep deprivation and its physiological and psychological effects. The World Health Organization is not a subjective body.

What is more, even so-called subjective responses can yield objective results. If I experience repeated psychic stress because of my "subjective" distaste for a given noise, that stress can still harm me.

You're just complaining because you don't like (my kind of) music.

This is like saying that a person who opposes rape doesn't like sex. In fact, there is convincing anecdotal evidence to suggest that people who truly love music are the most likely to resent having it forced on them.

I have just as much right to make noise as you have to be quiet.

You are absolutely correct—in your own dwelling, you do. But when your noise enters my dwelling, your "right" stays at the door. In a public as opposed to a private space we each have a right to expect, and if need be to negotiate, our fair share of its enjoyment, but no one has the right to co-opt the entire space, including the entire acoustical space, for his or her own purposes.

Nobody else is complaining.

A lack of complaint doesn't mean a lack of distress. Studies have shown that a "falling off" of complaints sometimes correlates to an increase in noise annoyance. Often people stop complaining, or don't complain to begin with, because they feel helpless.

Noise is the price we pay for progress.

Usual meaning: Noise is the price you should be willing to pay for my progress. In other words, you should be willing to lose some of your property value or the enjoyment of your home so that I can maximize the profits of my property or the enjoyment of my home. Robbing Peter to pay Paul is not progress. It's robbery.

A noise of short duration has no lasting effect.

The same might be said for a slap in the face. In each case, the effects are contingent. If a repeated noise is sporadic and unpredictable, a person's concentration, sleep, or sense of well-being will be affected during the quiet intervals of anticipation. Even a single, nonrepeatable noise event can have effects that last far beyond its duration. If, after a fifteen-second noise event, you miss the last twenty-five words of a performance or lecture you've been attending for the past two hours, how long does the effect last? If you paid for your attendance, how much did that brief interruption cost you?

You're just an elitist.

Unlike people who throw around words like *elitist*. Elitism is a strange charge coming from someone with the means—usually the pricey mechanical or electronic means—to produce a substantial noise annoyance. Elitist relative to what? The elitist in an armed robbery is the person holding the gun.

Okay, so what's your solution?

A person who calls an ambulance is not obliged to prescribe his or her own medical treatment; much less is he or she obliged to design a national health care program. Private citizens who complain about an established or proposed noise source are not obliged to provide remedies for its mitigation, much less "solutions" for the larger environmental or economic problems (climate change, unemployment, what people are supposed to do with the dune buggies they got for Christmas) that a given project is touted as addressing.

What if everybody in the world said "Not in my backyard"?

The world would be a better place. First, because we'd have achieved a level of distributive justice that permitted everyone to have a backyard. Second, because we'd have reached a level of culture that encouraged people to love their backyards. Third, because we'd have reached a level of environmental sustainability in which people spent more time in their backyards. And, finally, because we'd have reached a stage of political maturity in which everybody had to come to the table and discuss what they wanted and what they were willing to pony up to get it. In other words, we would have reached a stage in which one person's peace was not purchased at the price of another person's noise.

STRATEGIES FOR THOSE SUFFERING FROM NOISE

1. Always consider, and as much as possible respect, the degree to which someone is personally invested in his or her noise. For example, is the noise source a tool the person uses to make his or her living? An object that serves to identify the person in some way (as a biker, a musician)? A member of the person's family (a noisy child or a barking dog)? Calling a noise "stupid," "obnoxious," or "entirely unnecessary" is sometimes close to calling a person stupid, obnoxious, or entirely unnecessary. Not a good way to go.

2. If the noise is amplified music, focus on the volume of the music as opposed to the type of music. If you happen to like the type of music being played, say so!

3. Get to know people before a noise dispute erupts. Knowing your neighbors will determine not only how you hear their noise but also how your neighbors will hear you if you complain about it. In addition to the likelihood of being ineffective, confronting a perfect stranger about noise also has the potential to be dangerous. Someone who "doesn't care" whom he disturbs might not care whom he hurts either.

4. Know the noise code for your town or city. It is a mistake to assume that the police will know the code or will necessarily take action if they do. Police are likely to be just as annoyed by repeated calls from the same complainant as you are by the source of your complaint. But if you show a knowledge of the code from the get-go, you're likely to be taken more seriously should you make a complaint.

5. To stand and fight a noise is a braver course than to cut and run, but ultimately it may not be the most successful. Weigh your costs and benefits, and talk to people who've fought similar battles. Especially in cases of noise that will last only for a limited duration (graduation parties, civic celebrations, etc.), consider the possibility of getting away.

STRATEGIES FOR THOSE LIKELY TO BE MAKING NOISE

1. Beware of using noise as a weapon against people who irritate you. They may decide to retaliate in kind, or to escalate the conflict with a more lethal weapon.

2. Letting people know about a potential noise ahead of time is a good way to reduce its annoyance. As with number 3 above, knowing your neighbors is probably the best sound-mitigation tool in existence. If you know your neighbors, you are likely to know if there is sickness in their household, a new baby, somebody who works a night shift and sleeps during the day. Also, if you can tell people when a noise event is likely

to end, that may reduce stress. "Not knowing when [or that] the noise will end" has been cited as a factor for those who suffer from it.

3. Inviting people to a noisy celebration often makes it more tolerable to them—even if they decide not to come.

4. In the case of a proposed business or development project, recognize that misrepresenting the noise potential is only going to increase the noise annoyance once you go into operation. By then, of course, it may be "too late" for the affected parties to do anything, but it will also be too late for you to establish the credibility you'll want should you decide to expand your operation or open a new installation elsewhere. "They lied to us last time."

5. Studies show that noise annoyance is reduced whenever people (a) feel that a permitting process has been fair, (b) retain some control over the noise source, and (c) derive economic benefit from the noise source. To the extent that you allow a community an investment in your operating procedures and profits, you reduce the noise they hear.

THINGS FOR COMMUNITIES AND INDIVIDUALS TO KEEP IN MIND DURING A PERMITTING PROCESS

1. There are a number of sound metrics, none of which is entirely satisfactory for describing what you will actually hear and how it will affect you. Decibels, for example, can be A-weighted, B-weighted, and C-weighted, with the first usually used as an approximation of how the loudness of the sound will be perceived. (See Common Terms Used in Discussions of Noise.) A decibel is not a very helpful measure of sound unless considered along with the duration of the sound, the nature of the sound, and the distance at which the decibel level applies. Finally, beware of trying to compute relative decibel readings with simple arithmetic. Decibels are logarithmic, and it probably won't hurt your credibility to use that word in a hearing or discussion. The general rule of thumb is that perceived loudness will double for every increase of 10 decibels.

2. Be sure to ask who is paying the expert witness. Ask how many times that witness or his or her firm has represented individual citizens or communities as opposed to private corporations with the means to pay. If the developer receives tax subsidies, ask your political representatives why you, as a taxpayer, are obliged to pay for the developer's consultants as well as your own. These questions are not likely to gain you any change in how the game is played, but they will make the game clearer to media reporters and to undecided citizens.

3. A noise code is only as good as the willingness and ability to enforce it. A civil authority can impose all kinds of acoustical restrictions, and a developer can agree to every one of them. Until the sound source is in operation, these assurances mean about as much as promises made during an act of seduction. What citizen groups need

to ask is this: Will those regulations be enforceable once the facility is in place? For example, who will be willing or able to shut down an airport runway if its decibel levels exceed the DNL promised by the developer or required by the licensing authority? The obvious answer is no one. Furthermore, once the source is in operation, is it even possible to determine whether the sound levels meet the restrictions? Is the proper equipment available? Does anyone have the authority and the knowledge to use it?

4. Beware of percentages and averages. For example: You are told that a noise event will take place "no more than 2 percent of the nighttime hours." If a night is taken as eight hours long, 2 percent is a little under ten minutes. Bearable, right? Unless, after ten continuous minutes of ear-shattering noise, it takes you two hours to get back to sleep. Or, unless there are twenty separate noise events, roughly two to three per hour, each only thirty seconds in duration, but each loud enough to wake you up. A night that is 2 percent noisy can be 100 percent sleepless!

5. Beware of promises to soundproof your home. First of all, soundproofing is not uniformly effective with all types of sound; it is notably less effective with low-frequency noise. Second, recognize that soundproofing can work only within the interior of your dwelling and that a "soundproofed home" can amount to a quiet prison. There is also the ethical question of where noise ought to be mitigated—at the boundary of the source or at the boundary of the receiver. Who's creating the problem in the first place?

A PERSONAL NOISE CODE

1. I will turn off all noise sources over which I have immediate control. If the TV is driving me nuts, why do I have it on?

2. I will not patronize businesses that subject their customers or employees to loud or manipulative "soundtracks." I will make it clear why I am taking my business elsewhere. I will do so quietly.

3. I will do as much as I can to reduce the impact of my own noise on others in hotels, restaurants, neighborhoods, and public spaces. In other words, I will do my best to express my awareness that others exist besides myself.

4. As much as possible, I will do physical chores with bodily powered tools. I will rake my leaves. If someone invents an electric fork, I will do without one. I will keep the machines I do use in good working order.

5. I will play music and make conversation in the same way that a healthy person eats food. In other words, I will do so with intention, attention, hospitality, and a sense of pleasure. I will always have a better aim than "filling space" or "killing time."

6. I will not own an animal whose needs for companionship, exercise, and diversion I cannot meet.

7. I will travel less, or at least less wastefully. I will walk or ride a bike whenever I can. I will prefer public to private transportation. I will reduce the number of times I fly.

8. I will buy more locally raised food. I will also support initiatives to provide locally produced, owned, and consumed energy—even at the cost of hearing more of my own noise.

9. I will vote for political candidates who support quiet diplomacy. I will always insist that blowing apart human beings shall be the option of last resort, or no option at all.

10. I will get to know my neighbors and invite more of them to my celebrations.

11. In both my social interactions and my exercise of political power I will strive to give my fellow citizens compelling evidence that they are recognized and valued. In other words, I will reduce their need to make noise in order to gain attention.

12. I will not take the sounds I love for granted. I will cherish and protect them. To the best of my ability, I will try to listen to them.

NOTES

CHAPTER 1: NOISE IS INTERESTED IN YOU

3 *Jazz Age roar* Raymond W. Smilor, "American Noise, 1900–1930," in *Hearing History: A Reader*, ed. Mark M. Smith (Athens: University of Georgia Press, 2004), 319–330. See also Emily Thompson, *The Soundscape of Modernity: Architectural Acoustics and the Culture of Listening in America, 1900–1933* (2002; Cambridge, MA: MIT Press, 2004), 148–157, as well as Karin Bijsterveld, *Mechanical Sound: Technology, Culture, and Public Problems of Noise in the Twentieth Century* (Cambridge, MA: MIT Press, 2008), 161–171.

3 *New York Noise Abatement Commission* Thompson, *Soundscape of Modernity*, 157–168.

4 *Norse myth* H. R. Ellis Davidson, *Gods and Myths of Northern Europe* (Harmondsworth, UK: Penguin, 1964), 34.

4 *Your list will include children* Gary Evans and others, "Community Noise Exposure and Stress in Children," *Journal of the Acoustical Society of America* 109, no. 3 (March 2001): 1023–1027. See also William Hal Martin and others, "Noise-Induced Hearing Loss in Children: Preventing the Silent Epidemic," *Journal of Otology* 1 (2006): 11–21. Martin and his co-authors note that "children from families with low poverty-to-income ratios have more high-frequency hearing loss than children from the middle and high poverty-to-income ratios, suggesting that class disparities may exist" (12). See note for *the poor* below.

4 *the elderly* Birgitta Berglund, Thomas Lindvall, and Dietrich H. Schwela, eds., *Guidelines for Community Noise* (Geneva: World Health Organization, 1999), 53. See also "Older Adults' Speech-Processing Difficulties May Stem from 'Fast, Noisy Talk,' Not Deafness," *Science Daily*, http://www.sciencedaily.com/releases/1998/10/981023073930.htm.

4 *the physically ill* Berglund, Lindvall, and Schwela, *Guidelines for Community Noise*, xiii, 53. See also James E. West and others, "Characterizing Noise in Hospitals" (paper presented at Inter-Noise, Honolulu, 3–6 December 2006).

5 *racial minorities* U.S. Census "American Housing Survey" figures cited by Les Blomberg in "Acoustical Slums, Green Buildings, and the Acoustics of Sustainability" (paper presented at "Noise! Design, Health and the Urban Soundscape," Graduate School of Architecture, Planning and Preservation, Columbia University, New York, 21 September 2009).

5 *people with autism* See Roy Richard Grinker, *Unstrange Minds: Remapping the World of Autism* (New York: Basic Books, 2007), 277, 279; Laura Schreibman, *The Science and Fiction of Autism* (Cambridge, MA: Harvard University Press, 2005), 41–42; and Temple Grandin, *Thinking in Pictures: And Other Reports from My Life with Autism* (1995; New York: Vintage, 2006), 107–108. See also Tony Attwood, *The Complete Guide to Asperger's Syndrome* (London: Jessica Kingsley Publishers, 2007), 271–279. An estimated 70–85 percent of children with Asperger's syndrome have an "extreme sensitivity" to certain sounds. Other senses can be similarly affected. "Some adults with Asperger's syndrome consider their sensory sensitivity has a greater impact on their daily lives than problems with making friends, managing emotions and finding appropriate employment."

5 *the poor* For a sample of the research, see Robin R. Sobotta, Heather E. Campbell, and Beverly Owens, "Aviation Noise and Environmental Justice: The Barrio Barrier," *Journal of Regional Science* 47, no. 1 (2007): 125–154. See also Yvonne de Kluizenaar and others, "Hypertension and Road Traffic Noise Exposure," *Journal of Occupational and Environmental Medicine* 49 (2007): 484–492. A poignant depiction of the perceived auditory entitlements of different social classes is found in Elizabeth Strout's acutely class-conscious novel, *Amy and Isabelle* (New York: Random House, 1998), 19. "A dog, chained outside in the cold, sometimes barked at the reindeer all night, but no one thought to call the owner or the police, as they would certainly have done across the river, where people expected, or demanded, a good night's sleep."

5 *laborers* Pierre Deshaies and others, "Noise as an Explanatory Factor in Work-Related Fatality Reports: A Descriptive Study" (paper presented at 9th International Congress on Noise as a Public Health Problem, ICBEN, Foxwoods, CT, 2008). See also Charlie Fidelman, "Noise Can Kill, Health Report Finds," *Montreal Gazette*, 24 November 2007, as well as Thomas L. Bean, "Noise on the Farm Can Cause Hearing Loss," Agriculture and Natural Resources, Ohio State University, 2008, http://ohioline.osu.edu/aex-fact/pdf/AEX_590_08.pdf.

5 *prisoners* "You try to read, and find you've been reading the same paragraph for hours. The noise level is high. You can't think or concentrate. . . . When the lights go out you lie there, and relief comes only between midnight and breakfast. You stay up all night enjoying the tremendous relief. The noise which literally vibrates your brain is gone." Jack Henry Abbott, *In the Belly of the Beast: Letters from Prison* (New York: Random House, 1981), 66.

5 *able to kill fish* See Jack Boulware, "Feel the Noise," *Wired 8.10*, http://www.wired.com/wired/archive/8.10/stereocar_pr.html.

5 *footnote, safety standards . . . allow toys* Jane E. Brody, "All That Noise Is Damaging Children's Hearing," *New York Times*, 9 December 2008, http://www.nytimes.com/2008/12/09/health/09brod.html?_r=1.

5 *weapons fire* Sharon M. Abel, "Hearing Loss in Military Aviation and Other Trades: Investigation of Prevalence and Risk Factors," *Aviation, Space, and Environmental Medicine* 76, no. 12 (December 2005): 1128.

7 *"learned helplessness"* Gary W. Evans and others, "Community Noise Exposure and Stress in Children," *Journal of the Acoustical Society of America* 109, no. 3 (March 2001): 1026.

7 *normal human speech* Karl D. Kryter, *The Handbook of Hearing and the Effects of Noise: Physiology, Psychology and Public Health* (New York: Academic Press, 1994), 291–295.

7 *footnote, loudness of average American conversation* Cited in Richard Mahler, "The Human Cost of Silence Lost: How a Noisy Environment Hurts Our Health," in *Thrillcraft: The Environmental Costs of Motorized Recreation*, ed. George Wuerthner (White River Junction, VT: Chelsea Green, 2007), 38.

8 *cost . . . due to sleeplessness* "You're Getting Very Sleepy: More Sleep Would Make Most Americans Happier, Healthier and Safer," American Psychological Association, APA Online, www.psychologymatters.org/sleep.html, 12 May 2004.

8 *prescriptions* Bruce L. Lambert, Ken-Yu Chang, and Swu-Jane Lin, "Effect of Orthographic and Phonological Similarity on False Recognition of Drug Names," *Social Science & Medicine* 52 (2001): 1843–1857. See also Bruce L. Lambert and others, "Frequency and Neighborhood Effects on Auditory Perception of Drug Names in Noise" (paper presented at Noise-Con, Minneapolis, 17–19 October 2005).

8 *ovenbirds* Lucas Habib, Erin M. Bayne, and Stan Boutin, "Chronic Industrial Noise Affects Pairing Success and Age Structure of Ovenbirds *Seiurus aurocapilla*, *Journal of Applied Ecology* 44, no. 1 (February 2007): 176–184.

8 *ground squirrels* Lawrence Rabin, Richard G. Coss, and Donald H. Owings, "The Effects of Wind Turbines on Antipredator Behavior in California

Ground Squirrels (*Spermophilus beecheyi*), *Biological Conservation* 131, no. 3 (August 2006): 410–420.

8 *effects of noise on a variety of animals* See Gordon Hempton and John Grossman, *One Square Inch of Silence: One Man's Search for Natural Silence in a Noisy World* (New York: Free Press, 2009), especially Chapter 8, "Nature's Symphony in Decline," in which the authors note: "The avian choir is not just shrinking, but forgetting its repertoire"—due to noise, 185. The authors also recount the possibly fatal effects of noise on the Yantze River dolphin, once known as the "Goddess of the Yangtzee," a creature that was virtually sightless and relied on its "sonar-based sensory system to navigate and feed." Its seeming disappearance may represent "the first mammal extinction in 50 years" (186).

8 *desert kangaroo rat* Autumn Lyn Radle, "The Effect of Noise on Wildlife: A Literature Review," March 2007, http://interact.uoregon.edu/MediaLit/wfae/library/articles/radle_effect_noise_wildlife.pdf, 8.

9 *Tennessee snail darter* John Fleischman, "Counting Darters, Endangered Fish Species," *Audubon* 98, no. 4 (July 1996): 84. See also Jeffrey St. Clair, "Glory Boy and the Snail Darter," *CounterPunch*, 3–4 March 2007, http://www.counterpunch.org/stclair03032007.html.

10 *footnote, the number of hours we work* Bill McKibben, *Deep Economy: The Wealth of Communities and the Durable Future* (New York: Henry Holt, 2007), 114–115.

11 *anti-noise activists of the nineteenth and early twentieth centuries* See sources for "Jazz Age roar" above.

13 *Kathy Jackson* Kevin Cole, "Woman Shot in Battle over Loud Music Dies," *Omaha World Herald*, 30 May 2008.

13 *James Eckenrode* Alex Branch, "Man, 22, Slain After Argument over Noise," *Fort Worth Star-Telegram*, 10 February 2008, B-1.

14 *"a tragic story"* Karin Bijsterveld, *Mechanical Sound*, 234.

15 *adverse and well-documented physiological consequences* Berglund, Lindvall, and Schwela, *Guidelines for Community Noise*, 39–42 (on hearing loss), 47–48 (on cardiovascular and other physical effects), 140 (on lower birth weight). See also Evans and others, "Community Noise Exposure," as well as Hildegard Niemann and Christian Maschke, *WHO LARES* [World Health Organization Large Analysis and Review of European Housing and Health Status]: *Final Report* (Berlin: Berlin Center of Public Health, 2004), EUR/04/5047477.

16 *Adolf Eichmann* See David Cesarani, *Becoming Eichmann: Rethinking the Life, Crimes, and Trial of a "Desk Murderer"* (New York: Da Capo, 2006), 36, 210.

See also Roger A. Salerno, *Beyond the Enlightenment: Lives and Thoughts of Social Theorists* (Westport, CT: Greenwood, 2004), 144.

16 *Ted Bundy . . .* Peter Vronksy, *Serial Killers: The Method and Madness of Monsters* (New York: Berkley Books, 2004), 133.

16 *David Berkowitz* David Montaldo, "David Berkowtiz—The Son of Sam," http://crime.about.com/od/murder/p/sonofsam.htm.

16 *Ted Kaczynski* Gary Greenberg, "In the Kingdom of the Unabomber," *McSweeney's* 3 (1999): 70–71. A shorter, more recent version of Greenberg's essay appears in his *The Noble Lie: When Scientists Give the Right Answers for the Wrong Reasons* (New York: Wiley, 2008), 103–126.

16 *"There's a little bit of the Unabomber"* Robert Wright, cited in Greenberg, "In the Kingdom of the Unabomber" (78).

17 *Hitler . . .* He made the statement in 1939 in his *Manual of German Radio.* Diane Ackerman, *A Natural History of the Senses* (New York: Vintage, 1995), 188.

20 *"the science of happiness"* Daniel Nettle, *Happiness: The Science Behind Your Smile* (Oxford: Oxford University Press, 2005), 83–85.

CHAPTER 2: THE UNWANTED SOUND OF EVERYTHING WE WANT

21 *"It was a sound you were obliged to take personally"* Ian McEwan, *Atonement* (2001; New York: Anchor, 2003), 222.

21 *people in the ancient world were capable of reading silently* See G. L. Hendrikson, "Ancient Reading," *The Classical Journal* 25, no. 3 (December 1929): 182–196. See also Alberto Manguel, *History of Reading* (New York: HarperCollins, 1996), 41–53.

22 *Augustine of Hippo* Hendrikson, "Ancient Reading," 185–186. The passage is found in the sixth book of Augustine's *Confessions.*

22 *footnote, as a provincial . . . Augustine* Bernard M. W. Knox, "Silent Reading in Antiquity," *Greek, Roman, and Byzantine Studies* 9, no. 1–4 (Spring 1968): 422.

23 *paper mill in central Maine* I am indebted to Tony Lyons of NewPage in Rumford, Maine.

23 *"most prevalent irreversible occupational hazard"* Birgitta Berglund, Thomas Lindvall, and Dietrich H. Schwela, eds., *Guidelines for Community Noise* (Geneva: World Health Organization, 1999), 39.

24 *footnote, "Transportation noise . . . environmental noise pollution"* Berglund, Lindvall, and Schwela, *Guidelines for Community Noise,* 24.

25 *Biju* Kiran Desai, *The Inheritance of Loss* (New York: Grove Press, 2006), 255, 55.

25 *Lenny* Harold Pinter, *The Homecoming* (New York: Grove Press, 1965), 28.

25 *Death Cab for Cutie* Reference to "Marching Bands of Manhattan," *Plans*, Atlantic Records, 2005.

25 *people who suffer from tinnitus* American Tinnitus Association, "How Many People Have Tinnitus," in Frequently Asked Questions, http://www.ata.org/about-tinnitus/patient-faq1.

25 *a ringing or buzzing . . . or a sound like crickets* American Tinnitus Association, FAQ. See also Elizabeth Willingham, "Tinnitus," Baylor College of Medicine website, http://ww.bcm.edu/oto/grand/07_22_04.htm, and Groopman below.

25 *tinnitus was believed to indicate mystical awareness* Bernard Dan, "Titus's Tinnitus," *Journal of the History of the Neurosciences* 14, no. 3 (September 2005): 210.

26 *a 64-year-old retired machine repairman* Jerome Groopman, "That Buzzing Sound: The Mystery of Tinnitus," *New Yorker*, 9 and 16 February 2009, 43.

26 *Tinnitus has . . . many causes* Groopman, "That Buzzing Sound," 42–44. See also American Tinnitus Association, FAQ.

26 *Department of Veterans Affairs* Groopman, "That Buzzing Sound," 46.

26 *derives from the same Latin root* "Noise," *Oxford English Dictionary*, 2nd ed. (New York: Oxford University Press, 1989).

27 *neonatal hospital units . . . and "preemies"* Matt Kryger, "A Placid Place for Preemies," *Indianapolis Star*, 2 February 2008.

27 *"sound out of place"* Karin Bijsterveld, *Mechanical Sound: Technology, Culture, and Public Problems of Noise in the Twentieth Century* (Cambridge, MA: MIT Press, 2008), 240.

27 *tinnitus can be triggered in some people . . . by placing them in a silent environment* Groopman, "That Buzzing Sound," 42.

27 *Herman Gombiner* Isaac Bashevis Singer, "The Letter Writer," in *The Collected Stories* (New York: Farrar, Straus and Giroux, 1982), 263.

27 *"the sounds of other people"* Catherine Guastavino, "The Ideal Urban Soundscape: Investigating the Sound Quality of French Cities," *Acta Acustica United with Acustica* 92 (2006): 947.

28 *BBC agreed to install a noise machine* Kevin Maguire, "BBC Cheers Up Lonely Staff with the Chit-Chat Machine," *Guardian*, 14 October 1999, http://www.guardian.co.uk/media/1999/oct/14/bbc.uknews.

28 *"colonization of silence"* Andrew Waggoner, "The Colonization of Silence," 2007, http://www.newmusicbox.org.

28 *Les Blomberg* See www.nonoise.org.

29 *footnote, how we define sensitivity* Researchers Irene van Kamp and Hugh Davies give the figure of 12–15 percent for that segment of the population

"extremely sensitive to noise." See "Environmental Noise and Mental Health: Five-Year Review and Future Directions" (paper presented at 9th International Congress on Noise as a Public Health Problem, ICBEN, Foxwoods, CT, 2008). See also Wolfgang Ellermeier, Monika Eigenstetter, and Karin Zimmer, "Psychoacoustic Correlates of Individual Noise Sensitivity," *Journal of the Acoustical Society of America* 109, no. 4 (April 2001): 1464–1473. For a critique of noise-sensitivity as a tenable idea, see Dylan M. Jones and D. R. Davies, "Individual and Group Differences in the Response to Noise," in *Noise and Society*, ed. Dylan M. Jones and Antony Chapman (New York: John Wiley & Sons, 1984), 125–154.

29 *the Occupational Safety and Health Administration requires . . . hearing protection* See "Occupational Noise Exposure—1910.95," OSHA website. http://www.osha.gov/pls/oshaweb/owadisp.show_document?p_table=standards&p_id=9735.

29 *footnote, The Who* Tudor Vieru, "Searching for the Loudest Band in the World," *Softpedia*, 21 February 2009, http://news.softpedia.com/news/Searching-for-the-Loudest-Band-in-the-World-105105.shtml.

29 *he wants his MP3 player to be loud* Jane E. Brody, "All That Noise Is Damaging Children's Hearing," *New York Times*, 9 December 2008, http://www.nytimes.com/2008/12/09/health/09brod.html?_r=1.

30 *"sonic abuse"* Jamie C. Kassler, "Musicology and the Problem of Sonic Abuse," in *Music, Sensation, and Sensuality*, ed. Linda Phyllis Austern (London: Routledge, 2002), 321–333.

30 *60-year-old man* Karl Kryter, *The Handbook of Hearing and the Effects of Noise: Physiology, Psychology and Public Health* (New York: Academic Press, 1994), 272.

30 *"unlearned sensation of unwantedness"* Kryter, *The Handbook*, 53.

30 *neural processing centers* Kryter, *The Handbook*, 24.

31 *"a man of humanity in Europe"* Adam Smith, *Theory of Moral Sentiments*, 1st American edition (1759; Philadelphia: Anthony Finley, 1817), 215.

31 *Noise causes hearing loss* Office of Noise Abatement and Control, *Noise Effects Handbook: A Desk Reference to Health and Welfare Effects of Noise* (Fort Walton Beach, FL: National Association of Noise Control Officials, 1981), 2–7. For a succinct overview, see also Bart Kosko, *Noise* (New York: Viking, 2006), 48–55.

31 *new-style titanium clubs* Malcolm A. Buchanan and others, "Is Golf Bad for Your Hearing?" *British Medical Journal* 337 (20–27 December 2008): 1437–1438.

32 *16,000 fine hairs* Kosko, *Noise*, 52.

32 *footnote, hair cell regeneration* Brenda Ryals, "Hair Cell Regeneration: How It Works and What It Means for Human Beings," *ASHA Leader*, 5 May 2009, 14–16.

32 *120 million people* Berglund, Lindvall, and Schwela, *Guidelines for Community Noise*, 39. See also Kryter, *The Handbook*, 111–112.

32 *cross-cultural studies of elderly populations indicate that . . . presbycusis* Henry Still, *In Quest of Quiet* (Harrisburg, PA: Stackpole Books, 1970), 38–44.

32 *eliminating* sociocusis *from the industrialized world* Kryter, *The Handbook*, 273.

32 *Schafer's prediction* R. Murray Schafer, *The Soundscape: Our Sonic Environment and the Tuning of the World* (1977; Rochester, VT: Destiny Books, 1994), 181.

32 *noise is a stressor* Berglund, Lindvall, and Schwela, *Guidelines for Community Noise*, 47–48. For one example of the use of noise to test anti-stress medications, see Rajan Ravindran and others, "Noise-Stress-Induced Brain Neurotransmitter Changes and the Effect of *Ocimum sanctum* (Linn) Treatment on Albino Rats," *Journal of Pharmacological Sciences* 98, no. 4 (2005): 354–360.

32 *ancient Chinese text* Robert Alex Baron, *The Tyranny of Noise* (New York: St. Martin's, 1970), 113. Baron also cites more recent instances of noise torture. "A young Greek told a news conference that he had seen a man accused of being a Communist, but who maintained his innocence, tortured for three months by . . . intolerable reverberations from a bell outside his cell. The Russian Communists in turn expose their political prisoners to a novel form of modern noise torture, nothing as primitive as beatings bells or gongs. They simply place them in a noisy factory in a Siberian labor camp" (114).

32 *a form of torture today* Suzanne G. Cusick, "Music as Torture/Music as Weapon," *Transcultural Music Review* 10 (2006), 1, http://www.sibetrans.com/trans/trans10/cusick_eng.htm.

32 *stress hormones* Hugh Davies and Irene van Kamp, "Environmental Noise and Cardiovascular Disease: Five Year Review and Future Directions" (paper presented at 9th International Congress on Noise as a Public Health Problem, ICBEN, Foxwoods, CT, 2008).

33 *"A hunter-gatherer's sensitivity"* Kosko, *Noise*, 56.

33 *sawmill workers* Hind Sbihi, Hugh Davies, and Paul Demers, "Hypertension in Noise-Exposed Sawmill Workers: A Cohort Study," *Occupational and Environmental Medicine* 65 (2008): 643–646.

33 *Austrian schoolchildren* Gary Evans and others, "Community Noise Exposure and Stress in Children, *Journal of the Acoustical Society of America* 109, no. 3 (March 2001): 1023–1027.

33 *four major European airports* Reported by Sora Song, "Nighttime Noise and Blood Pressure, *Time*, 13 February 2008, http://www.time.com/time/health/article/0,8599,1713178,00.html. The study Song refers to: Alexandros S. Haralabidis and others, "Acute Effects of Night-Time Noise Exposure on Blood Pressure in Populations Living Near Airports," *European Heart Journal* 29, no. 5 (2008): 658–664.

33 *lower birth weights in Japan* Kryter, *The Handbook*, 549–553.

33 *number-one "burden of disease"* Eveline Maris, "The Social Side of Noise Annoyance" (doctoral dissertation, University of Leiden, 17 December 2008), 11.

33 *"Sleep deprivation . . . silent killer"* Michael Chee, "Sleep Deprivation Leads to Impaired Risky Decision Making," National University of Singapore, 20 January 2008, http://www.nus.edu.sg/research/rg124.php.

33 *Lack of sleep not only taxes the immune system* Lea Winerman, "Brain, Heal Thyself," *Monitor on Psychology* 37, no. 1 (January 2006), http://www.apa.org/monitor/jan06/brain.html.

33 *100,000 motor vehicle accidents* "You're Getting Very Sleepy," *American Psychological Association Online*, 12 May 2004, http://psychologymatters.org/sleep.html.

33 *17 percent increase* "You're Getting Very Sleepy."

34 *Arline Bronzaft* "The Effect of Elevated Train Noise on Reading Ability," *Environment and Behavior* 7, no. 4 (1975): 517–528.

34 *Subsequent studies have confirmed her findings* Arline Bronzaft, "The Effect of a Noise Abatement Program on Reading Ability," *Journal of Environmental Psychology* 1 (1981): 215–222. See also Berglund, Lindvall, and Schwela, *Guidelines for Community Noise*, 50, which notes that in studies of airport noise "the adverse effects were larger in children with lower school achievement." The term *jet-pause teaching* is sometimes used to refer to interruptions of classroom activity by exterior noise sources. For a recent study on the educational effects of a cell phone ringing *within* the classroom, see Jill T. Shelton and others, "The Distracting Effects of a Ringing Cell Phone: An Investigation of the Laboratory and the Classroom Setting," *Journal of Environmental Psychology* 29, no. 4 (December 2009): 513–521.

34 *noise and mental illness* van Kamp and Davies, "Environmental Noise and Mental Health."

34 *solitary confinement* Atul Gawande, "Hellhole: Is Solitary Confinement Torture?" *New Yorker*, 30 March 2009, 40.

34 *quiet places . . . mentally restorative* van Kamp and Davies, "Environmental Noise and Mental Health."

34 *Attempts to prove objective connections . . . generate controversy* For a brief and thoughtful analysis of the difficulties involved, see R. F. Soames Job and

Chika Sakashita, "Conceptual Differences Between Experimental and Epidemiological Approaches to Assessing the Causal Role of Noise in Health Effects" (paper presented at 9th International Congress on Noise as a Public Health Problem, ICBEN, Foxwoods, CT, 2008).

35 *Sound is a vibration* See Gilles A. Daigle, "Atmospheric Acoustics," in *McGraw-Hill Encyclopedia of Science and Technology* (New York: McGraw Hill, 2000). See also Kryter, *The Handbook*, 1–15.

35 *"One could imagine an alien species"* Daniel J. Levitin, *This Is Your Brain on Music: The Science of a Human Obsession* (New York: Dutton, 2006), 41–42.

35 *medieval tapestries* C. M. Woolgar, *The Senses in Late Medieval England* (New Haven: Yale University Press, 2006), 67.

35 *nineteenth-century ladies* Emily Thompson, *The Soundscape of Modernity: Architectural Acoustics and the Culture of Listening in America 1900–1933* (Cambridge, MA: MIT Press, 2002), 374.

36 *Greek amphitheater* Schafer, *The Soundscape*, 220.

36 *medieval cathedrals* Woolgar, *The Senses*, 66. "The new Gothic style of the thirteenth century changed the proportions of churches, so that the largest buildings had naves much higher than they were wide, increasing the amount of reverberation within the building. It is exactly at this moment that one finds the development of polyphony."

36 *Mayan ball courts* David Lubman, "Acoustics of the Great Ball Court at Chichen Itza," *Journal of the Acoustical Society of America* 120, no. 5 (November 2006): 3279.

36 *today's commercial jets* See Bijsterveld, "A Booming Business: The Search for a Practical Aircraft Noise Index," in *Mechanical Sound*, 193–232. See also International Civil Aviation Organization, *Review of Noise Abatement Procedure Research and Development and Implementation Results: Discussion of Survey Results*, ICAO, 2007, http://www.icao.int/icao/en/env/ReviewNADRD.pdf.

36 *phenomenal growth in air travel* Les Blomberg and David Morris, "Sound Decisions," *New Rules*, Winter 1999, 16. For "air cargo travel" the authors cite a 2,156 percent increase since 1960.

36 *European countries have been paving their highways* European Environment Agency, *Transport at a Crossroads* (Copenhagen: Office for Official Publications of the European Communities, 2009), 24–25. See also Forum of European National Highway Research Laboratories, *Sustainable Road Surfaces for Traffic Noise Control Project* (SILVIA), 2008, http://www.trl.co.uk/silvia/.

36 *Equipment . . . allows bar owners to monitor noise* See www.acopacific.com. I do not endorse the products of this company or any others mentioned in this book.

36 *who pays them* The vexed connection between technical expertise and the means to employ it struck me numerous times in the preparation of this book but never more explicitly than during a round-table discussion with experts from several prominent acoustical-engineering firms. With refreshing candor and obvious frustration they discussed such issues as the federal regulation of U.S. airspace and the plight of disadvantaged communities near major transportation routes. An executive at one of the firms, who had been participating by speaker phone and who seemed to feel the discussion had gone too far in a critical direction, finally broke in and offered these remarks: "Everybody who works at [his firm] is interested in transportation and its development and in doing it in a responsible way. *There's nobody in our office at all who works against these programs.* We're all trying to fit them into a complex social situation where there can be very strong feelings about it. . . . *It isn't as though we take the part of the community against these projects.* Transportation is a critical part of social need. Doing that in a responsible open way that builds on trust—I think that's a crucial part of what we try to bring to projects, pointing out both sides of the issue, *but in an objective enough way so that the decision makers can see the benefit of the project, and the community can also see the benefit,* and both sides can see the impacts and work towards minimizing those. *We are transportation advocates*" (emphasis added).

37 *U.S. Office of Noise Abatement and Control* Phil Wisman, "EPA History (1970–1985)," http://www.epa.gov/history/tipics/epa/15b.htm.

37 *footnote, fear of crashes* Kryter, *The Handbook,* 516.

38 *lack of control . . . a major factor in noise annoyance* Sheldon Cohen and Shirlynn Spacapan, "The Social Psychology of Noise," in *Noise and Society,* ed. Dylan M. Jones and Antony J. Chapman (New York: John Wiley & Sons, 1984), 238–239. See also Maris, "The Social Side," 15.

38 *the "fair process effect"* Maris, "The Social Side," 20–23. The bulk of Maris's dissertation is devoted to an experimental exploration of the fair process effect as it relates to noise.

39 *Swedish and Dutch scientists have recently completed a study* Eja Pedersen and others, "Reponse to Noise from Modern Wind Farms in the Netherlands," *Journal of the Acoustical Society of America* 126, no. 2 (August 2009): 634.

39 *Tampopo* Directed by Juzo Itami, 1985.

39 *"eat more noisily"* Andrew Horvat, *Japanese Beyond Words: How to Walk and Talk Like a Native Speaker* (Berkeley, CA: Stone Bridge Press, 2000), 78.

39 *ongoing noise dispute* Donald Richie, "Hisako Shiraishi," in *The Donald Richie Reader: 50 Years of Writing on Japan,* ed. Arturo Silva (Berkeley, CA: Stone Bridge Press, 2001), 109–115.

39 *one person driving a motor scooter* Kosko, *Noise,* 55.

40 *"the wailing noise"* David Lister, "Unionists Protest Against Building of Ulster Mosque," *The Times,* 14 January 2003, http://www.timesonline.co.uk/tol/news/uk/article812068.ece.

40 *traditional music of Chinese laborers* Krystyn R. Moon, *Yellowface: Creating the Chinese in American Popular Music and Performance, 1850s–1920s* (New Brunswick, NJ: Rutgers University Press, 2005), 3, 7, 9–13, 16–17, etc.

40 *footnote, jazz was . . . dismissed* Emily Thompson, *The Soundscape of Modernity,* 130–132. See also Jacques Attali, *Noise* (1977; Minneapolis: University of Minnesota Press, 1985), p. 104. A 1920 review in the French *Revue Musicale* stated: "Jazz is cynically the orchestra of brutes with nonopposable thumbs and still prehensile toes, in the forest of Voodoo. It is entirely excess, and for that reason more than monotone: the monkey is left to his own devices, without morals, without discipline, thrown back to all the groves of instinct, showing his meat still more obscene. These slaves must be subjugated, or there will be no more master. Their reign is shameful. The shame is ugliness and its triumph."

40 *Rai* Joan Gross, David McMurray, and Ted Swedenburg, "Arab Noise and Ramadan Nights: *Rai,* Rap and Franco-Maghrebi Identities," in *The Anthropology of Globalization: A Reader,* ed. Jonathan Xavier Inda and Renato Rosaldo (Malden, MA: Blackwell Publishing, 2002), 198–222.

41 *"our music foretells our future"* Attali, *Noise,* 11.

42 *Hawthorne* Leo Marx, *The Machine in the Garden: Technology and the Pastoral Ideal in America* (New York: Oxford University Press, 1964), 12–14.

42 *Thoreau* Leo Marx, *The Machine,* 260.

43 *Sudanese folktale* Ali Lufti Abdallah, *The Clever Shiekh of the Butana and Other Stories: Sudanese Folktales* (New York: Interlink Books, 1999), 97ff.

43 *ancient Egyptian papyri* S.D.G. Stephens, "The Treatment of Tinnitus—a Historical Perspective," *The Journal of Laryngology and Otology* 98 (October 1984): 963.

43 *first record of a noise complaint* In *The Epic of Gilgamesh,* trans. N. K. Sandars (Harmondsworth, UK: Penguin, 1972), 108.

43 *knowledge that noise can damage hearing* Bijsterveld, *Mechanical Sound,* 71.

43 *Queen Elizabeth I* Bruce R. Smith, *The Acoustic World of Early Modern England: Attending to the O-Factor* (Chicago: University of Chicago Press, 1999), 52–53.

43 *women of Aurillac* Alain Corbin, *Village Bells: Sound and Meaning in the 19th-Century French Countryside,* trans. Martin Thom (1994; New York: Columbia University Press, 1998), 304.

43 *"Tomorrow, and tomorrow. . . . " Macbeth* 5.5.19–28.

44 *"all is signal"* Kosko, *Noise,* xviii.

44 *Hegel* Georg Wilhelm Friedrich Hegel, "Reciprocal Recognition, Spirit, and the Concept of Right," in *Lectures on the Philosophy of Spirit 1827–8,* trans. Robert P. Williams (New York: Oxford University Press, 2007), 22–25.

45 *Since the 1880s* Thompson, *Soundscape of Modernity,* 122.

45 *when we listen to noise "we find it fascinating"* John Cage, *Silence* (Middleton, CT: Wesleyan University Press, 1961), 3.

45 *Barack Obama* In *Dreams of My Father: A Story of Race and Inheritance* (New York: Random House, 1995), 269.

CHAPTER 3: THE NOISE OF POLITICAL ANIMALS

47 *"Wherever Noise is granted immunity"* R. Murray Schafer, *The Soundscape: Our Sonic Environment and the Tuning of the World* (1977; Rochester, VT: Destiny Books, 1994), 76.

47 *"man as political animal"* H. D. F. Kitto, *The Greeks* (Harmondsworth, UK: Penguin, 1951), 111.

47 *Aristotle's teacher Plato* Kitto, *The Greeks,* 65.

47 *acrobats and stuttering orators* See Andrew G. P. Lang and Hugh D. Amos, *These Were the Greeks* (Uwchland, PA: Dufour, 1991), 145, and "Stuttering Isn't Fun," *One Voice,* http://www.stutterisa.org/OneVoice/OV26.pdf.

48 *tenor bell of Notre Dame* Alain Corbin, *Village Bells: Sound and Meaning in the 19th-Century French Countryside,* trans. Martin Thom (1994; New York: Columbia University Press, 1998), 10. The bell was smashed in 1792.

48 *a "Cockney," for example* Bruce R. Smith, *The Acoustic World of Early Modern England: Attending to the O Factor* (Chicago: University of Chicago Press, 1999), 53.

48 *bigger bells or more peals were sounded for boys* Corbin, *Village Bells,* 164–167. "In Publier the nuptials of 'a girl who had gone astray' were celebrated 'noiselessly.'" In the towns of Cuvat, Publier, and Ballaison, there was no ringing at the baptisms of children who entered 'the world before the sixth or seventh month of marriage'" (166).

49 *masculine power* Recent studies show that men prefer louder music than women and consume more alcohol when listening to loud music than women do. See Barry Blesser and Linda-Ruth Salter, "The Unexamined Rewards for Excessive Loudness" (Communications at 9th International Conference on Noise as a Public Health Problem, ICBEN, Foxwoods, CT, 2008), 3–4.

49 *"sound imperialism"* Schafer, *The Soundscape,* 77–78.

49 *David Owen* In *Noise,* directed and written by Henry Bean, 2007. If you see the film you might also want to read David Denby's insightful review, "The Unquiet Life," in *New Yorker,* 19 May 2008.

49 *"learned helplessness"* Gary W. Evans and others, "Community Noise Exposure and Stress in Children," *Journal of the Acoustical Society of America* 109, no. 3 (March 2001): 1026.

50 *altruism and sense of self-determination* Sheldon Cohen and Shirlynn Spacapan, "The Social Psychology of Noise," in *Noise and Society,* ed. Dylan M. Jones and Antony J. Chapman (New York: John Wiley & Sons, 1984), 222–227.

50 *to "unmake the world"* Elaine Scarry, *The Body in Pain: The Making and Unmaking of the World* (New York: Oxford University Press, 1985), 45.

50 *Russian Gulag* Scarry, *Body in Pain,* 40.

51 *trash-can lids* Rosemary Sales, *Women Divided: Gender, Religion and Politics in Northern Ireland* (London: Routledge, 1997), 72. On the August 9 commemoration, see David Ramsbotham, "Why I Will Vote Against the 42-Day Law," *Guardian,* 13 October 2008, http://www.guardian.co.uk/commentisfree/2008/oct/13/terrorism-lords.

51 *"If I had a hammer"* The Weavers, "If I Had a Hammer," Charter Records, 1949.

51 *"Because they sell the righteous for silver"* Amos 1:2.

51 *"I'm black and I'm proud"* James Brown, "Say It Loud (I'm Black and Proud)," Universal Motown, 1968.

51 *"stochastic resonance"* Bart Kosko, *Noise* (New York: Viking, 2006), 145–166.

52 *Mass of the Ass* Gary Shapiro, *Alcyone: Nietzsche on Gifts, Noise, and Women* (Albany: SUNY Press, 1991), 106.

52 *"rough music"* Smith, *Acoustic World,* 154–157. See also David Underdown, *Revel, Riot, and Rebellion: Popular Politics and Culture in England 1603–1660* (Oxford: Clarendon Press, 1985), 100–103.

52 *"skimmington"* Thomas Hardy, *The Mayor of Casterbridge* (1886; New York: Macmillan, 1965), Chapter 39. Hardy, who was in closer historical earshot to "rough music" than we, refers to the skimmington as "the Demonic Sabbath." The moment when Lucetta realizes she is its object is pitiful: "'She's me—she's me—even to the parasol—my green parasol!' cried Lucetta, with a wild laugh as she stepped in. She stood motionless for one second—then fell heavily to the floor. Almost at the instant of her fall the rude music of the skimmington ceased." Mission accomplished.

52 *"Charivari can be interpreted. . . . "* Smith, *Acoustic World,* 156.

52 *the more anarchic sounds of Carnival* Jacques Attali, *Noise* (1977; Minneapolis: University of Minnesota Press, 1985), 21–23.

53 *1618 English skimmington* Barbara Ehrenreich, *Dancing in the Streets: A History of Collective Joy* (New York: Holt, 2006), 109. Smith, *Acoustic World*, 155. See also the source in the next note, also cited by Ehrenreich.

54 *Agnes Mills* Martin Ingram, "Ridings, Rough Music and the 'Reform of Popular Culture' in Early Modern England," *Past and Present*, no. 105 (1984): 82.

54 *For something purportedly so "subversive," noise can inspire* "When a culture accepts loudness as being a legitimate right in recreational sound venues, that acceptance tends to legitimize all forms of noise pollution. . . . We believe that acceptance of loudness in entertainment then carries over to a tolerance of disruptive noise from airplanes, jackhammers, powered garden equipment, and so on." Blesser and Salter, "The Unexamined Rewards," 2.

54 *"AURAL GLUT"* Liz Was, "knoise pearl #1," in *Sounding Off! Music as Subversion/Resistance/Revolution*, ed. Ron Sakolsky and Fred Wei-han Ho (Brooklyn: Autonomedia, 1995), 65.

54 *"Let them eat cake"* Consider also the following statement by Tricia Rose: "In the postindustrial urban context of dwindling low-income housing, a trickle of meaningless jobs for young people, mounting police brutality, and increasingly draconian depictions of young inner city residents, hip hop style *is* black urban renewal." *Black Noise: Rap Music and Black Culture in Contemporary America* (Hanover: Wesleyan University Press, 1994), 61. As a stark and celebratory summation of what has and has not happened in urban America, Rose's statement is impeccable. Still, I cannot read it without imagining a chorus of parsimonious (and racist) middle-class taxpayers responding with an enthusiastic "Amen!" In other words, *Let them eat rap.*

54 *footnote, Karl Marx* "The Civil War in France," in *The Marx-Engels Reader*, ed. Robert C. Tucker, 2nd ed. (New York: Norton, 1978), 647.

54 *John Rawls* In *A Theory of Justice* (Cambridge, MA: Harvard University Press, 1971), 3.

55 *In a recent British survey* Roger P. Murphy, *Antisocial Housing* (London: UK Noise Association, 2004), 5, 12–13. Sources I spoke to in London attributed some of the noise problems in that city's public housing to cheap construction in the building boom that followed the Second World War. Yet, according to the "American Housing Survey" of the U.S. Census (Table 2-8), 46 percent, or 1.4 million units, of the multifamily housing units constructed in the United States in the last ten years report noise problems; 15 percent, or 460,000 units, report "significant" noise problems. Figures cited by Les

Blomberg in "Acoustical Slums, Green Buildings, and the Acoustics of Sustainability" (paper presented at "Noise! Design, Health and the Urban Soundscape," Graduate School of Architecture, Planning and Preservation, Columbia University, New York, 21 September 2009).

55 *footnote, From a British Housing Association tenant* Murphy, *Antisocial Housing,* 4.

55 *One of the saddest noise stories* I heard this story from New York City noise researcher and consultant Arline Bronzaft.

56 *environmental "disamenities" are distributed by race and class* Robin R. Sobotta, Heather E. Campbell, and Beverly Owens, "Aviation Noise and Environmental Justice: The Barrio Barrier," *Journal of Regional Science* 47, no. 1 (2007): 125–126.

56 *Cornell researchers* Gary W. Evans and Lyscha A. Marcynyszyn, "Environmental Justice, Cumulative Environmental Risk, and Health Among Low- and Middle-Income Children in Upstate New York," *American Journal of Public Health* 94, no. 11 (November 2004): 1942–1944.

56 *"barrio barrier"* Sobotta, Campbell, and Owens, "The Barrio Barrier," 125–154.

57 *whether people travel to the . . . disamenity* Sobotta, Campbell, and Owens, "The Barrio Barrier," 127–128.

57 *African Americans . . . behind bars* Marc Mauer and Ryan Scott King, "Schools and Prisons: Fifty Years After *Brown v. Board of Education,*" The Sentencing Project, http://www.sentencingproject.org/doc/publications/rd_brownvboard.pdf. See also Chapter 7 of this book.

57 *footnote, Clinton administration's Executive Order* Environmental Justice Resource Center, Executive Order No. 12898, http://www.ejrc.cau.edu/execordr.html.

58 *Blomberg thinks that noise is an issue of "the commons"* Les Blomberg, "The Nature of Noise: Civility, Sovereignty, Community, Reciprocity, Power, Tyranny, and Technology," *The Quiet Zone* (Noise Pollution Clearinghouse newsletter), Fall 2006, 4–8.

58 *landlords evicted their tenant farmers* Karl Marx, "Expropriation of the Agricultural Population from the Land," in *Capital,* ed. Friedrich Engels, *The Great Books of the Western World,* Vol. 50 (Chicago: Britannica, 1952), 355–364.

59 *When the British were determining their bombing strategy* Robert L. O'Connell, *Of Arms and Men: A History of War, Weapons, and Aggression* (New York: Oxford University Press, 1989), 284.

59 *"the unmaking of civilization"* Scarry, *Body in Pain,* 77.

60 *HMMH* Harris Miller Miller & Hanson.

60 *the DNL . . . that the Federal Aviation Administration uses* Paul Schomer, "Criteria for Assessment of Noise Annoyance," *Noise Control Engineering Journal* 53, no. 4 (July-August 2005): 132–144.

61 *Environmental Protection Agency* Schomer, "Criteria for Assessment," 139.

61 *a policy recommended by the World Health Organization* Birgitta Berglund, Thomas Lindvall, and Dietrich H. Schwela, eds., *Guidelines for Community Noise* (Geneva: World Health Organization, 1999), 66.

62 *the only feasible redress is home buyout* Christian Broer, a sociology professor at the University of Amsterdam, has recently suggested that aircraft noise complaints are "a product of [noise] policy." "Policy Annoyance: How Policies Shape the Experience of Aircraft Sound." *Aerlines, e-zine edition,* http://www.scribd.com/doc/12475160/38-Broer-Noise-Annoyance. As an example he cites how aircraft noise complaints in Switzerland not only are less frequent than in the Netherlands but also tend to be expressed in different terms. The noise is presumably the same, but different policies provide different frameworks for how the noise is perceived. His analysis is more nuanced than I have made it sound, and I could not agree more with his concluding recommendation that noise be addressed in tandem with other issues "like global warming, consumption critique, local autonomy and democratization." As for his main thesis, it is no doubt correct, but correct in the same way as saying that the Americans with Disabilities Act has prompted disabled people in the United States to be more vocal about "access." I would hope so.

62 *footnote, "on how to construct an aluminum helmet"* See Leonard Vintiñi, "Tracing the 'Hum,'" *The Epoch Times,* 19 July 2008, http://en.epochtimes.com/n2/content/view/7074/. See also the photograph of the "Faraday cage," which is "designed to prevent the passage of electromagnetic waves," in Maria Blondeel, "Listening to Acoustic Energy and Not Hearing," *Soundscape: The Journal of Acoustic Ecology* 6, no. 1 (Spring/Summer 2005): 23.

63 *"soft-sound sensitivity"* See the note for *how we define sensitivity* above.

63 *teenagers hear higher frequencies* Acoustician Rein Pirn tells the following story: "We are summoned to the local elementary school to investigate a noise that, the teachers swear, exists only in the kids' imagination. Upon entering the room, I thought I heard a faint hiss—nothing to get excited about. But the sound level meter told otherwise. Through 8000 Hz, everything was normal, but at 16,000 Hz, where we grownups begin to experience hearing loss, the level skyrocketed by more than 30 dB! A faulty radiator

valve, it turned out." Deborah Melone and Eric W. Wood, *Sound Ideas: Acoustical Consulting at BNN and Acentech* (Cambridge, MA: Acentech, 2005), 168.

63 *"like a dentist's drill"* Temple Grandin, *Thinking in Pictures: And Other Reports from My Life with Autism* (1995; New York: Vintage, 2006), 63.

63 *governess* Temple Grandin and Margaret M. Scariano, *Emergence: Labeled Autistic* (Novato, CA: Arena Press, 1989), 25.

63 *electric cars* Rick Rojas, "Hybrid Cars May Pose Silent Threat to the Blind: Advocates Want Vehicles to Make Some Sound," *Courier-Journal* (Louisville, KY), 8 July 2008.

64 *least Bell's vireo* Douglas E. Barrett, "Traffic-Noise Impact Study for Least Bell's Vireo Habitat Along California State Route 83," *Transportation Research Record 1559*, National Research Council (1996): 3–7.

64 *kennels* "Findings," *Harper's Magazine*, October 2006, 96.

64 *the U.S. Supreme Court ruled . . . that Navy sonar exercises* "U.S. Supreme Court Rules in Favor of Navy over Whales," posted by Chris V. Thangham, *Digital Journal*, www.digitaljournal.com/article/262302. See also Jim Roach, "Military Sonar May Give Whales the Bends, Study Says," *National Geographic News*, 8 October 2003, http://news.nationalgeographic.com/news/2003/10/1008_031008_whalebends.html.

64 *biophany* Bernie Krause, "Anatomy of the Soundscape: Evolving Perspectives," *Journal of the Audio Engineering Society* 56, no. 1/2 (January/February 2008): 73.

65 *Kornhauser's experiment* Karl D. Kryter, *The Handbook of Hearing and the Effects of Noise: Physiology, Psychology and Public Health* (New York: Academic Press, 1994), 422. Kryter discusses "acoustic perfume" on the same page and the effect of noise on the sensation of time's passage on 402. His summary of the conflicting evidence regarding the effects of noise on task performance is on 427.

65 *controversy . . . over leaf-blowers* Carolyn Zinko, "Gardeners in California City to Protest Leaf Blower Ban, Claiming Ban Is Racist," *San Francisco Chronicle*, 3 March 1998, A13. See also Daniel Olmos, "A Ban on a Noisy Existence: The Los Angeles Leaf Blower Ban, Spatialized Whiteness and the Gardeners' Struggle for Dignity" (paper presented at the annual meeting of The American Studies Association, Philadelphia, 11 October 2007).

66 *footnote, Americans . . . work 350 hours more per year* David E. Shi, *The Simple Life: Plain Living and High Thinking in American Culture* (1985; Athens: University of Georgia Press, 2007), 278.

66 *footnote, the Japanese Samurai class* Noel Perrin, *Giving Up the Gun: Japan's Reversion to the Sword, 1543–1879* (1979; Jaffrey, NH: David R. Godine, 1989).

67 *Into Great Silence* Directed by Philip Gröning, 2005. Silence is an entire subject unto itself. In addition to John Cage's *Silence*, cited later in this chapter, see Susan Sontag's essay "The Aesthetics of Silence" in *Styles of Radical Will* (New York: Farrar, Straus and Giroux, 1969), pp. 3–34, in which she notes that "[t]he mystical tradition has always recognized, in Norman Brown's phrase, 'the neurotic character of language'" (22). R. Murray Schafer sees that recognition realized most fully in the religion of Taoism. "No philosophy or religion catches the positive felicity of stillness better than Taoism. It is a philosophy that would make all noise abatement legislation unnecessary." *The Soundscape*, 258. A dense and extended philosophical treatment of silence is to be found in Bernard P. Dausenhauer, *Silence: The Phenomenon and Its Ontological Significance* (Bloomington: University of Indiana Press, 1980). Regarding the ethical implications of deliberate silence, Dausenhauer says that "silence is never an act of unmitigated autonomy. Rather . . . silence involves a yielding following upon an awareness of finitude and awe" (24). Within Dausenhauer's framework, silence is justifiably viewed as the opposite of noise, which very often *is* "an act of unmitigated autonomy." For a fictional treatment of silence and our perception of it, see Kevin Brockmeier, "The Year of Silence," in *The Best American Short Stories 2008*, ed. Salman Rushdie and Heidi Pitlor (Boston: Houghton Mifflin, 2008), 22–35.

67 *video arcade* Jane E. Brody, "All That Noise Is Damaging Children's Hearing," *New York Times*, 9 December 2008.

67 *most of us would or should be outraged* The imposition of silence on women and the vilification of "noisy" women are hallmarks of patriarchal cultures. "Silence was valued very strongly and was considered an indication of modesty and good upbringing," writes Monia Hejaiej of traditional Tunisian culture in *Behind Closed Doors: Women's Oral Narratives in Tunis* (New Brunswick, NJ: Rutgers, 1996), 64. Hejaiej illustrates the point by citing the story "Sabra," in which a wife "never uttered a word of complaint" after her husband kills their children one by one. "You were well mannered," he says in praise of her, "you never raised your voice, you never complained, may the womb that bore you be blessed." Readers inclined to see this as a "Muslim thing" are directed to "The Clerk's Tale" in Chaucer's *Canterbury Tales*, in which a husband puts his wife to the same test (though he only pretends to murder the children). Like Sabra, she passes.

67 *forced inmates . . . to sing a song* Joseph J. Moreno, "Orpheus in Hell: Music in the Holocaust," in *Music and Manipulation: On the Social Uses and Social Control of Music,* ed. Stephen Brown and Ulrik Volgsten (New York: Berghahn Books, 2006), 267–268.

68 *Amnesty International* Scarry, *Body in Pain,* 49.

68 *introverts . . . more likely to gravitate toward quietness* D. M. Jones and D. R. Davies, "Individual and Group Differences in the Response to Noise," in *Noise and Society,* ed. Dylan M. Jones and Antony J. Chapman (New York: John Wiley and Sons, 1984), 134–136.

68 *"one's flesh and blood"* James Baldwin, *No Name in the Street,* in *Collected Essays* (New York: Library of America, 1998), 468.

69 *U.S. Noise Control Act* United States Code 42 USC 4901. See http://www.nonoise.org/lawlib/usc/42/4901.htm.

70 *"complaining in style"* Karin Bijsterveld, *Mechanical Sound: Technology, Culture, and Public Problems of Noise in the Twentieth Century* (Cambridge, MA: MIT Press, 2008), 260.

71 *the republic of one* It is interesting to note that the SONY corporation, which used to market car stereo speakers with the motto "Disturb the Peace," now markets "Noise Canceling Headphones"—not unlike the computer companies that simultaneously market "spyware" and "firewalls." The audio products and the ads that promote them are aimed at different markets, but the antisocial subtext is similar. Just as the young dude-nik was invited by the speaker ads to see himself as separate from and superior to the staid grownups he annoyed, the young, neatly dressed (usually) male professional sporting the headphones is shown on an airplane aloofly unperturbed by loquacious dowdy women and their fecund younger sisters, all of whose heads have been replaced with megaphones. In this connection, see Tocqueville's statement on despotism on page 175 of Chapter 7, "Loud America."

71 *footnote, the Branch Davidian compound in Waco, Texas* Gore Vidal, "The Meaning of Timothy McVeigh," in *The Best American Essays 2002,* ed. Stephen Jay Gould and Robert Atwan (Boston: Houghton Mifflin, 2002), 312.

71 *"threshold of pain"* Sources differ on the designation, with some (e.g., Kosko, *Noise,* 50) placing it at 120 dB; others (e.g., Kryter, *The Handbook,* 53) place it at 140 dB, while noting 120 as the threshold of "discomfort." Obviously there is a continuum of pain, which varies among individuals and according to physiological and subjective circumstances.

71 *John Cage* In *Silence* (Middleton, CT: Wesleyan University Press, 1961), 93.

CHAPTER 4: WHAT THE PYTHON SAID: PREHISTORY TO THE EVE OF THE INDUSTRIAL AGE

75 *"My dear friend," said the python to the squirrel . . . "* Godfrey Igwebuike Onah, "The Meaning of Peace in African Traditional Religion and Culture," *Afrika World.net*, http://www.afrikaworld.net/afrel/goddionah.htm.

75 *Big Bang* David L. Chandler, "Universe Started with Hiss, Not Bang," *New Scientist*, 12 June 2004, http://www.newscientist.com/article/dn5092-universe-started-with-hiss-not-bang.html; Marcus Chown, "Big Bang Sounded Like a Deep Hum," *New Scientist*, 30 October 2003, http://www.newscientist.com/article/dn4320-big-bang-sounded-like-a-deep-hum.html; Brandon Keim, "Listening to the Big Bang," *Wired*, 28 September 2008, http://www.wired.com/wiredscience/2008/09/listening-to-th/; Maggie McKee, "Big Bang Waves Explain Galaxy Clustering," *New Scientist*, 12 January 2005, http://www.newscientist.com/article/dn6871.

76 *Krakatoa* John M. Picker, *Victorian Soundscape* (Oxford and New York: Oxford University Press, 2003), 4–5.

76 *Laetoli Footprints* See entry by that name at "Evolution Library," *PBS.org.*, http://www.pbs.org/wgbh/evolution/library/07/1/l_071_03.html.

76 *the sound of fingernails pulled over a blackboard* Diane Ackerman, *A Natural History of the Senses* (1990; New York: Vintage, 1995), 188.

77 *Native Americans* My source is the novelist Howard Frank Mosher, who drew this conclusion after talking to amateur archaeologists in the American West. While the areas along quiet streambeds often yield good finds of arrowheads and other artifacts, the ground adjacent to fast or cascading water is seldom as rich.

77 *Àjàpá is repeatedly awakened by . . . the Dawn Bird* Oyekan Owomoyela, *Yoruba Trickster Tales* (Lincoln: University of Nebraska Press, 1997), 57–68.

78 *python tries in vain to quiet a chattering squirrel* See the epigraph for this chapter.

78 *"driving pigs"* Francis Grose, *Provincial Glossary*, 1787, cited in Jeffrey Kacirk, *Forgotten English: 2006 Calendar* (Petaluma, CA: Pomegranate, 2006). "*He is driving his hogs over Swarston Bridge,* this is a saying used in Derbyshire when a man snores in his sleep. Swarston Bridge is very long, and not very wide, which causes the hogs to be crowded together, in which situation they always make a loud grunting noise."

79 *"the place of noises"* Samuel Adams Drake, *A Book of New England Legends and Folk Lore* (Boston: Little, Brown, 1906), 427–429.

79 *Sirens* See Book 12 of *The Odyssey.*

79 *Howard Frank Mosher* He wrote and published the story he told me in "Thunder from a Cloudless Sky," *Washington Post Magazine*, 12 July 1998, 23–24.

80 *D. H. Lawrence* In *Twilight in Italy* (1916; New York: Viking, 1958), 197.

80 *English settlers* Drake, *New England Legends*, 428.

80 *village exorcisms* James Frazer, *The New Golden Bough*, edited and abridged by Theodor H. Gaster (1890; New York: New American Library, 1964), 599–609. In an Egyptian myth a mother's "great scream of terror" is sufficient to break a spell that would otherwise have made her son immortal. Isis, the goddess who was weaving the spell even as she was mourning her dead brother, Osiris, soon lets out "an ear-piercing shriek" of her own, which kills the boy. Joyce Tyldesley, *Tales from Ancient Egypt* (Bolton, UK: Rutherford Press, 2004), 19.

80 *and animals* In China "Mao identified sparrows as one of the 'four pests' (rats, flies, and mosquitoes were the others) because they ate grain in the fields, in effect stealing from the workers. He commanded people to kill the birds, which they did, often by banging gongs whenever the birds tried to land, until they dropped dead from exhaustion." Jeff Goodell, *Big Coal: The Dirty Secret Behind America's Energy Future* (Boston: Houghton Mifflin, 2006), 238.

80 *footnote, "hums"* Leonard Vintiñi, "Tracing the 'Hum,'" *The Epoch Times* 19 July 2008, http://en.epochtimes.com/n2/content/view/7074/.

80 *the hero Gilgamesh* N. K. Sandars, trans. *The Epic of Gilgamesh* (Harmondsworth, UK: Penguin, 1972), 108.

81 *if the Sumerians were the first* Abba Eban, *Heritage: Civilization and the Jews* (New York: Summit Books, 1984), 8–12.

82 *"The Lord saw"* Genesis 6:5.

82 *"[v]iolence shall no more be heard"* Isaiah 60:18.

82 *history of weaponry* Robert L. O'Connell, *Of Arms and Men: A History of War, Weapons, and Aggression* (New York: Oxford University Press, 1989), 47.

82 *cry of Achilles* Homer, *The Iliad*, trans. Robert Fitzgerald (Garden City, NY: Anchor-Doubleday, 1974), 442. See also Victor Davis Hanson, *The Western Way of War: Infantry Battle in Classical Greece* (Berkeley: University of California Press, 2000), 149. "The greatest noise was produced not by individuals talking and shouting, but by the collective war-cry that the army uttered in unison—the ancient equivalent of the rebel yell, which was designed to make each soldier forget his own fear as he sent a message of terror to the enemy."

83 *Scottish war cry* Samuel Taylor Coleridge, *The Notebooks of Samuel Taylor Coleridge: Notebooks 1819–1826*, ed. Kathleen Coburn and Merton Christensen (London: Routledge, 1990), 168.

83 *"rebel yell"* D. Scott Hartwig, "'It's All Smoke and Dust and Noise': The Face of Battle at Gettysburg," in *Battle: The Nature and Consequences of Civil War Combat*, ed. Kent Gramm (Tuscaloosa: University of Alabama Press, 2008), 22.

83 *an "association" . . . between noise and religion* R. Murray Schafer, *The Soundscape: Our Sonic Environment and the Tuning of the World* (1977; Rochester, VT: Destiny Books, 1994), 50.

83 *"Day of Noise and Clamour"* The Qur'an, 101:1–11.

83 *Christian equivalent* 1 Corinthians 15:52. Revelation 1:10.

83 *Mount Sinai* Exodus 19:16–23; 20:18–21.

84 *"You shall love your neighbor"* Leviticus 19:18. Les Blomberg emphasizes that *lack* of reciprocity is an essential element in noise as an abuse of power. See "The Nature of Noise: Civility, Sovereignty, Community, Reciprocity, Power, Tyranny, and Technology," *The Quiet Zone* (Noise Pollution Clearinghouse newsletter), Fall 2006, 4–8.

84 *Sabbath* Exodus 20:8–11. See also Liz Harris, *Holy Days: The World of a Hasidic Family* (New York: Summit Books, 1985), 54–76. "What happens when we stop working and controlling nature?" asks Harris's Hasidic host. "When we don't operate machines, or pick flowers, or pluck fish from the sea, or change darkness to light, or turn wood into furniture? When we cease interfering with the world we are acknowledging that it is God's world" (69). In addition, see Gershom G. Scholem, ed., *Zohar: The Book of Splendor* (1949; New York: Schocken Books, 1963), 36. "And with the incoming of the Sabbath, it behooves them [Torah students] to rejoice their wives, to the honor of the heavenly union."

85 *spatial approach of zoning* See Karin Bijsterveld, *Mechanical Sound: Technology, Culture, and Public Problems of Noise in the Twentieth Century* (Cambridge, MA: MIT Press, 2008), 68–69.

85 *noise policy in the European Union involves . . . sonic mapping* See http://ec.europa.eu/environment/noise/.

85 *Medieval records* C. M. Woolgar, *The Senses in Late Medieval England* (New Haven: Yale University Press, 2006), 97.

85 *schools and hospitals* Bijsterveld, *Mechanical Sound*, 68–69. See also Raymond W. Smilor, "American Noise 1900–1930," in Mark M. Smith, *Hearing History: A Reader* (Athens: University of Georgia Press, 2004), 324.

86 *efforts in New York City . . . to make certain streets automobile-free* For a different approach to the same problem, see Felix Salmon, "How Driving a Car into Manhattan Costs $160," *Reuters.com*, 3 July 2009, http://blogs.Reuters.com/felix-salmon/2009/07/03/how-driving-a-car-into-manhattan-costs-160/.

86 *footnote, a threshold of 100 dB for "ceremonies, festivals and entertainment events"* Birgitta Berglund, Thomas Lindvall, and Dietrich H. Schwela, *Guidelines for Community Noise* (Geneva: World Health Organization, 1999), 65.

86 *Japanese tea ceremony* Okakura Kakuzo, *The Book of Tea* (1906; Rutland, VT, and Tokyo, Japan: Charles E. Tuttle, 1956), 33, 61.

86 *Elijah is witness to a clamor of meteorological noises* 1 Kings 19:11–12.

86 *the chosen servant of God "will not cry or lift up his voice"* Isaiah 42:2.

87 *"quietness and confidence"* Isaiah 30:15.

87 *Buddha* Karen Armstrong, *Buddha* (New York: Viking Penguin, 2001), 122. See also Bhikkhu Nayanatusita, "Buddhism and Sound Pollution," http://www.bps.lk/other_library/buddhism_and_sound_pollution.pdf.

87 *Sri Ramakrishna* In *The Gospel of Sri Ramakrishna*, trans. Swami Nikhilananda (New York: Ramakrishna-Vivekananda Center, 1977), 735.

87 *For contemplatives* According to the Confucian teacher Chu Hsi, "In his [the sage's] activity the principle of quietude is not yet done away with and in his quietude the source of activity is not yet stopped." Rodney S. Taylor, *The Confucian Way of Contemplation: Okaha Takehiko and the Tradition of Quiet Sitting* (Columbia: University of South Carolina Press, 1988), 39. Similarly, the traditional Quaker emphasis on silence was not restricted to "meeting" but was supposed to extend to all phases of life. "They recommended silence by their example, having very few words upon all occasions." William Penn, "For of Light Came Sight," in *The Quaker Reader*, ed. Jessamyn West (1962; Wallingford, PA: Pendle Hill, 1992), 111.

87 *Meister Eckhart* Raymond B. Blakney, ed. *Meister Eckhart: A Modern Translation* (New York: Harper and Row, 1941), 243.

87 *American transcendentalists* "But what is classification but the perceiving that these objects [of nature] are not chaotic, and are not foreign, but have a law which is also a law of the human mind? . . . [N]ature is the opposite of the soul, answering for it part for part." Ralph Waldo Emerson, "The American Scholar," in *The Romantic Movement in American Writing*, ed. Richard Harter Fogle (New York: Odyssey Press, 1966), 28.

88 *Julius Caesar* Schafer, *The Soundscape*, 189.

88 *"They make a wilderness . . ."* Tacitus is ostensibly quoting Calgacus, leader of the Britons, prior to engaging the Romans at the Battle of the Grampians. See John Bartlett, *Familiar Quotations*, 13th ed. (Boston: Little, Brown, 1955), 63.

88 *Juvenal* Schafer, *The Soundscape*, 190.

89 *Seneca's approach to the noise from the gym* "On Noise," in *The Art of the Personal Essay*, ed. Phillip Lopate (New York: Doubleday, 1994), 5–8.

89 *a Roman law prohibiting coppersmiths* Bijsterveld, *Mechanical Sound,* 56.

89 *Ordered to commit suicide* Alain de Botton, *The Consolations of Philosophy* (New York: Pantheon, 2000), 76–77.

90 *getting used to . . . noise* Karl Kryter, *The Handbook of Hearing and the Effects of Noise: Physiology, Psychology and Public Health* (New York: Academic Press, 1994), 630, 637.

90 *Roman libraries* Lionel Casson, *Libraries in the Ancient World* (New Haven: Yale University Press, 2001), 80–108. See also Scott Douglas, *Quiet Please: Dispatches from a Public Librarian* (New York: Da Capo, 2008), 75, as well as Alberto Manguel, *History of Reading* (New York: HarperCollins, 1996), 43–44. Manguel reports that we have no record of Roman library patrons complaining about the noise.

91 *continuous noise is . . . less disturbing to people than impulsive noise* Kryter, *The Handbook,* 85.

91 *"the massive noise" of a traditional yeshiva* Jacob Neusner, *Judaism's Theological Voice: The Melody of the Talmud* (Chicago: University of Chicago Press, 1995), 14.

91 *the year 1253* Woolgar, *The Senses,* 91.

91 *approval to the use of torture* Charles Williams, *The Descent of the Dove* (Grand Rapids, MI: Eerdmans 1939), 239.

91 *screaming* Hillel Schwartz, "Noise and Silence: The Soundscape of Spirituality" (address to the Inter-Religious Federation for World Peace, Seoul, South Korea, August 1995), http://www.noisepollution.org/library.noisesil/noisesil.htm.

92 *"Tongues mixed and mingled"* Dante Alighieri, *Hell,* trans. Dorothy L. Sayers (Harmondsworth, UK: Penguin, 1949), canto 3, lines 25–27, page 86.

92 *Genghis Khan* Trevor Nevitt Dupuy, *The Military Life of Genghis: Khan of Khans* (New York: Franklin Watts, 1969), 106. The extent of Genghis's use of gunpowder to hurl projectiles is an open question. See John Joseph Saunders, "Did the Mongols Use Guns?" appendix 2 in *The History of Mongol Conquests* (Philadelphia: University of Pennsylvania Press, 2001), 196–199.

92 *gunnery . . . as the invention of hell* John Milton, *Paradise Lost,* in *Complete Poems and Major Prose,* ed. Merritt Y. Hughes (Indianapolis: Bobbs-Merrill, 1957), book IV, lines 482–491, page 335.

93 *Thomas Dekker* Bruce R. Smith, *The Acoustic World of Early Modern England: Attending to the O-Factor* (Chicago: University of Chicago Press, 1999), 62.

93 *castle kitchens also carried hellish associations* Woolgar, *The Senses,* 68, 76.

93 *against the law . . . for a man to disturb the peace by beating his wife* Eveline Maris, "The Social Side of Noise Annoyance" (doctoral dissertation, University of Leiden, 17 December 2008), 10.

93 *town fathers of New Orleans moved the jail* Dell Upton, *Another City: Urban Life and Urban Spaces in the New American Republic* (New Haven: Yale University Press, 2008), 65.

93 *rape of Hispaniola* Bartolomé de las Casa, "1542: Hispaniola," from *A Brief Account of the Destruction of the Indies,* in *Lapham's Quarterly,* Winter 2008, 122–124.

93 *"the people of that island "are by nature very Clamourous . . . "* Bruce R. Smith, *Acoustic World,* 305.

94 *"Indian Sagomore"* Bruce R. Smith, *Acoustic World,* 315.

94 *Shaker communities* Information obtained at Canterbury Shaker Village in Canterbury, NH. See http://www.shakers.org/.

94 *It bore the nickname of Bedlam* "Bedlam: The Hospital of St. Mary of Bethlehem," British Broadcasting Corporation, http://www.bbc.co.uk/dna/h2g2/A2554157?s id=1.

95 *"ugliness was very rare"* John Wain, *Samuel Johnson* (New York: Viking, 1975), 43.

95 *engraving by William Hogarth* Emily Thompson, *The Soundscape of Modernity: Architectural Acoustics and the Culture of Listening in America 1900–1933* (Cambridge, MA: MIT Press, 2002), 116.

95 *Liza Picard* Cited in Jeffrey Meyers, *Samuel Johnson: The Struggle* (New York: Basic Books, 2008), 92.

95 *the majority of these noises were "organic"* Thompson, *Soundscape of Modernity,* 116–117.

95 *Jonathan Swift* "A Description of the Morning," in *Jonathan Swift: Poems Selected by Derek Mahon* (London: Faber and Faber, 2006), 3. Mahon writes in his introduction: "A townsman, if not always urbane, Swift delighted in the noise and squalor of city life" (xi).

97 *English Peasant Revolt* Woolgar, *The Senses,* 90.

97 *nineteenth- and twentieth-century diatribes* See note for Chapter 1 in this book on "Jazz Age roar." See also Peter Bailey, "Breaking the Sound Barrier," Chapter 9 in *Popular Culture and Performance in the Victorian City* (Cambridge, UK: Cambridge University Press, 1998), 203–210.

98 *"London"* In *Samuel Johnson: Selected Poetry and Prose,* ed. Frank Brady and W. K. Wimsatt (Berkeley: University of California Press, 1977), 53, lines 228–235.

98 *Bramble's litany of complaints* Tobias Smollett, *The Expedition of Humphry Clinker* (1771; New York: Rinehart, 1950), 29–34.

98 *Mohocks* Margaret Lane, *Samuel Johnson and His World* (London: Hamish Hamilton, 1975), 76.

99 *footnote, survey in nineteen urban "open spaces"* Lei Yu and Jian Kang, "Effects of Social, Demographical and Behavioral Factors on the Sound Level Evaluation in Urban Open Spaces, *Journal of the Acoustical Society of America* 123, no. 2 (February 2008): 782.

99 *other inventions would have to be heard* See Bill McKibben, *Deep Economy: The Wealth of Communities and the Durable Future* (New York: Holt, 2007), 5, on the steam engine. See Schafer, *The Soundscape*, 72, for a list of eighteenth-century mechanical inventions.

CHAPTER 5: WHAT LAURA HEARD: THE INDUSTRIAL AGE TO THE PRESENT

101 *"They had never heard such a racket"* Laura Ingalls Wilder, *Little House in the Big Woods* (1932; New York: Harper and Row, 1971), 224.

101 *"a great invention!"* Wilder, *Little House in the Big Woods*, 227–228.

102 *"flat line" sound* R. Murray Schafer, *The Soundscape: Our Sonic Environment and the Tuning of the World* (1977; Rochester, VT: Destiny Books, 1994), 78.

102 *"acoustic ecology"* Barry Truax, *Acoustic Ecology* (Norwood, NJ: Ablex, 1984), 74.

103 *In medieval Wales . . . a plowing team consisted of two men* Wendell Berry, *The Unsettling of America: Culture and Agriculture* (1977; San Francisco: Sierra Club Books, 1986), 139.

103 *Lewis Mumford* Cited in Schafer, *The Soundscape*, 64.

103 *"beaten with the strap"* Schafer, *The Soundscape*, 75.

103 *Mary "Mother" Jones* Betsy Harvey Kraft, *Mother Jones: One Woman's Fight for Labor* (New York: Clarion Books, 1995), 54–55.

103 *"Boilermakers' disease"* Schafer, *The Soundscape*, 76.

103 *industrial hearing loss* Karin Bijsterveld, *Mechanical Sound: Technology, Culture, and Public Problems of Noise in the Twentieth Century* (Cambridge, MA: MIT Press, 2008), 71.

103 *"A nail boy wearing earplugs"* Bijsterveld, *Mechanical Sound*, 75.

104 *"pickles and 'rusty' bacon"* Peter Bailey, "Breaking the Sound Barrier," Chapter 9 in *Popular Culture and Performance in the Victorian City* (Cambridge, UK: Cambridge University Press, 1998), 208.

104 *footnote, Karl Marx* On "the noise and turmoil of the new system of production," see *Capital*, ed. Frederick Engels, *The Great Books of the Western World*

50 (Chicago: Britannica, 1952), 134. On factory work exhausting "the nervous system," see *Capital*, 207.

104 *French observer* Leo Marx, *The Machine in the Garden: Technology and the Pastoral Ideal in America* (New York: Oxford University Press, 1964), 208.

104 *higher wages* Bijsterveld, *Mechanical Sound*, 78.

105 *"from the idiocy of rural life"* Karl Marx and Friedrich Engels, *Manifesto of the Communist Party*, in *The Great Books of the Western World*, Vol. 50, 421.

105 *elevated trains* Emily Thompson, *The Soundscape of Modernity: Architectural Acoustics and the Culture of Listening in America 1900–1933* (Cambridge, MA: MIT Press, 2002), 120.

105 *Duncan Sandys* Bijsterveld, *Mechanical Sound*, 198.

105 *Charles Dickens and Emile Zola* Schafer, *The Soundscape*, 75.

105 *John Ruskin* John D. Rosenberg, *The Darkening Glass: A Portrait of Ruskin's Genius* (New York: Columbia University Press, 1961), 183–185.

105 *vexing issue of street musicians* John M. Picker, "The Soundproof Study: Victorian Professional Identity and Urban Noise," in *Victorian Soundscapes* (Oxford: Oxford University Press, 2003), 41–81.

106 *Dickens and twenty-six other signatories* Picker, "Soundproof Study," 60–61.

106 *Period cartoons* Picker, "Soundproof Study," 64–76.

106 *LaGuardia* Michael Pollak, "Silence of the Cranks," *New York Times*, 13 February 2005, http://www.nytimes.com/2005/02/13/nyregion/thecity/13fyi.html?_r=1.

106 *new class of home-based artists* See Picker, "Soundproof Study," 52–55.

106 *Goethe* Thompson, *Soundscape of Modernity*, 116.

106 *Schopenhauer* Arthur Schopenhauer, "On Noise," in *Great Essays*, ed. Houston Peterson (New York: Washington Square Press, 1960), 156–161.

107 *Thomas Carlyle's failed attempt . . . to make a soundproof study* Marcel Proust also attempted to make a soundproof study by filling its walls with cork. Schafer, *The Soundscape*, 228. William Gladstone referred to his study as "The Temple of Peace." Anne Fadiman, *Ex Libris: Confessions of a Common Reader* (New York: Farrar, Straus and Giroux, 1998), 144.

107 *"Chelsea is unfashionable"* Carlyle's House Memorial Trust, *Carlyle's House Catalog* (London: Chiswick Press, 1995), 29.

107 *"the houses are cheap and excellent"* Fred Kaplan, *Thomas Carlyle* (Berkeley: University of California Press, 1983), 205.

107 *From the backyard* In *Carlyle's House Catalog*, 27–28.

107 *"hard battle against fate"* In *Carlyle's House Catalog*, 38.

107 *"All summer"* Picker, "Soundproof Study," 43.

107 *"demon foul"* In *Carlyle's House Catalog*, 63.

108 *"a practical Liverpool railway man"* In *Carlyle's House Catalog*, 64.

108 *Jane . . . seemed rather to enjoy it* Picker, "Soundproof Study," 55–56.

108 *"a flattering delusion"* Carlyle Memorial Trust, House Papers.

108 *"the silent room is the noisiest"* Carlyle Memorial Trust, House Papers.

108 *History of Frederick the Great* Kaplan, *Thomas Carlyle*, 384.

108 *"One hour spent"* Virginia Woolf, "Great Men's Houses," in *The London Scene: Five Essays by Virginia Woolf* (New York: Random House, 1982), 23.

108 *"two of the most nervous"* Woolf, "Great Men's Houses," 24.

108 *footnote, Tennyson* Kaplan, *Thomas Carlyle*, 258.

108 *some disadvantaged, invisible Fanny whom we ignore* Carlyle Memorial Trust, House Papers.

109 *"how loud the battle was"* D. Scott Hartwig, "'It's All Smoke and Dust and Noise': The Face of Battle at Gettysburg," in *Battle: The Nature and Consequences of Civil War Combat*, ed. Kent Gramm (Tuscaloosa: University of Alabama Press, 2008), 20.

110 *General John Gibbon* Hartwig, "' It's All Smoke and . . . Noise,'" 20–21.

110 *medical records of Union Army soldiers* D. Scott Mcllwain, Kathy Gates, and Donald Ciliax, "Heritage of Army Audiology and the Road Ahead: The Army Hearing Program," *American Journal of Public Health* 98, no. 12 (December 2008): 2167.

110 *footnote, The General Law of 1862 and the Disability Act of 1890* Mcllwain, Gates, and Ciliax, "Heritage of Army Audiology," 2167.

110 *Dixon Irwin* Eric T. Dean, "'The Awful Shock and Rage of Battle': Rethinking the Meaning and Consequences of Combat in the Civil War," in Gramm, *Battle*, 100–101.

111 *"quiet . . . of plantation life"* Mark M. Smith, *Listening to Nineteenth-Century America* (Chapel Hill: University of North Carolina Press, 2001), 19.

111 *"streets of New-York"* Smith, *Listening*, 3.

111 *"the cracking of that whip"* Mark M. Smith, "Listening to the Heard Worlds of Antebellum America," in *Hearing History: A Reader* (Athens: University of Georgia Press, 2004), 370.

111 *American theater audiences . . . underwent a change* John F. Kasson, *Rudeness & Civility: Manners in Nineteenth-Century Urban America* (New York: Hill and Wang, 1990), 217–260. These changes were not confined to America. See James H. Johnson, "Listening and Silence in Eighteenth- and Nineteenth-Century France," in Mark M. Smith, *Hearing History*, 178–180. Johnson writes: "The golden rule of bourgeois decency was not to bother others" (178), by which he presumably means other bourgeoisie.

111 *guides to etiquette* Kasson, *Rudeness & Civility*, 44.

111 *Theodore Thomas* Kasson, *Rudeness & Civility*, 234–239. Thomas had a continental parallel in Gustav Mahler, "who hated all extraneous noise, threw out singers' fan clubs, cut short applause between numbers, glared icily at talkative concertgoers, and forced late-comers to wait in the lobby." Alex Ross, *The Rest Is Noise: Listening to the Twentieth Century* (New York: Farrar, Straus and Giroux, 2007), 19–20.

112 *footnote, "one wonders how anything less can ever suffice"* Kasson, *Rudeness & Civility*, 260.

113 *Yiddish theater* Kasson, *Rudeness & Civility*, 251.

113 *"guerrilla fighters"* Kasson, *Rudeness & Civility*, 250.

113 *"spoilsport"* Johan Huizinga, *Homo Ludens: A Study of the Play Element in Culture* (Boston: Beacon, 1955), 11–12.

113 *3,000 blasts* Thompson, *Soundscape of Modernity*, 121.

113 *the landmark Bennet Act* Raymond Smilor, "American Noise, 1900–1930," in Mark M. Smith, *Hearing History*, 53.

113 *footnote, noise ordinances* Smilor, "American Noise," 325.

114 *Society for the Suppression of Unnecessary Noise* Thompson, *Soundscape of Modernity*, 121.

114 *footnote, Chicago peddlers* Thompson, *Soundscape of Modernity*, 124–125.

114 *opinions of boat captains* Thompson, *Soundscape of Modernity*, 121.

114 *Mark Twain* Smilor, "American Noise," 324.

114 *"I offer up this sacrifice"* Hollis Godfrey, "The City's Noise," *Atlantic Monthly*, November 1909, 609–610.

114 *"It is no exaggeration"* Thompson, *Soundscape of Modernity*, 126.

115 *Arline Bronzaft* "The Effect of Elevated Train Noise on Reading Ability," *Environment and Behavior* 7, no. 4 (1975): 517–528.

115 *Theodor Lessing* Lawrence Baron, "Noise and Degeneration: Theodor Lessing's Crusade for Quiet," *Journal of Contemporary History* 17 (1982): 165–178.

115 *Wallace Sabine* Thompson, *Soundscape of Modernity*, 13–45.

116 *mixed blessing* See Bijsterveld, *Mechanical Sound*, 199. "Mechanical objectivity originally aimed at eliminating the subjectivity of aesthetic judgment and scholarly dogmas from the registration of natural phenomena, and it ended up dismissing the human mind and body as trustworthy witnesses in favor of machines."

116 *audiometer* Thompson, *Soundscape of Modernity*, 146.

116 *adoption of the decibel* Bijsterveld, *Mechanical Sound*, 105.

116 *"out of the hands of cranks"* Thompson, *Soundscape of Modernity*, 146.

117 *Koji Nagahata* "What Do Citizens Imagine Is a Level of 80 DB?: A Basic Study of Environmental Communication on Soundscape Issues" (paper presented at Inter-Noise, Honolulu, 3–6 December 2006). See also Sonoko Kuwano, Seiichiro Namba, and Tohru Kato, "Perception and Memory of Loudness of Various Sounds" from the same conference. After conducting several experiments, the authors concluded that "the loudness of recalled sounds was judged on the basis of memorized sounds, not on the basis of the subjective image of the sound sources."

117 *footnote, Credit for inventing the radio* See "Who Invented Radio?" http://www.pbs.org/tesla/ll/ll_whoradio.html.

118 *the world before artificial lighting* John M. Staudenmaier, "Denying the Holy Dark: The Enlightenment Ideal and the European Mystical Tradition," in *Progress: Fact or Illusion?* ed. Leo Marx and Bruce Mazlish (Ann Arbor: University of Michigan Press, 1996), 183.

118 *"It is time"* Staudenmaier, *Denying the Holy Dark*, 175.

118 *"Time and space died"* Fillipo Tommaso Marinetti, *The Founding and Manifesto of Futurism 1909*, in *Futurist Manifestos*, ed. Umbro Apollonio (Boston: MFA Publications, 2001), 21–22.

119 *footnote, Nietzsche's assertion* Friedrich Nietzsche, *Twilight of the Idols*, in *The Portable Nietzsche*, ed. Walter Kaufman (New York: Viking, 1968), 542.

119 *a curious document* Luigi Russolo, *The Art of Noises* (extracts) *1913*, in Apollonio, *Futurist Manifestos*, 74–76; 85–88.

119 *concert in Milan* Thompson, *Soundscape of Modernity*, 136–138.

119 *Helmholtz's often-cited distinction* Thompson, *Soundscape of Modernity*, 132.

120 *footnote, gray area* Richard Strauss's first opera, *Guntram*, for example, "was a flop at its 1894 premiere, mainly because the orchestration drowned out the singers." Ross, *The Rest Is Noise*, 16.

120 *Luftwaffe would equip its bombers* Bailey, "Breaking the Sound Barrier," 198.

120 *Official tastes of the Reich would influence a sweeping postwar reaction against tonality* Ross, *The Rest Is Noise*, 305–339.

120 *footnote, liberal politics and . . . conservative style* Ross, *The Rest Is Noise*, 317.

121 *Hiroshima railway station* John Hersey, *Hiroshima* (1946; New York: Bantam, 1959), 11–13, 16–19, 84.

122 *feature article* S. Smith Stevens, "The Science of Noise," *Atlantic Monthly*, July 1946, 98–100.

123 *U.S. Interstate* AA Roads Interstate Guide, http://www.interstate-guide.com/.

123 *Heathrow Airport* Bijsterveld, *Mechanical Sound*, 195.

123 *Robert Sarnoff* David E. Shi, *The Simple Life: Plain Living and High Thinking in American Culture* (1985; Athens: University of Georgia Press, 2007), 249.

123 *Muzak* Bijsterveld, *Mechanical Sound*, 86.

123 *high school teachers* Christopher Lasch, *The Culture of Narcissism: American Life in an Age of Diminishing Expectations* (New York: Norton, 1979), 309–310.

124 *right-wing propaganda* David Brock, *The Republican Noise Machine: Right-Wing Media and How It Corrupts Democracy* (New York: Crown, 2004).

124 *environmentalist mindset* Thompson, *Soundscape of Modernity*, 123.

124 *Jimmy Carter* Shi, *The Simple Life*, 270.

124 *UNESCO* Schafer, *The Soundscape*, 97.

124 *Noise Control Act* See Noise Pollution Clearinghouse, "Noise Control Act, Law Library. http://www.nonoise.org/epa/act.htm. See also Sidney A. Shapiro, *The Dormant Noise Control Act and Options to Abate Noise Pollution*, Administrative Conference of the United States, November 1991.

125 *"Levels Document"* Paul Schomer, "Criteria for Assessment of Noise Annoyance," *Noise Control Engineering Journal* 53, no. 4 (July–August 2005): 139.

125 *demonstrations at Heathrow* Bijsterveld, *Mechanical Sound*, 198. See also www.hacan.org and www.planestupid.com.

125 *"Let's Be Gentle with Each Other"* Bijsterveld, *Mechanical Sound*, 184–186, 191.

125 *Ronald Reagan* Shapiro, *Dormant Noise Control Act*. See also Phil Wisman, "EPA History (1970–1985)," http://www.epa.gov/history/tipics/epa/15b.htm.

126 [*911*] *oral history* http://911research.wtc7.net/wtc/evidence/eyewitnesses.html. See also http://911research.wtc7.net/wtc/evidence/oralhistories/explosions.html.

126 *siege of Fallujah* Suzanne G. Cusick, "Music as Torture/Music as Weapon," *Transcultural Music Review* 10 (2006), 1, http://www.sibetrans.com/trans/trans10/cusick_eng.htm.

126 *Manuel Noriega* Cusick, "Music as Torture," 1.

126 *Binyam Mohamed* "Pull the Plug on Torture Music: Binyam Mohamed," http://www.reprieve.org.uk.

127 *other songs reportedly used to break down prisoners* Cusick, "Music as Torture," 3.

127 *earworms* Oliver Sacks, *Musicophilia: Tales of Music and the Brain* (New York: Alfred A. Knopf, 2007), 41–48.

128 *Meinkinder* See http://www.myspace.com/meinkinder.

130 *talking into his cell phone* For a passage that ought to accompany any cell phone placed in a time capsule, see Philip Roth, *Everyman* (Boston: Houghton Mifflin, 2007), 64–65. "Everywhere I walked, somebody was approaching me talking on a phone and someone was behind me talking on a phone. Inside the cars, the drivers were on the phone. When I took a taxi,

the cabbie was on the phone. For one who frequently went without talking to anyone for days at a time, I had to wonder what . . . that had previously held them up had collapsed in people to make incessant talking into a telephone preferable to walking about under no one's surveillance. . . . For me it made the streets appear comic and the people ridiculous."

CHAPTER 6: THEIR WORLD TOO: NOISE TODAY

131 *Beirut* Alexander Besant and Willy Lowry, "Beirut's Nightlife Is Back—and Not Everyone Is Happy," *Daily Star* (Beirut, Lebanon), 16 June 2008.

131 *Thailand* "Court Dismisses Cases Concerning Suvarnabhumi Airport," *Bangkok Post*, 24 June 2009, http://www.bangkokpost.com/mail/147036/.

131 *A recent editorial* "Reducing the Level of Noise," *China Daily*, 23 November 2007, http://www.china.org.cn/english/China/232916.htm.

131 *Ghana* "National Noise Awareness Day Launched," *ModernGhana.com*,16 April 2008, http://www.modernghana.com/news/162814/1/national-noise-awareness-day-launched.html.

131 *Moscow* Chloe Arnold, "Russia: Moscow's Noise Pollution Reaches Dangerous Levels," *Radio Free Europe*, 19 September 2007, http://www.rferl.org/content/article/1078719.html.

131 *Arctic Circle* R. Murray Schafer, *The Soundscape: Our Sonic Environment and the Tuning of the World* (1977; Rochester, VT: Destiny Books, 1994), 85.

131 *One out of six Australians* Mike Safe, "Bad Vibrations," *The Australian*, 31 January 2009, http://www.theaustralian.com.au/news/features/bad-vibrations/story-e6frg8h6-1111118667268.

132 *Stichting BAM* See http://www.stichtingbam.nl.

132 *densely populated* "The World's Most Densely Populated Countries," *Telegraph.co.uk*, 23 October 2009, http://www.telegraph.co.uk/news/worldnews/6413308/The-worlds-most-densely-populated-countries.html. See also Adriana Stuijt, "Noise Pollution Kills 600 Dutch a Year," *Digital Journal*, 23 February 2009, http://www.digitaljournal.com/article/267835.

132 *neighbor noise* "[N]otably by the sounds from the radio, stereo, or television." Karin Bijsterveld, *Mechanical Sound: Technology, Culture, and Public Problems of Noise in the Twentieth Century* (Cambridge, MA: MIT Press, 2008), 160.

133 *noise levels above 50 A-weighted decibels* Netherlands Environmental Assessment Agency, "Area and Dwellings in the Netherlands Exposed to Noise Levels in Excess of 50 db(A), 2002," Environmental Data Compendium, http://www.mnp.nl/mnc/i-en-0295.html.

133 *pigs* My source is Frits van den Berg of the Amsterdam Municipal Health Service, cited extensively in Chapter 8.

133 *geographical features* Stuijt, "Noise Pollution." See also Martin Dunford, Phil Lee, and Suzanne Morton Taylor, "The Battle with the Sea," in *The Rough Guide to The Netherlands* (New York: Rough Guides, 2007), unpaginated insert between 168 and 169.

133 *footnote, ocean sonar* Stuijt, "Noise Pollution."

134 *Pliny* Dunford, *Rough Guide,* 357.

135 *right-wing backlash* Dunford, *Rough Guide,* 370–371.

136 *DALYS per year* Anne B. Knol, *Trends in the Environmental Burden of Disease in the Netherlands 1980–2020,* RIVM report 500029001/2005, especially sections on noise 16–17, 40–46, 60–65, and "The DALY Debate," 80ff. Contact Anne.Knol@RIVM.nl.

136 *"Give us back our bikes!"* Dunford, *Rough Guide,* 369.

137 *87 percent of Dutch travel* European Environment Agency, *Transport at a Crossroads* (Copenhagen: Office for Official Publications of the European Communities, 2009), 15, 44.

137 *between 1 and 2.5 kilometers* Dutch Ministry of Transport, Public Works and Water Management, *Nationale Mobiliteitsmonitor 2008,* http://www.verkeerenwaterstaat.nl/kennisplein/3/7/375943/Nationale_Mobiliteitsmonitor_2008_compleet.pdf, 15–16. I am grateful to Frits van den Berg for translating these pages.

137 *rising automobile ownership* EEA, *Transport,* 4.

137 *120 million Europeans* Birgitta Berglund, Thomas Lindvall, and Dietrich H. Schwela, eds., *Guidelines for Community Noise* (Geneva: World Health Organization, 1999), viii.

137 *67 million* EEA, *Transport,* 5.

137 *Road noise has three main sources* EEA, *Transport,* 24. Les Blomberg pointed out the additional factor of rattles. See also Associated Press, "Automakers Obsess over Reducing Wind Noise," MSNBC, 18 June 2008, http://www.msnbc.msn.com/id/25248338/.

137 *trucks . . . 90 percent quieter European Commission Green Paper,* 1 (full citation at *footnote, "Green Paper"* below). On quieter pavements see EEA, *Transport,* 24–25. See also Forum of European National Highway Research Laboratories, *Sustainable Road Surfaces for Traffic Noise Control Project* (SILVIA), 2008, http://www.trl.co.uk/silvia/.

137 *German study* EEA, *Transport,* 23.

137 *increase in heart attack risk* J. Selander and others, "Long-Term Exposure to Road Traffic Noise and Myocardial Infarction," *Epidemiology* 20, no. 2 (March 2009): 272–279.

138 *breakdown "is very similar for other Western European countries"* EEA, *Transport*, 29.

138 *Europe's population* European Environment Agency, "About Noise," http://www.eea.europa.eu/themes/noise/about-noise.

138 *"baked in" environmental effect* C. Vlek, "Could We Have a Little More Quiet, Please? A Behavioral-Science Commentary on *Research for a Quieter Europe in 2020*," *Noise & Health* 7, no. 26 (2005): 67.

138 *footnote, trips to work* Jean-Paul Rodrigue, "The Geography of Transport Systems," Hofstra University, www.people.hofstra.edu/geotrans/eng/ch6en/conc6en/ch6c3en.html.

138 *footnote, The Swiss have a car-sharing "Mobility Cooperative"* EEA, *Transport*, 30.

139 *"noise levels of aircraft have been reduced"* EEA, *Transport*, 26.

139 *the FAA has projected* Federal Aviation Administration, Forecasts FY 2009–2025, 31 March 2009, http://www.faa.gov/data_research/aviation/aerospace_forecasts/2009–2025/.

139 *The widely discussed RANCH . . . study* Charlotte Clark and others, "Exposure-Effect Relations Between Aircraft and Road Traffic Noise Exposure at School and Reading Comprehension: The RANCH Project," *American Journal of Epidemiology* 163, no. 1 (2006): 27–37.

140 *footnote, "Green Paper"* Commission of the European Communities, *Future Noise Policy: European Commission Green Paper* (Brussels, 11 April 1996, COM 96 final), http://europa.eu/documentation/official-docs/green-papers/index_en.htm.

140 *airports in France* EEA, *Transport*, 26–27.

140 *the UK and Spain* EEA, *Transport*, 26.

141 *the country's first, and one of Europe's last, national highway systems* National Roads Authority of Ireland, *Guidelines for the Treatment of Noise and Vibration in National Road Schemes* (Dublin: National Roads Authority, 2004).

141 *165 percent increase in their country's greenhouse emissions* EEA, *Transport*, 17.

141 *study conducted . . . in Dublin* Jan Battles, "Noise a Major Threat to Health of City Dwellers," *Sunday Times*, 12 October 2008.

141 *footnote, youthful country* Robin Gauldie, *Ireland* (London: New Holland, 2006), 28.

143 *"yoga, Iberian style"* Sara Miller Llana, "Sleepless in Spain: The Siesta Recedes," *Christian Science Monitor*, 19 January 2006, http://www.csmonitor.com/2006/0119/p20s01-woeu.html.

143 *mikro ipno* See "Rhodes Culture: Social Conventions," *Monarch Holidays*, http://holidays.monarch.co.uk/destinations/Greece/Rhodes/culture.

143 *Syria* Raya Mamarbachi, "Cultural Differences and Taboos in Syrian Business Situations," *Going Global*, http://old.goinglobal.com/hot_topics/syria_business_mamarb.asp.

144 *Carnival at Santa Cruz* See http://www.spanish-fiestas.com/spanish/festivals/carnival.htm.

144 *festival of San Isidro* See http://www.spanish-fiestas.com/Madrid/san-isidro.htm.

144 *Basra and Silk Road Festivals in Syria* See http://travel.mapsofworld.com/syria/syria-festivals/.

144 *pre-Lenten carnivals held throughout Greece* Lance Chilton and others, *The Rough Guide to the Greek Islands* (New York: Rough Guides, 2004), 60.

144 *siesta has come under fire* Llana, "Sleepless in Spain."

144 *national limits were not instituted until 1993* Department for Environment, Food & Rural Affairs (DEFRA, UK), "Noise and Nuisance Policy: A Review of National and European Practices," http://www.defra.gov.uk/environment/noise/research/climate/nannexb.htm.

144 *"a man's thing"* Llana, "Sleepless in Spain."

144 *Spanish Civil War* Llana, "Sleepless in Spain."

145 *a fable . . . about an owl* Russell Hoban, *Riddley Walker* (1980; New York: Touchstone, 1992), 85–86.

146 *Albert Camus* In *The Rebel: An Essay on Man in Revolt*, trans. Anthony Bower (1951; New York: Vintage, 1956), p. 13.

146 *Keiko Torigoe* Kozo Hiramatsu, "A Review of Soundscape Studies in Japan," *Acta Acustica United with Acustica* 92 (2006): 857–858.

146 *Best Hundred Sounds* I am indebted to Dr. Katsuaki Terasawa for his kindness in translating some of the sounds from the Japanese Ministry of the Environment website. See also Keiko Torigoe, "Insights Taken from Three Visited Soundscapes in Japan," *Soundscape: The Journal of Acoustic Ecology* 6, no. 2 (Fall/Winter 2005): 9–12.

146 *Basho composed a famous haiku* In *The Sound of Water: Haiku by Bashō, Buson, Isssa, and Other Poets*, translated by Sam Hamill (Boston: Shambala, 2000), 37.

147 *Finland* Helmi Järviluoma, "Listen! They Are Still Threshing: One Hundred Finnish Soundscapes Selected," *Soundscape: The Journal of Acoustic Ecology* 6, no. 2 (Fall/Winter 2005): 28–30.

147 *Taiko drumming* Chiaki Ishiwata and Toshiko Fukuchi, "Results of Field Measurements of Generated Sound of Japanese Drum 'Taiko' and Consideration Towards Its Sound Insulation" (paper presented at Inter-Noise, Honolulu, 3–6 December 2006).

147 *"riotous laughter"* For a glimpse (through Western eyes) of traditional rural mores in post–Second World War Japan, see Richard K. Beardsley, John W. Hall, and Robert E. Ward, *Village Japan* (Chicago: University of Chicago Press, 1959), 66. "The sense of quiet tranquility in Niiike comes in good part from the restraint put on all expressions of emotion. Exuberance is repressed; so is every form of violence. The people smile readily and often; they laugh not infrequently, but gently, never in boisterous guffaws. Noisy argument or quarreling is as rare as riotous laughter, and physical violence among adults is shocking even to think about."

147 *"average Japanese bar"* Andrew Horvat, *Japanese Beyond Words: How to Walk and Talk Like a Native Speaker* (Berkeley, CA: Stone Bridge Press, 2000), 81.

147 *uyokusha* Alice Gordenker, "Sound Trucks," *Japan Times Online,* 18 April 2006, http://search.japantimes.co.jp/cgi-bin/ek20060418wh.html.

147 *bylaws enacted against them* Hiramatsu, "A Review of Soundscape Studies," 862.

147 *"recent day trip"* Mary Miyamoto, "Speaking Up on Noise," *Kansai Scene Magazine,* 3 February 2009, http://www.kansaiscene.com/2009_01/html/update.shtml.

148 *Metropolitan Assembly of Accra* "AMA Slams Churches for Excessive Noise," *HappyGhana.com,* 18 April 2008, http://www.happyghana.com/newsdetails.asp?id=3314&cat_id=1.

148 *Durga Pooja festival* Sudhakar B. Ogale, "South-East Asian Region," in Berglund, Lindvall, and Schwela, *Guidelines for Community Noise,* 131.

148 *Sri Lanka* Bhikkhu Nayanatusita, "Buddhism and Sound Pollution," http://www.bps.lk/other_library/buddhism_and_sound_pollution.pdf.

148 *"the artillery of the church"* Nayanatusita, "Buddhism."

149 *Says one Buddhist text* Nayanatusita, "Buddhism."

150 *Pico Iyer* "The Eloquent Sounds of Silence," *Time,* 25 January 1993.

150 *Mexico City's 20 million inhabitants* Lisa J. Adams, "Silver Lining: Flu Brings Less Crime, Less Noise and Blue Sky to Mexico City," Associated Press, 30 April 2009.

151 *More and more cities of Latin America* Guillermo Fuchs, "Latin America," in Berglund, Lindvall, and Schwela, *Guidelines for Community Noise,* 113–114.

151 *bustling cities of India's economic boom* Phil Hazlewood, "Fighting the Noise in India's 'Maximum City,'" *ThingsAsian,* 16 February 2009, http://www.thingsasian.com/stories-photos/34842.

151 *Residents of Mumbai . . . are exposed to a constant 80–85 decibels* Hazlewood, "Fighting the Noise." See also Mark Jacobson, "Dharavi: Mumbi's Shadow

City," *National Geographic*, July 2007, 74. The beginning of Jacobson's article adds another nuance to our understanding of the relationship between poverty, development, and noise. "All cities in India are loud, but nothing matches the 24/7 decibel level of Mumbai, the former Bombay, where the traffic never stops and the horns always honk. Noise, however, is not a problem in Dharavi, the teeming slum of one million souls, where as many as 18,000 people crowd into a single acre. By nightfall, deep inside the maze of lanes too narrow even for the *putt-putt* of auto rickshaws, the slum is as still as a verdant glade. Once you get accustomed to sharing 300 square feet of floor with 15 humans and an uncounted number of mice, a strange sense of relaxation sets in—ah, at last a moment to think straight."

151 *motorized rickshaws* Hazlewood, "Fighting the Noise."

151 *to better navigate Mumbai's bustling narrow streets* "No Horn Day a Success: Mumbai Cops," *Times of India*, 8 April 2008, http://timesofindia.indiatimes.com/articleshow/2934032.cms.

151 *Yeshwant Oke* Hazlewood, "Fighting the Noise."

151 *35 percent of the population . . . had bilateral sensory neural hearing loss* Ogale, "South-East Asian Region," 129.

151 *Karachi* Shabih H. Zaidi, "Eastern Mediterranean Region," in Berglund, Lindvall, and Schwela, *Guidelines for Community Noise*," 124–125.

152 *2009 report* Saadia Qamar, "Noise Pollution Creating Problems for Karachiites," *The Nation*, 26 April 2009, http://www.nation.com.pk/pakistan-news-newspaper-daily-english-online/Regional/Karachi/26-Apr-2009/Noise-pollution-creating-problems-for-Karachiites.

152 *noise and heat can be combated* Zaidi, "Eastern Mediterranean Region," 127.

152 *certain court rulings out of India* Sairam Bhat, "Noise and the Law," *India Together*, November 2003, http://www.indiatogether.org/2003/nov/law-noise.htm.

154 *"the developing sector"* Etienne Grond, "South Africa," in Berglund, Lindvall, and Schwela, *Guidelines for Community Noise*, 121.

154 *"the rural sector"* Grond, "South Africa," 121.

155 *"non-addition" of Chapter 2 aircraft . . . and a phase-out of those aircraft* Paul Goldschagg, "Airport Noise and Environmental Justice in South Africa," *International Research in Geographical and Environmental Education* 11, no. 1 (2002): 72–75.

155 *footnote, ICAO resolved to phase out* Goldshagg, "Airport Noise," 72–73. See also EEA, *Transport*, 25–26.

156 *Gary Greenberg* Quoted with his kind permission.

157 *unanimously passed a thoroughgoing revision of its . . . noise code* See Richard Siegler and Eva Talel, "Impact of New York City's Amended Noise Control Code," *New York Law Journal,* 2 July 2008, 3, 6–7. See also Erich Thalheimer and Charles Shamoon, "New York City's New and Improved Construction Noise Regulation" (paper presented at Noise-Con, Reno, 22–24 October 2007).

158 *jacket for that iconic urban noisemaker, the jackhammer* According to Charles Shamoon, the jacket was developed by Cosmo Iannicco, an engineer with KeySpan Energy Utility of New York.

158 *model for other U.S. cities* Winnie Hu, "New York Leads Politeness Trend? Get Outta Here!" *New York Times,* 8 April 2008, http://www.nytimes.com/2006/04/16/nyregion/16conduct.html?_r=1.

159 *governor's hotline* Kathleen Lucadamo, "Since New Noise Code Began in July, 135,589 Noise Complaints Made to 311," *New York Daily News,* 4 December 2007.

161 *New Orleans* I owe much of my understanding of what happened to New Orleans during and after Katrina to four individuals I met there in 2007: The Rev. Walter Baer, John Biguenet, Dr. Paul Farmer, and Dana Land. On the retrieval of corpses, see "After the Flood," *The Nation,* 10 September 2007.

162 *China* Bill McKibben, *Deep Economy: The Wealth of Communities and the Durable Future* (New York: Henry Holt, 2007), 194.

162 *footnote, the use-level of automobile horns* Jacques Attali, *Noise* (1977; Minneapolis: University of Minnesota Press, 1985), 124.

CHAPTER 7: LOUD AMERICA

165 *"beneath the nearly invincible and despairing noise"* James Baldwin, *No Name in the Street,* in *Collected Essays* (New York: Library of America, 1998), 382.

166 *fourteen military engagements* Korea, Cuba, Vietnam, Dominican Republic, Lebanon, Grenada, Panama, Kuwait, Somalia, Haiti, Bosnia, Kosovo, Afghanistan, Iraq.

166 *"American's biggest exports are television programs and movies"* Bill McKibben, *Deep Economy: The Wealth of Communities and the Durable Future* (New York: Henry Holt, 2007), 196.

166 *footnote, cars* See http://www.plunkettresearch.com/Industries/AutomobilesTrucks/AutomobileTrends/tabid/89/Default.aspx. See also Emma Rothchild, "Can We Transform the Auto-Industrial Society?" *New York Review of Books,* 26 February 2009, 8–9. *televisions* See http://www.ce.org/Press/CurrentNews/press_release_detail.asp?id=10764. *guns* See Laura MacInnis, "U.S. Most Armed Country with 90 Guns per 100 People," *Reuters,*

28 August 2007, http://www.Reuters.com/article/wtMostRead/edUSL283 4893820070828.

166 *footnote, "enplanements"* Federal Aviation Administration. Forecasts FY 2009–2025, 31 March 2009, http://www.faa.gov/data_research/aviation/aerospace_forecasts/2009–2025/.

166 *footnote, "three broad categories of incivility"* Jonathan Schonsheck, "Rudeness, Rasp, and Repudiation," in *Civility and Its Discontents: Essays on Civic Virtue, Toleration, and Cultural Fragmentation,* ed. Christine Sistare (Lawrence: University Press of Kansas, 2004), 169. For an explicit linking of noise with issues of civility, see the chapter "Keep It Down (and Rediscover Silence)" in P. M. Forni, *Choosing Civility: The Twenty-Five Rules of Considerate Conduct* (New York: St. Martin's, 2002), 93–96.

167 *barbaric yawp* Walt Whitman, "Song of Myself," in *Leaves of Grass* (1855; New York: Mentor, 1954), section 52, page 96.

167 *the Saint John Will-I-Am Coltrane African Orthodox Church* See http://www.coltranechurch.org/.

168 *"the loudest country"* Oscar Wilde, "Impressions of America" (1882), cited in Henry Still, *In Quest of Quiet* (Harrisburg, PA: Stackpole Books, 1970), 22.

168 *Philip Larkin* Eric Nisenson, *Ascension: John Coltrane and His Quest* (New York: St. Martin's, 1993), 129–130.

168 *"Mozartean celebration"* Nisenson, *Ascension,* 94.

168 *"Some people say"* Bill Cole, *John Coltrane* (1976; New York: Da Capo, 2001), 11.

168 *footnote, Frank Lowe* Nat Hentoff, liner notes to *The Very Best of John Coltrane* (Rhino Records, 2000), 8.

169 *religious experience* Cole, *John Coltrane,* 23.

169 *Alice Coltrane* Cole, *John Coltrane,* 158.

169 *African and Asian musical traditions* Cole pays particular emphasis to Coltrane's global influences; see index headings for Africa, Eastern Considerations, Indian Music, and Indonesian Music.

169 *Einstein* Cole, *John Coltrane,* 16, 197.

169 *LSD* Nisenson, *Ascension,* 166.

169 *"one of the gentlest"* Hentoff, liner notes, 6.

169 *could never recall his having said a single malicious word* Lewis Porter, *John Coltrane: His Life and Music* (Ann Arbor: University of Michigan Press, 1998), 250–251. See also Hentoff above, 6.

169 *woman in a jazz club* Porter, *John Coltrane,* 252.

169 *the way he dressed* Cole, *John Coltrane,* 47.

169 *so as not to wake the rest of the hotel* Cole, *John Coltrane,* 120.

169 *producer quipped* Cole, *John Coltrane,* 189.

169 *tears in his eyes* Cole, *John Coltrane*, 157.

169 *ruptured the blood vessels* Cole, *John Coltrane*, 171.

169 *"It was part of his nature"* Cole, *John Coltrane*, 83.

169 *footnote, Coltrane played with, and . . . studied under* See Alex Ross, *The Rest Is Noise: Listening to the Twentieth Century* (New York: Farrar, Straus and Giroux, 2007), 478, 498, as well as Nisenson, *Ascension*, 223–224, and Cole, *John Coltrane*, 146. Ross reports that the American composer Steve Reich went to see Coltrane perform at least fifty times (498).

170 *jazz was influenced by . . . urban commotion* Emily Thompson, *The Soundscape of Modernity: Architectural Acoustics and the Culture of Listening in America, 1900–1933* (2002; Cambridge, MA: MIT Press, 2004), 119. "Jazz musicians and avant-garde composers created new kinds of music directly inspired by the noises of the modern world. By doing so they tested long-standing definitions of musical sound, and they challenged listeners to reevaluate their own distinctions between music and noise." Thompson quotes Duke Ellington commenting on his *Harlem Air Shaft*: "So much goes on in a Harlem air shaft. You get the full essence of Harlem in an air shaft. You hear fights, you smell dinner, you hear people making love. You hear intimate gossip floating down. You hear the radio. An airshaft is one great big loudspeaker." She adds: "The connection between jazz and urban noise that Ellington celebrated was, however, far more frequently invoked by those who condemned it" (131). Without detracting from the beauty of the music or the genius involved in making it from such sources, one wants to ask how many people forced to live next to a "Harlem air shaft" were pleased with the situation.

170 *Wynton Marsalis* In Ken Burns, *Jazz*, PBS, 2001, episode ten.

170 *Pianist McCoy Tyner* Cole, *John Coltrane*, 52–53, 117. Writing in 1993, Eric Nisenson called Tyner "one of the two or three most important pianists of the past thirty years" (*Ascension*, 88).

170 *"his true soul mate"* Gary Giddens, *Visions of Jazz: The First Century* (New York: Oxford University Press,1998), 482.

170 *devout Muslim* Cole, *John Coltrane*, 118, 157.

170 *"magic in sound"* Cole, *John Coltrane*, 60.

171 *Tyner explained* Porter, *John Coltrane*, 266.

171 *footnote, Among Coltrane's . . . innovations* Giddens, *Visions of Jazz*, 488.

171 *Jones's explanation* Porter, *John Coltrane*, 267.

171 *neither . . . had bad words* Porter, *John Coltrane*, 267, 296.

171 *England, in 1966* Greil Marcus, *Like a Rolling Stone: Bob Dylan at the Crossroads* (New York: PublicAffairs, 2005), 182,

172 *a few tomatoes* Cole, *John Coltrane*, 190–191.

172 *"Judas!"* Marcus, *Like a Rolling Stone*, 26.

172 *"most storied event"* Marcus, *Like a Rolling Stone*, 155. For Marcus's telling of the event, see his chapter "Three Stages."

172 *1963* Fred Goodman, excerpt from *The Mansion on the Hill*, in *Studio A: The Bob Dylan Reader*, ed. Benjamin Hedin (New York: Norton, 2004), 40.

172 *"face of folk music"* Mike Marqusee, *Wicked Messenger: Bob Dylan and the 1960s* (New York: Seven Stories Press, 2005), 150. For Marqusee's account, including Dylan's outfit for the 1965 performance, see his chapter "Little Boy Lost."

172 *Pete Seeger was heard shouting* Goodman, *The Mansion on the Hill*, in *Studio A*, 42. Accounts of what Seeger said vary but agree in substance.

172 *"pandering to the crowd"* Marcus, *Like a Rolling Stone*, 157.

172 *Oscar Brand* Marqusee, *Wicked Messenger*, 153.

172 *footnote, $50-a-day scale* Goodman, *The Mansion on the Hill*, in *Studio A*, 41, 43.

173 *"I can't put anybody down"* From interview with Nat Hentoff, *Playboy*, March 1966, in *Bob Dylan: The Essential Interviews*, ed. Jonathan Cott (New York: Wenner Books, 2006).

173 *can't be a good egg* C. S. Lewis, *Mere Christianity* (1952; San Francisco: Harper San Francisco, 2001), 199.

173 *"I was doing fine"* From interview with Nora Ephron and Susan Edmiston, *Positively Tie Dream*, August 1965, in *The Essential Interviews*, 52.

174 *footnote, jamming* See http://www.babylon.com/definition/jam_session/.

174 *coordinated clapping* Marcus, *Like a Rolling Stone*, 180. The use of noise to disrupt music perceived as noise has precedents. To cite one with a happier outcome than the Newport fiasco: Apparently some of Arnold Schoenberg's detractors brought whistles and noisemakers to his 1913 world premiere of *Gurre-Leider*, but ended up weeping and chanting the composer's name. Ross, *The Rest is Noise*, 54.

175 *"Despotism"* Alexis de Tocqueville, *Democracy in America*, Vol. 2, ed. Phillips Bradley, trans. Henry Reeve, and rev. Francis Bowen (New York: Vintage, 1945), 110 (beginning of Chapter 4).

175 *footnote, Plato* Cited in Jacques Attali, *Noise* (1977; Minneapolis: University of Minnesota Press, 1985), 34.

175 *As of 2009* "U.S. Prisons and Drug Laws," *Bill Moyers Journal*, PBS, 17 April 2009. See also David C. Fahti, "Lock 'Em Up? It Costs You," *Human Rights Watch*, http://www.hrw.org/en/news/2009/04/01/lock-em-it-costs-you.

175 *the United States has 25 percent of the world's prisoners* Atul Gawande, "Hellhole: Is Solitary Confinement Torture?" *New Yorker*, 30 March 2009, 41.

175 *44 percent of them black* "Prison Statistics," United States Department of Justice, 30 June 2009, http://www.ojp.usdoj.gov/bjs/prisons.htm.

176 *one out of three* Stan Wilson, "Behind the Scenes: Life After San Quentin," *CNN.com*, 26 February 2009, http://www.cnn.com/2009/LIVING/wayoflife/02/26/btsc.life.after.prison/index.html.

176 *quadrupled its number* Gawande, "Hellhole," 41.

176 *acoustical nightmare* Jack Evans, "What Should a Prison Sound Like?" *Texas Architect* 45, no. 1 (January-February 1995): 76.

176 *"an alarm clock"* Jeffrey Ian Ross and Stephen C. Richard, *Behind Bars: Surviving Prison* (New York: Alpha, 2002), 135. See also Sasha Abramsky, "Prison Breakdown: Overcrowding Has Pushed California's Prison System to the Brink," *Real Cost of Prisons.org*, 22 October 2007, http://realcostofprisons.org/blog/archives.2007/10/in_these-timesp.html.

176 *"It's when it gets quiet"* Ralph "Sonny" Barger with Keith and Lent Zimmerman *Hell's Angel: The Life and Times of Sonny Barger and the Hell's Angel Motorcycle Gang* (New York: William Morrow, 2000), 200.

177 *Abdulrahman Zeitoun* Frankie Martin, "Muslim Man Imprisoned During Rescue of Katrina Victims," *Journey Into America*, 16 March 2009, http://journeyintoamerica.wordpress.com/2009/03/16/muslim-imprisoned-during-rescue-of-katrina-victims/. Zeitoun's story is told at length in a recent book by Dave Eggers called *Zeitoun* (McSweeney's, 2009). In a radio interview prior to publication, Eggers made this interesting observation: "Because the canoe was so quiet, he was able to hear things that others weren't—and others going around in motorboats and fan boats that are so incredibly loud. He was in this quiet canoe and was able to sort of help animals and people, and see things and hear things that others weren't able to." Dave Eggers, interview with Guy Raz, *All Things Considered*, NPR, 25 July 2009.

177 *Kaczynski's "oversensitivity to sound"* Gary Greenberg, "In the Kingdom of the Unabomber," *McSweeney's* 3 (1999): 70–71 A shorter, more recent version of Greenberg's essay appears in his *The Noble Lie: When Scientists Give the Right Answers for the Wrong Reasons* (New York: Wiley, 2008), 103–126.

177 *footnote, Solitary confinement . . . is currently used* Gawande, "Hellhole," 40–44.

178 *"Even if we walked"* Stephen A. Carter, "Can You Hear Me Now?" *Correctionalnews.com*, 13 April 2004, http://correctionalnews.com/ME2/Audiences/dirmod.asp?sid=&nm=&type=Publishing&mod=Publications%3A%3AArticle&mid=8F3A7027421841978F18BE895F87F791&tier=4&id=00093BE8A7F74DEFA8D7B2E95D13EA6F&AudID=A8CD3887511441F7AA259DA5A2CCFA71.

178 *a survey conducted . . . in four states* Jerry P. Christoff and Knut A. Rostad, "Noise in Corrections and Detention Facilities: A Fresh Look at the ACA Noise Standard and the Impact of Noise on Staff and Inmates" (Bethesda, MD: Committee on Acoustics in Corrections, 2004), Appendix D.

178 *ACA . . . daytime standard limit* Carter, "Can You Hear Me Now?"

179 *"talking with people from different backgrounds"* Patrick Lincoln, "Knocking Down Walls: From Protest to Prison to Building Movement," http://www.marlboro.edu/resources/library/zines.php.

179 *George Catlin* Alfred Runte, *National Parks: The American Experience*, 3rd ed. (Lincoln, Nebraska: University of Nebraska Press, 1997), 26.

179 *Yellowstone became the first* See http://www.nps/gov.tpr/.

179 *visually stunning monuments* Runte, *National Parks*, 1–47.

179 *Everglades* Runte, *National Parks*, 17.

179 *footnote, Stegner was quoting . . . Bryce* See Runte, *National Parks*, xi. Senator Ken Salazar identified the speaker as Bryce in a speech before the National Parks Conservation Association in Washington, DC, 15 April 2008.

180 *"desolate fields"* National Park Foundation, *Mirror of America: Literary Encounters with the National Parks* (Boulder, CO: Roberts Rinehart, 1989), 108.

180 *Tall Grass Prairie* See http://www.nps.gov/tapr/.

180 *John Muir* Runte, *National Parks*, 31.

180 *Natural Sounds Program* The program website is at http://www.nature.nps.gov/naturalsounds/organization/. Most of my information on the parks came from interviews at the Fort Collins, Colorado, headquarters of the National Park Service's Natural Sounds Program and with managers of several parks, as noted below and in the acknowledgments of this book. See also Robert E. Manning, *Parks and Carrying Capacity: Commons Without Tragedy* (Washington, DC: Island Press, 2007). See also Steven M. Miller, "A Conversation with Nicholas Miller, November 2005," *Soundscape: The Journal of Acoustic Ecology* 7, no. 1 (Fall-Winter 2007): 19–22.

180 *1998 study* G. E. Haas and T. J. Wakefield, *National Parks and the American Public: A National Opinion Survey on the National Park System* (Washington, DC: National Parks and Conservation Association, 1998).

180 *footnote, Edward Abbey* See "*Polemic*: Industrial Tourism and the National Parks," in *Desert Solitaire: A Season in the Wilderness* (1968; New York: Ballantine, 1971), 48–73.

181 *Mount Rushmore* Thanks to Vicki McCusker and Frank Turina of the Natural Sounds Office.

182 *Zion National Park* Thanks to Jock Whitworth, supervisor of Zion.

182 *Muir Woods* Thanks to Ericka Pilcher of Colorado State University and the National Park Service. See also Manning, *Parks*, 155–166.

182 *"low flier" incidents involving military jets* Thanks to Vicki McCusker of the Natural Sounds Program and to Gregg Fauth, wilderness coordinator for Sequoia and Kings.

183 *the most intractable noise problem . . . comes from the sky* My information comes from Kurt Fristrup of the National Sounds Program, Les Blomberg of the Noise Pollution Clearinghouse, and Nicholas Miller of Harris Miller Miller & Hanson.

184 *the number . . . is projected to reach 1.1 billion* FAA, Forecasts FY 2009–2025.

185 *Grand Canyon controversy* A good summary can be found at the Grand Canyon National Park website, http://www.nps.gov/archives/grca/overflights/documents/chronology.htm.

186 *"thrillcraft"* George Wuerthner, Introduction to *Thrillcraft: The Environmental Consequences of Motorized Recreation*, ed. George Wuerthner (White River Junction, VT: Chelsea Green; Foundation for Deep Ecology, 2007), xxxv–xxxvi.

186 *Yellowstone dispute* "Court Rules to Protect Yellowstone," Greater Yellowstone Coalition, http://news.greateryellowstone.org/node/154. See also "2007–2008 Monitoring Reports and Other Studies," Yellowstone National Park website, http://www.nps.gov/yell/parkmgmt/winterusetechnicaldocuments.htm.

187 *Elk and bison* Barrie Gilbert, "No Wild, No Wildlife: The Threat from Motorized Recreation," in Wuerthner, *Thrillcraft*, 119–123.

187 *percent of public land use* Wuerthner, Introduction to *Thrillcraft*, xxxv.

187 *jet skis* Charles Komanoff and Howard Shaw, *Drowning in Noise: Noise Costs of Jet Skis in America* (Montpelier, VT: Noise Pollution Clearinghouse, 2000). *Jet ski* is used as both a proper name for a company and as a common term for so-called personal watercraft manufactured by any company. I am using the word only in the latter sense.

187 *peculiar noisemaking characteristics* Komanoff and Shaw, *Drowning*, 12–15.

187 *financial loss to beachgoers* Komanoff and Shaw, *Drowning*, 17–38.

187 *at least 200,000 more jet skis now* See http://www.bts.gov/publications/national_transportation_statistics/html/table_02_44.html. See also Komanoff and Shaw, *Drowning in Noise*, 1, as well as Sean Smith and Carl Schneebeck, "Troubled Waters: Protecting Our Communities and Ourselves from Jet Ski Noise," in Wuerthner, *Thrillcraft*, 148.

187 *footnote, emissions-per-hour rate of a jet ski* Komanoff and Shaw, *Drowning in Noise*, 54–55.

188 *political "jamming"* See David Havlick, "Smoke and Gears: Seeing Through the Off-Roaders' Demographic Mirage," in Wuerthner, *Thrillcraft*, 59–62. See also Scott Silver, "From Recreation to Wreckreation: Efforts to Commercialize, Privatize, and Motorize the Great Outdoors," in Wuerthner, *Thrillcraft*, 69–72.

188 *A new jet ski . . . costs between $6,000 and $12,000* See "U.S. Demographics," www.jetski.com.

188 *figures for snowmobile ownership* See http://www.snowmobile.org/facts_snfcts.asp. As of July 2008, the average household income for snowmobilers was $72,000.

188 *study of ORV users* Havlick, "Smoke and Gears," 61.

188 *support from ORV manufacturers and the timber industry* Havlick, "Smoke and Gears," 59.

188 *Mr. Apology* "Alan S. Bridge, 50, Conceptual Artist," obituary in *New York Times*, 11 August 1995. See also Lydia Nibley, "All Apologies," *Westword.com*, 16 December 2004, http://westword.com/2004-12-16/new-years-eve/all-apologies/.

188 *footnote, Under Ronald Reagan* Bethanie Walder, "Miles from Everywhere: Roads, Off-Road Vehicles, and Watershed Restoration on Public Lands," in Wuerthner, *Thrillcraft*, 208.

189 *population 6,400* This statistic and many others can be found at the rally website. http://www.sturgismotorcyclerally.com.

191 *modified exhaust systems* Emily Fredrix, "Noise Laws Could Muffle Motorcycle Noise," *Associated Press*, 22 August 2007.

191 *Noisy Dozen Award* "Oakland Police Department: Roaring into Illegality," Noise Free America Press Release, 1 July 2008, http://www.noisefree.org/newsroom/pressreleases.php.

191 *footnote, ads for modified pipes* Anamosa, IA: J & P Cycles, 2007. Without discounting the motive of showing off—and pissing off—I suspect that for many riders the matter of "exhaust note" is mainly aesthetic. As one enthusiast told me, "I have a passion for motors. To hear a healthy-sounding motor—it does something for you, absolutely. It's part of the pleasure." To dismiss this pleasure as "merely aesthetic" strikes me as similar to dismissing noise annoyance as "merely subjective." I find it interesting that those with a passion for quiet and those with a passion for motors (not necessarily different people) both feel obliged to produce the evidence of dead bodies ("Noise Kills," "Loud Pipes Save Lives") in order to justify their auditory pleasures. Perhaps by freeing the word *aesthetic* from the dismissive qualifier

merely, we can succeed in freeing ourselves from some noise, including the negative "jamming" that occurs when an issue like this is debated.

192 *the average American biker* Steven E. Alford, "Riding Your Harley Back into Nature: Hobbes, Rousseau, and the Paradox of Biker Identity," in *Harley-Davidson and Philosophy: Full-Throttle Aristotle,* ed. Bernard E. Rollin and others (Chicago: Open Court, 2006), 145.

192 *RUBs* Thomas G. Endres, *Sturgis Stories: Celebrating the People of the World's Largest Motorcycle Rally* (Minneapolis: Kirk House, 2002), 11.

193 *footnote, Hell's Angels* Barger, *Hell's Angel,* 28.

193 *footnote, Harley-Davidson stock* Endres, *Sturgis Stories,* 14.

195 *"the Angels shall be Kings!"* Barger, *Hell's Angel,* 255.

195 *triumphal journey* Barger, *Hell's Angel,* 7–8.

196 *Tom Petty* Reference to "It's Good to Be King," from *Wildflowers,* Warner Brothers Records, 1994.

196 *sacred to more than thirty Native American tribes* Tamra Brennan, "Sacred Sites vs. a Modern Day Disease, Greed," *Protect Bear Butte,* 24 May 2009, http://bearbutte.blogspot.com/2009/05/sacred-sites-vs-modern-day-disease.html.

196 *Jackpine Gypsies* Endres, *Sturgis Stories,* 9.

197 *Jay Allen* Jim Robbins, "For Sacred Site, New Neighbors Are Far from Welcome," *New York Times,* 4 August 2006, http://www.nytimes.com/2006/08/04/us/04sacred.html.

201 *"pure products of America"* William Carlos Williams, "To Elsie," in *Selected Poems* (1949; New York: New Directions, 1969), 28.

202 *John McCain* Peter Hamby, "McCain Makes the Rounds at Biker Rally," *CNN Politics.com,* 5 August 2008, http://politicalticker.blogs.cnn.com/2008/08/05/scenes-from-the-trail-mccain-makes-the-rounds-at-biker-rally/.

204 *Singleton . . . delivered one of the most interesting papers* Herb Singleton, "Community Perceptions and Their Role in Noise Control: A Tale of One City" (paper presented at Inter-Noise, Honolulu, 3–6 December 2006).

209 *one who'd stayed* Cole, *John Coltrane,* 125.

CHAPTER 8: SUSTAINABILITY AND CELEBRATION

211 *"You must change your life"* Rainer Maria Rilke, "Archaic Torso of Apollo," in *The Essential Rilke,* trans. Galway Kinnell and Hannah Liebmann (New York: Ecco Press, 1999), 33.

211 *we will need to live closer together* I am indebted to John Moyers for pointing out the relationship between the environmental desirability of denser communities and the corresponding importance of noise abatement.

211 *there are 6 billion of us now* United States Census Bureau, World POPclock Projection, http://www.census.gov/ipc/www/popclockworld.html. For projections of future population growth, see United Nations, *The World at Six Billion,* 1999, www.un.org/esa/population/publications/sixbillion/sixbil part1.pdf.

211 *footnote, it takes half a gallon of oil* Bill McKibben, *Deep Economy: The Wealth of Communities and the Durable Future* (New York: Henry Holt, 2007), 64. McKibben also reports that "the fastest growing part of the food business is shipment by airplane" (65).

212 *"P & Q"* The Irish consultant was Karl Searson of Dublin.

212 *If the people of China* Emma Rothchild, "Can We Transform the Auto-Industrial Society?" *New York Review of Books,* 26 February 2009, 8–9.

212 *the average Chinese . . . the average American's* George Monbiot, *Heat: How to Stop the Planet from Burning* (2007; Cambridge, MA: South End Press, 2009), xii–xiii.

212 *family in Tanzania* McKibben, *Deep Economy,* 196.

213 *every village . . . ought to have a refrigerator* McKibben, *Deep Economy,* 197.

213 *footnote, The average American* McKibben, *Deep Economy,* 184.

213 *learn to live more quietly* For an affecting essay on this theme, see W. D. Wetherell, "Quiet," in *North of Now: A Celebration of Country and the Soon to Be Gone* (Guilford, CT: Lyons Press, 2002), 130–147. Among other trenchant remarks, Wetherell says, "Not watching television . . . is probably the rarest, most extreme form of dissent left to anyone in our culture" (131), a culture in which "'silence has become, in the final dark irony, all but synonymous with sexual abuse" (136).

213 *footnote, "Big Money"* Wendell Berry, "The Agrarian Standard," in *The Essential Agrarian Reader: The Future of Culture, Community, and the Land,* ed. Norman Wirzba (Lexington: University Press of Kentucky, 2003), 26.

213 *"our distinctively human heritage"* Barbara Ehrenreich, *Dancing in the Streets: A History of Collective Joy* (New York: Holt, 2006), 260.

214 *London burned 24,000 tons of coal* Monbiot, *Heat,* 2. Monbiot also reports: "In 2025, according to the US government's Energy Information Administration, the United States will burn 40 percent more coal than it does today" (82).

214 *coal miners* Jeffrey S. Vipperman, Eric R. Bauer, and Daniel R. Babich, "Survey of Noise in Coal Preparation Plants," *Journal of the Acoustical Society of America* 121, no. 1 (January 2006): 206.

214 *coal-fueled power plants* Vipperman, Bauer, and Babich, "Survey of Noise in Coal Plants," 206–212. The authors found that overall noise levels in the eight coal plants they visited ranged from 75.9 to 115 dB.

214 *U.S. corn yields . . . agricultural energy consumption* McKibben, *Deep Economy,* 64.

214 *someone living next to a datacom substation* Patrick Thibodeau, "Data Center Noise Levels Rising," *Computerworld,* 3 September 2007, http://computer world.com/s/article/299947/Can_You_Hear_Me_Now. See also American Society of Heating, Refrigerating and Air-Conditioning Engineers, "Acoustical Noise Emissions," Chapter 9 of *ASHRAE Design Considerations for Datacom Equipment Centers* (ASHRAE, 2006), http://searchdatacenter .techtarget.com/searchDataCenter/downloads/Datacom_Bk_Ch9.pdf.

215 *a strong preference* Les Blomberg, *Wind, Noise and Energy* (Montpelier, VT: Noise Pollution Clearinghouse, 2007), 6.

215 *wishing . . . to move* Blomberg, *Wind, Noise and Energy,* 6.

215 *more prospective home-buyers were concerned* I am grateful to Helen Matthews of the British Department of the Environment, Food, and Rural Affairs (DEFRA) for this information.

216 *90 percent by 2030* Monbiot, *Heat,* xii.

216 *footnote, UK scientists* Monbiot, *Heat,* 15.

218 *Chekhov* The letter was written to Alexei Suvorin, 27 March 1894. Anton Pavlovich Chekhov, Simon Karlinsky, and Henry Heim, *Anton Chekhov's Life and Thought: Selected Letters and Commentary* (1973; Evanston, IL: Northwestern University Press, 1997), 261.

218 *"chicken suicide case"* Cited in Denny Hatch, *Cedarhurst Alley: A Lighter-Than-Air, Anti-Noise Novel* (New York: iUniverse, 2005), 195–197.

219 *The average American family . . . flew* McKibben, *Deep Economy,* 34.

219 *flying remains the most polluting form of transportation* Monbiot, *Heat,* 173.

219 *less than 500 miles* See National Resources Defense Council, *Flying Off Course: Environmental Impacts of America's Airports,* 1996, www.eltoroairport.org./issues/ nrdc-flying.html. For a more recent report, see "Air Travel Delays Expected to Rise with Improving Economy," *NBC News Channel,* 8 October 2009, www2.wsav.com/sav/news/national/article/air_travel_delays _expected_to_rise_with_improving_economy/55326/. Adie Tomar, a transportation research analyst with the Brookings Institution, says, "We found that the short-haul flights less than 500 miles make up about half of the flights in this country." In other words, they found that little has changed in over ten years, concerns about climate change notwithstanding.

219 *HACAN ClearSkies* The Heathrow Association for the Control of Aircraft Noise has merged with ClearSkies. See www.hacan.org.

219 *Plane Stupid* www.planestupid.com. This site contains posts pertaining to recent protests at British airports.

219 *footnote, number of people . . . affected by aircraft noise* Compare Gerald L. Dillingham, "Aviation and the Environment: Impact of Aviation Noise on Communities Presents Challenges for Airport Operations and Future Growth of the National Airspace System," United States Government Accountability Office, GAO-08–216T, 24 October 2007, 5; and Cahal Milmo, "Airline Emissions 'Far Higher Than Previous Estimates,'" *The Independent,* 6 May 2008, http://www.indepdendent.co.uk/environment/climate-change/airline-emissions-far-higher-than-previous-estimates-821598,html.

220 *barred from public trains* "From the Mother of All Injunctions to the Mother of All Setbacks," *HACAN ClearSkies,* 7 August 2007, http://www.hacan.org.uk/news/press_releases.php?id=195.

220 *Bishop of London* "It's a Sin to Fly, Says Church," *London Sunday Times,* 23 July 2006, http://www.timesonline.co.uk/tol/news/uk/article691423.ece.

220 *"If you fly"* Monbiot, *Heat,* 188.

220 *footnote, a planted tree* Monbiot, *Heat,* 211.

221 *"another necessity"* Charles Komanoff, "Whither Wind: A Journey Through the Heated Debate over Wind Power," *Orion Magazine,* September/October 2006, 5.

222 *"What had surprised me"* Godefridus Petrus van den Berg, "The Sound of High Winds: The Effect of Atmospheric Stability on Wind Turbine Sound and Microphone Noise," doctoral dissertation, University of Groningen, Netherlands, http://irs.ub.rug.nl/ppn/294294104, 5.

222 *Mr. Bellamy* van den Berg, "Sound of High Winds," 8.

222 *Other descriptions* Derived from interviews at Mars Hill (later in chapter) as well as with van den Berg. See also UK Noise Association, *Location, Location, Location: An Investigation into Wind Farms and Noise by the Noise Association* (London: UK Noise Association, July 2006), 8; van den Berg, "Sound of High Winds," 158; and Barbara J. Frey and Peter J. Hadden, *Noise Radiation from Wind Turbines Installed Near Homes: Effects on Health,* Vol. 1, February 2007, www.windturbinenoisehealthhumanrights.com, 10, 17, 26.

222 *persistent low frequency noise* Kerstin Persson Waye, "Health Aspects of Low Frequency Noise" (paper presented at Inter-Noise, Honolulu, 3–6 December 2006).

222 *"vibroacoustic disease"* UK Noise Association, *Location,* 16.

222 *footnote, "cocktail party effect"* Karl D. Kryter, *The Handbook of Hearing and the Effects of Noise: Physiology, Psychology and Public Health* (New York: Academic Press, 1994), 309.

223 *"A typical approach"* Nina Pierpont, *Wind Turbine Syndrome: A Report on a National Experiment* (Santa Fe, NM: K-Selected Books, 2009), p. 30.

223 *Les Blomberg notes the following* In *Wind, Noise and Energy* (Montpelier, VT: Noise Pollution Clearinghouse, 2007).

223 *Swedish researchers* Eja Pederson and Kerstin Persson Waye, "Wind Turbine Noise, Annoyance and Self-Reported Health and Well-Being in Different Living Environments," *Occupational and Environmental Medicine* 64 (2007): 485. See also Eja Pederson and Kerstin Persson Waye, "Perception and Annoyance Due to Wind Turbine Noise—a Dose-Response Relationship," *Journal of the Acoustical Society of America* 116, no. 6 (December 2004): 3460–3470.

224 *"For some informants"* UK Noise Association, *Location,* 9.

224 *"been screwed"* The consultant has asked that his remark not be attributed.

224 *mile minimums recommended* UK Noise Association, *Location,* 21; Frey and Hadden, *Noise Radiation,* 5. For an approach to siting based on sound level measurements, along with a good review of the pertinent literature, see George W. Kamperman and Richard R. James, "Sample Guidelines for Siting Wind Turbines to Prevent Health Risks" (paper presented at 9th International Congress on Noise as a Public Health Problem, ICBEN, Foxwoods, CT, 2008).

224 *footnote, siting wind turbines offshore* Monbiot, *Heat,* 105.

227 *"a new Dutch disease"* van den Berg, "Sound of High Winds," 6.

227 *"due to strong winds"* van den Berg, "Sound of High Winds," 2.

227 *"now reach higher"* van den Berg, "Sound of High Winds," 20.

227 *"positive ring"* van den Berg, "Sound of High Winds," 3.

227 *"biting the hand"* van den Berg, "Sound of High Winds," 7.

229 *He finds it "astounding"* van den Berg, "Sound of High Winds," 10.

232 *footnote, "Everyone has the right to freedom of movement"* United Nations Declaration of Human Rights, cited in Simon Blackburn, *Being Good: An Introduction to Ethics* (Oxford: Oxford University Press, 2001), 136–143.

232 *the local gov'ment* "In Rural New York, Windmills Can Bring a Whiff of Corruption," *New York Times,* 17 August 2008, www.nytimes.com/2008/08/18/nyregion/ 18windmills.html.

234 *propitious directions* This section begins with a line from an untitled poem that begins "Terence, this is stupid stuff": "Oh I have been to Ludlow Fair/And left my necktie God knows where." In A. E. Housman, *A Shropshire Lad* (1896; New York: Dover, 1990), 42.

235 *pilgrimage to Mecca* Malcolm X, *Autobiography* (1964; New York: Grove, 1966), 344.

239 *In Hebrew* See "Repentance" in *The Oxford Companion to the Bible,* ed. Bruce M. Metzger and Michael D. Coogan (New York and Oxford: Oxford University Press, 1993), 646.

239 in *Rilke's German* "Archaic Torso of Apollo," in *The Essential Rilke,* 32–33.

239 *avant-garde skirt* Supposedly one of the shepherd children insisted that the Blessed Mother had appeared in a short skirt, a detail regarded as spurious by the Church authorities who first heard it. See Susan Neville, "Mystic or Maniac," *Oprah.com,* May 2007, http://www.oprah.com/article/omagazine/200705_omag_mysticvsmaniac/3.

CHAPTER 9: THE MOST BEAUTIFUL SOUND IN THE WORLD

241 *"I frequently hear music in the very heart of noise"* Quoted in Alex Ross, *The Rest Is Noise: Listening to the Twentieth Century* (New York: Farrar, Straus and Giroux), 143.

241 *soccer thug's curse* See Bill Buford, *Among the Thugs* (New York and London: Norton, 1992).

242 *motto of that Dutch anti-noise campaign* Karin Bijsterveld, *Mechanical Sound: Technology, Culture, and Public Problems of Noise in the Twentieth Century* (Cambridge, MA: MIT Press, 2008), 185.

243 *World Health Organization* Birgitta Berglund, Thomas Lindvall, and Dietrich H. Schwela, eds., *Guidelines for Community Noise* (Geneva: World Health Organization, 1999), 84.

244 *goose-stepping parades* George Orwell, "The Lion and the Unicorn: Socialism and the English Genius" (1941), in *My Country Right or Left 1940–1943,* Vol. 2, *The Collected Essays, Journalism and Letters,* ed. Sonia Orwell and Ian Angus (New York: Harcourt Brace Jovanovich, 1968), 62. "Why is the goose-step not used in England? There are, heaven knows, plenty of army officers who would be only too glad to introduce some such thing. It is not used because the people in the street would laugh. Beyond a certain point, military display is only possible in countries where the common people dare not laugh at the army."

245 *Ward v. Rock Against Racism* 491 U.S. 781 (1989). See http://supreme.justia.com/us/491/781. See also Linda Greenhouse, "Supreme Court Accord: Rock Music Is Loud," *New York Times,* 28 February 1989.

246 *footnote, Justice Thurgood Marshall* See http://law.jrank.org/pages/12675/Ward-v-Rock-Against-Racism.html.

246 *Scarry, Shklar, Rorty* See Scarry's *Body in Pain*, which is cited several times above in connection with noise and power; Judith Shklar, "Putting Cruelty First," Chapter 1 in *Ordinary Vices* (Cambridge, MA: Harvard University Press, 1984); and Richard Rorty, *Contigency, Irony, and Solidarity* (Cambridge, UK: Cambridge University Press, 1989), 88–95, 189–192.

248 *carnival barkers* Emily Thompson, *The Soundscape of Modernity: Architectural Acoustics and the Culture of Listening in America, 1900–1933* (2002; Cambridge, MA: MIT Press, 2004), 123–124.

254 *"Why could not everyone choose his or her telephone signal?"* R. Murray Schafer, *The Soundscape: Our Sonic Environment and the Tuning of the World* (1977; Rochester, VT: Destiny Books, 1994), 242.

255 *celebrated recordings* See www.wildsanctuary.com.

256 *wolves of history* See Jon T. Coleman, *Vicious: Wolves and Men in America* (New Haven: Yale University Press, 2004).

256 *flaneur* The ideas in the lecture were attributed to the French philosopher Michel de Certeau (1925–1986) and in particular to his book, *The Practice of Everyday Life* (1984). I have since learned that the pilfering of office supplies is apparently one example of a broader category of subversion called *la perruque*, literally "the wig," which Certeau describes as a tactic of the politically weak.

258 *"Señor . . . "* Robertson Davies, *Fifth Business* (1970; New York: Penguin, 1977), 224, 228.

259 *McCoy Tyner* See http:www.mccoytyner.com.

NOTE: All papers from 9th International Congress on Noise as a Public Health Issue (ICBEN) can be accessed online at http://www.icben.org/proceedings2008/.

A TIME LINE OF NOISE HISTORY

I took my dates from a wide variety of sources but most often from Karin Bijsterveld's *Mechanical Sound*, John Picker's *Victorian Soundscapes*, Alex Ross's *The Rest Is Noise*, R. Murray Schafer's *The Soundscape*, Bruce M. Smith's anthology *Hearing History*, Bruce R. Smith's *The Acoustic World of Early Modern England*, and Emily Thompson's *The Soundscape of Modernity*.

COMMON TERMS USED IN DISCUSSIONS OF NOISE

I am grateful to Les Blomberg and Nicholas Miller for a review of this section and to a number of written sources, including Giles Daigle's "Review and Tutorial of Sound Propagation," Karl D. Kryter's *The Handbook of Hearing and the Effects of Noise*, and the World Health Organization's *Guidelines for Community Noise*.

DECIBELS IN EVERYDAY LIFE AND EXTRAORDINARY SITUATIONS

Decibel charts are a frequent feature of articles and books about noise; where possible I have tried to determine the entries in mine by consensus. None are the results of my own measurements. Typical decibel charts often do not specify whether the readings have been weighted; it is probably safe to assume that A-weighting is implied in most cases. Many decibel charts also do not specify the distance between sound source and sound receiver for a given reading. In short, all readings on a typical decibel chart should be taken as approximate.

PRACTICAL CONSIDERATIONS FOR NOISE DISPUTES

For another set of arguments and responses, see Barry Truax, *Acoustic Communication.* Helpful advice for noise disputes can also be found on the website for the Noise Pollution Clearinghouse and Noise Free America.

BIBLIOGRAPHY

Abbey, Edward. *Desert Solitaire: A Season in the Wilderness.* New York: Ballantine, 1971.

Abbot, Jack Henry. *In the Belly of the Beast: Letters from Prison.* New York: Random House, 1981.

Abdallah, Ali Lutfi. *The Clever Sheikh of the Butana and Other Stories: Sudanese Folktales.* New York: Interlink Books, 1999.

Abel, Sharon M. "Hearing Loss in Military Aviation and Other Trades: Investigation of Prevalence and Risk Factors." *Aviation, Space, and Environmental Medicine* 76, no. 12 (December 2005): 1128–1135.

Abramsky, Sasha. "Prison Breakdown: Overcrowding Has Pushed California's Prison System to the Brink." *Real Cost of Prisons.org,* 22 October 2007. http://realcostofprisons.org/blog/archives.2007/10/in_these-timesp.html.

Ackerman, Diane. *A Natural History of the Senses.* New York: Vintage, 1995.

Adams, Lisa J. "Silver Lining: Flu Brings Less Crime, Less Noise and Blue Sky to Mexico City." Associated Press, 30 April 2009.

"After the Flood." *The Nation,* 10 September 2007.

"Aircraft Overflights: Chronology of Significant Events. Grand Canyon National Park Management Information. http://www.nps.gov/archives/grca/overflights/documents/chronology.htm.

"Air Travel Delays Expected to Rise with Improving Economy." *NBC News Channel,* 8 October 2009. http://www2.wsav.com/sav/news/national/article/air_travel_delays_expected_to_rise_with_improving_economy/55326/.

"Alan S. Bridge, 50, Conceptual Artist." Obituary in *New York Times,* 11 August 1995.

Alford, Steven E. "Riding Your Harley Back into Nature: Hobbes, Rousseau, and the Paradox of Biker Identity." In Rollin, Gray, Mommer, and Pineo, *Harley-Davidson and Philosophy: Full-Throttle Aristotle.*

Alighieri, Dante. *Hell.* Trans. Dorothy L. Sayers. Harmondsworth, UK: Penguin, 1949.

"AMA Slams Churches for Excessive Noise." *HappyGhana.com,* 18 April 2008. http://www.happyghana.com/newsdetails.asp?id=3314&cat_id=1.

American Society of Heating, Refrigerating and Air-Conditioning Engineers. "Acoustical Noise Emissions." Chapter 9 of *ASHRAE Design Considerations for Datacom Equipment Centers* (ASHRAE, 2006). http://searchdatacenter.techtarget.com/searchDataCenter/downloads/Datacom_Bk_Ch9.pdf.

American Tinnitus Association. Frequently Asked Questions. http://www.ata.org/about-tinnitus/patient-faq1.

Apollonio, Umbro, ed. *Futurist Manifestos*. Boston: MFA Publications, 2001.

Armstrong, Karen. *Buddha*. New York: Viking Penguin, 2001.

Arnold, Chloe. "Russia: Moscow's Noise Pollution Reaches Dangerous Levels." *Radio Free Europe*, 19 September 2007. http://www.rferl.org/content/article/1078719.html.

Associated Press. "Automakers Obsess over Reducing Wind Noise." MSNBC, 18 June 2008. http://www.msnbc.msn.com/id/25248338/.

Attali, Jacques. *Noise: The Political Economy of Music.* Trans. Brian Massumi. 1977; Minneapolis: University of Minnesota Press, 1985.

Attwood, Tony. *The Complete Guide to Asperger's Syndrome*. London: Jessica Kingsley Publishers, 2007.

Bailey, Peter. "Breaking the Sound Barrier." Chapter 9 in *Popular Culture and Performance in the Victorian City.* Cambridge, UK: Cambridge University Press, 1998.

Baldwin, James. *Collected Essays*. New York: Modern Library, 1998.

Baliles, Gerald L. "Aircraft Noise: Addressing a Potential Barrier to Global Growth." *Virginia Lawyer*, June/July 2001.

Barger, Ralph "Sonny," with Keith and Lent Zimmerman. *Hell's Angel: The Life and Times of Sonny Barger and the Hell's Angels Motorcycle Gang.* New York: William Morrow, 2000.

Baron, Lawrence. "Noise and Degeneration: Theodor Lessing's Crusade for Quiet." *Journal of Contemporary History* 17 (1982): 165–178.

Baron, Robert Alex. *The Tyranny of Noise.* New York: St. Martin's, 1970.

Barrett, Douglas E. "Traffic-Noise Impact Study for Least Bell's Vireo Habitat Along California State Route 83." *Transportation Research Record 1559,* National Research Council (1996): 3–7.

Battles, Jan. "Noise a Major Threat to Health of City Dwellers." *Sunday Times*, 12 October 2008.

Bean, Thomas L. "Noise on the Farm Can Cause Hearing Loss." Agriculture and Natural Resources, Ohio State University, 2008. http://ohioline.osu.edu/aex-fact/pdf/AEX_590_08.pdf.

Beardsley, Richard K., John W. Hall, and Robert E. Ward. *Village Japan*. Chicago: University of Chicago Press, 1959.

"Bedlam: The Hospital of St. Mary of Bethlehem." British Broadcasting Corporation. http://www.bbc.co.uk/dna/h2g2/A2554157?s id=1.

Berglund, Birgitta, Thomas Lindvall, and Dietrich H. Schwela, eds. *Guidelines for Community Noise*. Geneva: World Health Organization, 1999.

Berry, Wendell. "The Agrarian Standard." In *The Essential Agrarian Reader: The Future of Culture, Community, and the Land*. Ed. Norman Wirzba. Lexington: University Press of Kentucky, 2003.

———. *The Unsettling of America: Culture and Agriculture*. 1977; San Francisco: Sierra Club Books, 1986.

Besant, Alexander, and Willy Lowry. "Beirut's Nightlife Is Back—and Not Everyone Is Happy." *Daily Star* (Beirut, Lebanon), 16 June 2008.

Bhat, Sairam. "Noise and the Law." *India Together*, November 2003. http://www.indiatogether.org/2003/nov/law-noise.htm.

Bijsterveld, Karin. *Mechanical Sound: Technology, Culture, and Public Problems of Noise in the Twentieth Century*. Cambridge, MA: MIT Press, 2008.

Birnbaum, Pierre. *Geography of Hope: Exile, the Enlightenment, Disassimilation*. Trans. Charlotte Mandell. Berkeley: Stanford University Press, 2008.

Blesser, Barry, and Linda-Ruth Salter. "The Unexamined Rewards for Excessive Loudness." Communications: 9th International Conference on Noise as a Public Health Problem (ICBEN), Foxwoods, CT, 2008.

Blomberg, Les. "Acoustical Slums, Green Buildings, and the Acoustics of Sustainability." Paper presented at "Noise! Design, Health and the Urban Soundscape." Graduate School of Architecture, Planning and Preservation, Columbia University, New York, 21 September 2009.

———. "The Nature of Noise: Civility, Sovereignty, Community, Reciprocity, Power, Tyranny, and Technology." *The Quiet Zone* (Noise Pollution Clearinghouse newsletter), Fall 2006.

———. *Wind, Noise and Energy*. Montpelier, VT: Noise Pollution Clearinghouse, 2007.

Blomberg, Les, and David Morris. "Sound Decisions." *New Rules*, Winter 1999.

Blondeel, Maria. "Listening to Acoustic Energy and Not Hearing." *Soundscape: The Journal of Acoustic Ecology* 6, no. 1 (Spring/Summer 2005): 22–23.

Boulware, Jack. "Feel the Noise." *Wired 8.10*. http://www.wired.com/wired/archive/8.10/stereocar_pr.html.

Branch, Alex. "Man, 22, Slain After Argument over Noise." *Fort Worth Star-Telegram*, 10 February 2008.

Brennan, Tamra. "Sacred Sites vs. a Modern Day Disease, Greed." *Protect Bear Butte*, 24 May 2009. http://bearbutte.blogspot.com/2009/05/sacred-sites-vs-modern-day-disease.html.

Brock, David. *The Republican Noise Machine: Right-Wing Media and How It Corrupts Democracy*. New York: Crown, 2004.

Brockmeier, Kevin. "The Year of Silence." In *The Best American Short Stories 2008*. Ed. Salman Rushdie and Heidi Pitlor. Boston: Houghton Mifflin, 2008.

Brody, Jane E. "All That Noise Is Damaging Children's Hearing." *New York Times*, 9 December 2008. http://www.nytimes.com/2008/12/09/health/09brod.html?_r=1.

Broer, Christian. "Policy Annoyance: How Policies Shape the Experience of Aircraft Sound." *Aerlines, e-zine edition*. http://www.scribd.com/doc/12475160/38-Broer-Noise-Annoyance.

Bronzaft, Arline. "The Effect of a Noise Abatement Program on Reading Ability." *Journal of Environmental Psychology* 1 (1981): 215–222.

———. "The Effect of Elevated Train Noise on Reading Ability." *Environment and Behavior* 7, no. 4 (1975): 517–528.

———. *Listen to the Raindrops*. New York: Department of Environmental Protection, 2009.

Buchanan, Malcolm B., P. R. Prinsley, J. M. Wilkenson, and J. E. Fitzgerald. "Is Golf Bad for Your Hearing?" *British Medical Journal* 337 (20–27 December 2008): 1437–1438.

Buford, Bill. *Among the Thugs*. New York: Norton, 1992.

Burns, Ken. *Jazz*. Public Broadcasting System, 2001.

Cage, John. *Silence*. Middleton, CT: Wesleyan University Press, 1961.

Campbell, Heather E., and Laura R. Peck. "Not So Much: A Policy Brief on Recent Research on Environmental Justice in the Phoenix Area." *Perspectives in Public Affairs* (Spring 2008): 3–6.

Camus, Albert. *The Rebel: An Essay on Man in Revolt*. Trans. Anthony Bower. 1951; New York: Vintage, 1956.

Carlyle's House Memorial Trust. *Carlyle's House Catalog*. London: Chiswick Press, 1995.

Carter, Stephen. "Can You Hear Me Now?" *Correctionalnews.com*, 4 April 2004. http://correctionalnews.com/ME2/Audiences/dirmod.asp?sid=&nm=&type=Publishing&mod=Publications%3A%3AArticle&mid=8F3A7027421841978F18BE895F87F791&tier=4&id=00093BE8A7F74DEFA8D7B2E95D13EA6F&AudID=A8CD3887511441F7AA259DA5A2CCFA71.

Casson, Lionel. *Libraries in the Ancient World*. New Haven: Yale University Press, 2001.

Cesarani, David. *Becoming Eichmann: Rethinking the Life, Crimes, and Trial of a "Desk Murderer."* New York: Da Capo, 2006.

Chandler, David L. "Universe Started with Hiss, Not Bang." *New Scientist*, 12 June 2004. http://www.newscientist.com/article/dn5092-universe-started-with-hiss-not-bang.html.

Chekhov, Anton Pavlovich, Simon Karlinsky, and Henry Heim. *Anton Chekhov's Life and Thought: Selected Letters and Commentary*. 1973; Evanston, IL: Northwestern University Press, 1997.

Chilton, Lance, Marc S. Dubin, Don Bapst, Mark Ellingham, John Fisher, and Natania Nansc. *The Rough Guide to the Greek Islands*. New York: Rough Guides, 2004.

Chown, Marcus. "Big Bang Sounded Like a Deep Hum." *New Scientist*, 30 October 2003. http://www.newscientist.com/article/dn4320-big-bang-sounded-like-a-deep-hum.html.

Christoff, Jerry P., and Knut A. Rostad. "Noise in Corrections and Detention Facilities: A Fresh Look at the ACA Noise Standard and the Impact of Noise on Staff and Inmates." Bethesda, MD: Committee on Acoustics in Corrections, 2004.

Clark, Charlotte, Rocio Martin, Elsie van Kempen, Tamuno Alfred, Jenny Head, Hugh W. Davies, Mary M. Haines, Isabel Lopez Barrio, Mark Matheson, and Stephen Stansfeld. "Exposure-Effect Relations Between Aircraft and Road Traffic Noise Exposure at School and Reading Comprehension: The RANCH Project." *American Journal of Epidemiology* 163, no. 1 (2006): 27–37.

Cohen, Sheldon, and Shirlynn Spacapan. "The Social Psychology of Noise." In Jones and Chapman, *Noise and Society*.

Cole, Bill. *John Coltrane*. 1976; New York: Da Capo, 2001.

Cole, Kevin. "Woman Shot in Battle over Loud Music Dies." *Omaha World Herald*, 30 May 2008.

Coleman, Jon T. *Vicious: Wolves and Men in America*. New Haven: Yale University Press, 2004.

Commission of the European Communities. *Environmental Noise Directive 2002*. http://ec.europa.eu/environment/noise/directive.htm.

———. *Future Noise Policy: European Commission Green Paper*. Brussels: Commission of the European Communities, 11 April 1996. http://europa.eu/documentation/official-docs/green-papers/index_en.htm.

Corbin, Alain. *Village Bells: Sound and Meaning in the 19th-Century French Countryside*. Trans. Martin Thom. 1994; New York: Columbia University Press, 1998.

Cott, Jonathan, ed. *Bob Dylan: The Essential Interviews*. New York: Wenner Books, 2006.

"Court Dismisses Cases Concerning Suvarnabhumi Airport." *Baangkok Post,* 24 June 2009. http://www.bangkokpost.com/mail/147036/.

"Court Rules to Protect Yellowstone." Greater Yellowstone Coalition. http://news.greateryellowstone.org/node/154.

Cusick, Suzanne G. "Music as Torture/Music as Weapon." *Transcultural Music Review* 10 (2006). http://www.sibetrans.com/trans/trans10/cusick_eng.htm.

Daigle, Gilles A. "Atmospheric Acoustics." In *McGraw-Hill Encyclopedia of Science and Technology.* New York: McGraw Hill, 2000. Also www.AccessScience.com.

———. "Sound Propagation: Review and Tutorial." Paper presented at Inter-Noise, Honolulu, 3–6 December 2006.

Daley, Suzanne. "Spain Rudely Awakened to Workaday World." *New York Times,* 26 December 1999.

Dan, Bernard. "Titus's Tinnitus." *Journal of the History of the Neurosciences* 14, no. 3 (September 2005): 210–213.

Dauenhauer, Bernard P. *Silence: The Phenomenon and Its Ontological Significance.* Bloomington: University of Indiana Press, 1980.

Davidson, H. R. Ellis. *Gods and Myths of Northern Europe.* Harmondsworth, UK: Penguin, 1964.

Davies, Hugh, and Irene van Kamp. "Environmental Noise and Cardiovascular Disease: Five Year Review and Future Directions." Paper presented at 9th International Congress on Noise as a Public Health Problem, ICBEN, Foxwoods, CT, 2008.

Davies, Robertson. *Fifth Business.* 1970; New York: Penguin, 1977.

Dean, Eric T. "'The Awful Shock and Rage of Battle': Rethinking the Meaning and Consequences of Combat in the Civil War." In Gramm, *Battle.*

De Botton, Alain. *The Consolations of Philosophy.* New York: Pantheon, 2000.

De las Casa, Bartolomé. "1542: Hispaniola." Excerpt from *A Brief Account of the Destruction of the Indies.* In *Lapham's Quarterly,* Winter 2008.

Denby, David. "The Unquiet Life" (a review of the film *Noise*). *New Yorker,* 19 May 2008.

Department for Environment, Food & Rural Affairs (DEFRA, UK), "Noise and Nuisance Policy: A Review of National and European Practices." http://www.defra.gov.uk/environment/noise/research/climate/nannexb.htm.

Desai, Kiran. *The Inheritance of Loss.* New York: Grove Press, 2006.

Deshaies, Pierre, Richard Martin, Danny Belzile, Pauline Fortier, Chantal Laroche, Serge-André Girard, Tony Leroux, Hughes Nélisse, Robert Arcand, Maurice Poulin, and Michel Picard. "Noise as an Explanatory Factor in Work-Related Fatality Reports: A Descriptive Study." Paper presented at 9th International Congress on Noise as a Public Health Problem, ICBEN, Foxwoods, CT, 2008.

Dickens, Charles. Letter to House of Commons. Carlyle Memorial Trust.

Dillingham, Gerald L. "Aviation and the Environment: Impact of Aviation Noise on Communities Presents Challenges for Airport Operations and Future Growth of the National Airspace System." United States Government Accountability Office, GAO-08–216T. 24 October 2007.

Douglas, Scott. *Quiet, Please: Dispatches from a Public Librarian.* New York: Da Capo Press, 2008.

Drake, Samuel Adams. *A Book of New England Legends and Folk Lore.* Boston: Little, Brown, 1906.

Dunford, Martin, Phil Lee, and Suzanne Morton Taylor. *The Rough Guide to The Netherlands.* New York: Rough Guides, 2007.

Dupuy, Trevor Nevitt. *The Military Life of Genghis: Khan of Khans.* New York: Franklin Watts, 1969.

Dutch Ministry of Transport, Public Works and Water Management, *Nationale Mobiliteitsmonitor 2008.* http://www.verkeerenwaterstaat.nl/kennisplein/3/7/375943/Nationale_Mobiliteitsmonitor_2008_compleet.pdf.

Eban, Abba. *Heritage: Civilization and the Jews.* New York: Summit Books, 1984.

Eckhart, Meister. *Meister Eckhart: A Modern Translation.* Trans. Raymond Bernard Blakney. New York: Harper and Row, 1941.

Eggers, Dave. Interview with Guy Raz. *All Things Considered,* NPR, 25 July 2009.

Ehrenreich, Barbara. *Dancing in the Streets: A History of Collective Joy.* New York: Holt, 2006.

Ellermeier, Wolfgang, Monika Eigenstetter, and Karin Zimmer. "Psychoacoustic Correlates of Individual Noise Sensitivity." *Journal of the Acoustical Society of America* 109, no. 4 (April 2001): 1464–1473.

Emerson, Ralph Waldo. "The American Scholar." In *The Romantic Movement in American Writing.* Ed. Richard Harter Fogle. New York: Odyssey Press, 1966.

Endres, Thomas G. *Sturgis Stories: Celebrating the People of the World's Largest Motorcycle Rally.* Minneapolis: Kirk House, 2002.

Environmental Justice Resource Center. Executive Order No. 12898. http://www.ejrc.cau.edu/execordr.html.

European Environment Agency. "About Noise." http://www.eea.europa.eu/themes/noise/about-noise.

———. *Transport at a Crossroads.* Copenhagen: Office for Official Publications of the European Communities, 2009.

Evans, Gary W., Peter Lercher, Markus Meis, Hartmut Ising, and Walter W. Kofler. "Community Noise Exposure and Stress in Children." *Journal of the Acoustical Society of America* 109, no. 3 (March 2001): 1023–1027.

Evans, Gary W., and Lyscha A. Marcynyszyn. “Environmental Justice, Cumulative Environmental Risk, and Health Among Low- and Middle-Income Children in Upstate New York.” *American Journal of Public Health* 94, no. 11 (November 2004): 1942–1944.

Evans, Jack. “What Should a Prison Sound Like?” *Texas Architect* 45, no. 1 (January–February 1995): 76ff.

Fadiman, Anne. *Ex Libris: Confessions of a Common Reader.* New York: Farrar, Straus and Giroux, 1998.

Falzone, Kristin. “Airport Noise Pollution: Is There a Solution in Sight?” *Boston College Environmental Affairs Law Review* 26 (1999): 769–807.

Fathi, David C. “Lock ’Em Up? It Costs You.” *Human Rights Watch,* 1 April 2009. http://www.hrw.org/en/news/2009/04/01/lock-em-it-costs-you.

Federal Aviation Administration. Forecasts FY 2009–2025. 31 March 2009. http://www.faa.gov/data_research/aviation/aerospace_forecasts/2009–2025/.

Fidelman, Charlie. “Noise Can Kill, Health Report Finds.” *Montreal Gazette,* 24 November 2007.

“Findings.” *Harper’s Magazine,* October 2006.

Fleischman, John. “Counting Darters, Endangered Fish Species.” *Audubon* 98, no. 4 (July 1996): 84–89.

Folmer, Robert L., Susan Griest, and William Hal Martin. “Hearing Conservation Education Programs for Children.” *Journal of School Health* 72, no. 2 (February 2002): 51–57.

Forni, P. M. *Choosing Civility: The Twenty-Five Rules of Considerate Conduct.* New York: St. Martin’s, 2004.

Forum of European National Highway Research Laboratories. *Sustainable Road Surfaces for Traffic Noise Control Project* (SILVIA), 2008. http://www.trl.co.uk/silvia/.

Fraser, James George. *The New Golden Bough.* Ed. Theodor H. Gaster. 1890; New York: New American Library, 1964.

Fredrix, Emily. “Noise Laws Could Muffle Motorcycle Noise.” *Associated Press,* 22 August 2007.

Frey, Barbara J., and Peter J. Hadden. *Noise Radiation from Wind Turbines Installed Near Homes: Effects on Health,* Vol. 1. February 2007. www.windturbinenoise healthhumanrights.com.

“From the Mother of All Injunctions to the Mother of All Setbacks.” *HACAN ClearSkies,* 7 August 2007. http://www.hacan.org.uk/news/press_releases .php?id=195.

Fuchs, Guillermo. “Latin America.” In Berglund, Lindvall, and Schwela, *Guidelines for Community Noise,* 113–114.

Gauldie, Robin. *Ireland.* London: New Holland, 2006.

Gawande, Atul. "Hellhole: Is Solitary Confinement Torture?" *New Yorker,* 30 March 2009.

Giddins, Gary. *Visions of Jazz: The First Century.* New York: Oxford University Press, 1998.

Gilbert, Barrie. "No Wild, No Wildlife: The Threat from Motorized Recreation." In Wuerthner, *Thrillcraft.*

Godfrey, Hollis. "The City's Noise." *Atlantic Monthly,* November 1909.

Goldschagg, Paul. "Airport Noise and Environmental Justice in South Africa." *International Research in Geographical and Environmental Education* 11, no. 1 (2002): 72–75.

Goodell, Jeff. *Big Coal: The Dirty Secret Behind America's Energy Future.* Boston: Houghton Mifflin, 2006.

Goodman, Fred. *The Mansion on the Hill.* Excerpted in *Studio A: The Bob Dylan Reader.* Ed. Benjamin Hedin. New York: Norton, 2004.

Gordenker, Alice. "Sound Trucks." *Japan Times Online,* 18 April 2006. http://search.japantimes.co.jp/cgi-bin/ek20060418wh.html.

Gramm, Kent, ed. *Battle: The Nature and Consequences of Civil War Combat.* Tuscaloosa: University of Alabama Press, 2008.

Grandin, Temple. *Thinking in Pictures: And Other Reports from My Life with Autism.* 1995; New York: Vintage, 2006.

Grandin, Temple, and Margaret M. Scariano. *Emergence: Labeled Autistic.* Novato, CA: Arena Press, 1989.

Green, Penelope. "The Dream of Absolute Quiet." *New York Times,* 17 May 2007.

Greenberg, Gary. "In the Kingdom of the Unabomber." *McSweeney's* 3, 1999. Revised and reprinted in *The Noble Lie: When Scientists Give the Right Answers for the Wrong Reasons.* New York: Wiley, 2008.

Greenhouse, Linda. "Supreme Court Accord: Rock Music Is Loud." *New York Times,* 28 February 1989.

Grinker, Roy Richard. *Unstrange Minds: Remapping the World of Autism.* New York: Basic Books, 2007.

Grond, Etienne. "South Africa." In Berglund, Lindvall, and Schwela, *Guidelines for Community Noise,* 119–122.

Groopman, Jerome. "That Buzzing Sound: The Mystery of Tinnitus." *New Yorker,* 9 and 16 February 2009.

Gross, Joan, David McMurray, and Ted Swedenburg. "Arab Noise and Ramadan Nights: *Rai,* Rap and Franco-Maghrebi Identities." In *The Anthropology of Globalization: A Reader.* Ed. Jonathan Xavier Inda and Renato Rosaldo. Malden, MA: Blackwell Publishing, 2002.

Guastavino, Catherine. "The Ideal Urban Soundscape: Investigating the Sound Quality of French Cities." *Acta Acustica United with Acustica* 92 (2006): 945–951.

Haas, G. E., and T. J. Wakefield. *National Parks and the American Public: A National Opinion Survey on the National Park System*. Washington, DC: National Parks and Conservation Association, 1998.

Habib, Lucas, Erin M. Bayne, and Stan Boutin. "Chronic Industrial Noise Affects Pairing Success and Age Structure of Ovenbirds *Seiurus aurocapilla*. *Journal of Applied Ecology* 44, no. 1 (February 2007): 176–184.

Hamby, Peter. "McCain Makes the Rounds at Biker Rally." *CNNPolitics.com*, 5 August 2008. http://politicalticker.blogs.cnn.com/2008/08/05/scenes-from-the-trail-mccain-makes-the-rounds-at-biker-rally/.

Hamill, Sam, trans. *The Sound of Water: Haiku by Bashō, Buson, Issa, and Other Poets.* Boston: Shambhala, 2000.

Hanson, Victor Davis. *The Western Way of War: Infantry Battle in Classical Greece.* Berkeley: University of California Press, 2000.

Haralabidis, Alexandros S., Konstantina Dimakopoulou, Federica Vigna-Taglianti, Matteo Giampaolo, Alessandro Borgini, Marie-Louise Dudley, Göran Pershagen, Gösta Bluhm, Danny Houthuijs, Wolfgang Babisch, Manolis Velonakis, Klea Katsouyanni, and Lars Jarup. "Acute Effects of Night-Time Noise Exposure on Blood Pressure in Populations Living Near Airports. *European Heart Journal* 29, no. 5 (2008): 658–664.

Hardy, Thomas. *The Mayor of Casterbridge*. 1886; New York: Macmillan, 1965.

Harris, Liz. *Holy Days: The World of a Hasidic Family*. New York: Summit Books, 1985.

Hartwig, D. Scott. "'It's All Smoke and Dust and Noise': The Face of Battle at Gettysburg." In Gramm, *Battle*.

Hatch, Denny. *Cedarhurst Alley: A Lighter-Than-Air, Anti-Noise Novel*. New York: iUniverse, 2005.

Havlick, David. "Smoke and Gears: Seeing Through the Off-Roaders' Demographic Mirage." In Wuerthner, *Thrillcraft*.

Hazlewood, Phil. "Fighting the Noise in India's 'Maximum City.'" *ThingsAsian*, 16 February 2009. http://www.thingsasian.com/stories-photos/34842.

Hegel, Georg Wilhelm Friedrich. *The Philosophy of History*. Trans. J. Sibree. North Chemsford, MA: Courier Dover, 2004.

———. "Reciprocal Recognition, Spirit, and the Concept of Right." In *Lectures on the Philosophy of Spirit 1827–8*. Trans. Robert P. Williams. New York: Oxford University Press, 2007.

Hejaiej, Monia. *Behind Closed Doors: Women's Oral Narratives in Tunis*. New Brunswick, NJ: Rutgers University Press, 1996.

Hempton, Gordon, and John Grossmann. *One Square Inch of Silence: One Man's Search for Natural Silence in a Noisy World.* New York: Free Press, 2009.

Hendrikson, G. L. "Ancient Reading." *The Classical Journal* 25, no. 3 (December 1929): 182–196.

Hersey, John. *Hiroshima.* 1946; New York: Bantam, 1959.

Hiramatsu, Kozo. "A Review of Soundscape Studies in Japan." *Acta Acustica United with Acustica* 92 (2006): 857–864.

Hoban, Russell. *Riddley Walker.* 1980; New York: Touchstone, 1992.

Homer. *The Iliad.* Trans. Robert Fitzgerald. Garden City, NY: Anchor, 1975.

Horvat, Andrew. *Japanese Beyond Words: How to Walk and Talk Like a Native Speaker.* Berkeley, CA: Stone Bridge Press, 2000.

Housman, A. E. "Terence, this is stupid stuff." In *A Shropshire Lad.* 1896; New York: Dover, 1990.

Hu, Winnie. "New York Leads Politeness Trend? Get Outta Here!" *New York Times,* 8 April 2008. http://www.nytimes.com/2006/04/16/nyregion/16conduct.html?_r=1.

Huizinga, Johan. *Homo Ludens: A Study of the Play Element in Culture.* Boston: Beacon, 1955.

Ingram, Martin. "Ridings, Rough Music and the 'Reform of Popular Culture' in Early Modern England. *Past and Present,* no. 105 (1984): 77–113.

"In Rural New York, Windmills Can Bring a Whiff of Corruption." *New York Times,* 17 August 2008. www.nytimes.com/2008/08/18/nyregion/ 18windmills.html.

International Civil Aviation Organization. *Review of Noise Abatement Procedure Research and Development and Implementation Results: Discussion of Survey Results.* ICAO, 2007. http://www.icao.int/icao/en/env/ReviewNADRD.pdf.

Ishiwata, Chiaki, and Toshiko Fukuchi. "Results of Field Measurements of Generated Sound of Japanese Drum 'Taiko' and Consideration Towards Its Sound Insulation." Paper presented at Inter-Noise, Honolulu, 3–6 December 2006.

"It's a Sin to Fly, Says Church." *London Sunday Times,* 23 July 2006. http://www.timesonline.co.uk/tol/news/uk/article691423.ece.

Iyer, Pico. "The Eloquent Sounds of Silence." *Time,* 25 January 1993.

Jacobson, Mark. "Dharavi: Mumbai's Shadow City." *National Geographic,* May 2007.

J & P Cycles Catalog. Anamosa, IA: J & P Cycles, 2007.

Järviluoma, Helmi. "Listen! They Are Still Threshing: One Hundred Finnish Soundscapes Selected." *Soundscape: The Journal of Acoustic Ecology* 6, no. 2 (Fall/Winter 2005): 28–30.

Johnson, James H. "Listening and Silence in Eighteenth- and Nineteenth-Century France." In Mark M. Smith, *Hearing History.*

Johnson, Samuel. "London." In *Samuel Johnson: Selected Poetry and Prose.* Ed. Frank Brady and W. K. Wimsatt. Berkeley: University of California Press, 1977.

Jones, Dylan M., and Antony J. Chapman, ed. *Noise and Society.* New York: John Wiley & Sons, 1984.

Jones, Dylan M., and D. R. Davies. "Individual and Group Differences in the Response to Noise." In Dylan and Chapman, *Noise and Society.*

Kacirk, Jeffrey. *Forgotten English: A 365-Day Calendar of Vanishing Vocabulary and Folklore for 2006.* Petaluma, CA: Pomegranate, 2005.

Kahn, Douglas. *Noise, Water, Meat.* Cambridge, MA: MIT Press, 1999.

Kakuzo, Okakura. *The Book of Tea.* 1906; Rutland, VT: Charles E. Tuttle, 1956.

Kamperman, George W., and Richard R. James. "Sample Guidelines for Siting Wind Turbines to Prevent Health Risks." Paper presented at 9th International Congress on Noise as a Public Health Problem, ICBEN, Foxwoods, CT, 2008.

Kaplan, Fred. *Thomas Carlyle.* Berkeley: University of California Press, 1983.

Kassler, Jamie C. "Musicology and the Problem of Sonic Abuse." In *Music, Sensation, And Sensuality.* Ed. Linda Phyllis Austern. London: Routledge, 2002.

Kasson, John F. *Rudeness & Civility: Manners in Nineteenth-Century Urban America.* New York: Hill and Wang, 1990.

Keim, Brandon. "Listening to the Big Bang." *Wired,* 28 September 2008. http://www.wired.com/wiredscience/2008/09/listening-to-th/.

Keizer, Garret. "Preserving Silence in National Parks. *Smithsonian.org,* 6 August 2008. http://www.smithsonianmag.com/science-nature/sounds-in-parks.html.

———. "Sound and Fury: The Politics of Noise in a Loud Society." *Harper's Magazine,* March 2001.

Kitto, H. D. F. *The Greeks.* Harmondsworth, UK: Penguin, 1951.

Kluizenaar, Yvonne de, Ronald T. Gansevoort, Henk M. E. Miedema, and Paul E. de Jong. "Hypertension and Road Traffic Noise Exposure." *Journal of Occupational and Environmental Medicine* 49 (2007): 484–492.

Knol, Anne B. *Trends in the Environmental Burden of Disease in the Netherlands 1980–2020.* RIVM report 500029001/2005. Contact Anne.Knol@RIVM.nl.

Knox, Bernard. M. W. "Silent Reading in Antiquity." *Greek, Roman, and Byzantine Studies* 9, no. 1–4 (Spring 1968): 421–435.

Komanoff, Charles. "Whither Wind: A Journey Through the Heated Debate over Wind Power." *Orion Magazine,* September/October, 2006.

Komanoff, Charles, and Howard Shaw. *Drowning in Noise: Noise Costs of Jet Skis in America.* Montpelier, VT: Noise Pollution Clearinghouse, 2000.

Kosko, Bart. *Noise.* New York: Viking, 2006.

Kraft, Betsy Harvey. *Mother Jones: One Woman's Fight for Labor.* New York: Clarion Books, 1995.

Krause, Bernie. "Anatomy of the Soundscape: Evolving Perspectives." *Journal of the Audio Engineering Society* 56, no. 1/2 (January/February 2008): 73–88.

———. *Into a Wild Sanctuary*. Berkeley, CA: Heyday Books, 1998.

Kryger, Matt. "A Placid Place for Preemies. " *Indianapolis Star*, 2 February 2008.

Kryter, Karl D. *The Handbook of Hearing and the Effects of Noise: Physiology, Psychology and Public Health*. New York: Academic Press, 1994.

Kuwano, Sonoko, Seiichiro Namba, and Tohru Kato. "Perception and Memory of Loudness of Various Sounds." Paper presented at Inter-Noise, Honolulu, 3–6 December 2006.

"Laetoli Footprints." Evolution Library. *PBS.org*. http://www.pbs.org/wgbh/evolution/library/07/1/l_071_03.html.

Lambert, Bruce L., Ken-Yu Chang, and Swu-Jane Lin. "Effect of Orthographic and Phonological Similarity on False Recognition of Drug Names." *Social Science & Medicine* 52 (2001): 1843–1857.

Lambert, Bruce L., Paul A. Luce, William M. Fisher, Laura Walsh Dickey, and John W. Senders. "Frequency and Neighborhood Effects on Auditory Perception of Drug Names with Noise." Paper presented at Noise-Con, Minneapolis, 17–19 October 2005.

Lane, Margaret. *Samuel Johnson and His World*. London: Hamish Hamilton, 1975.

Lang, Andrew G. P., and Hugh D. Amos. *These Were the Greeks*. Uwchland, PA: Dufour, 1991.

Lasch, Christopher. *The Culture of Narcissism: American Life in an Age of Diminishing Expectations*. New York: Norton, 1979.

Lawrence, D. H. *Twilight in Italy*. 1916; New York: Viking, 2008.

Levitin, Daniel J. *This Is Your Brain on Music: The Science of a Human Obsession*. New York: Dutton, 2006.

Lewis, C. S. *Mere Christianity*. 1952; San Francisco: Harper San Francisco, 2001.

Lincoln, Patrick. "Knocking Down Walls: From Protest to Prison to Building Movement." http://www.marlboro.edu/resources/library/zines.php.

Lister, David. "Unionists Protest Against Building of Ulster Mosque." *The Times*, 14 January 2003. http://www.timesonline.co.uk/tol/news/uk/article 812068.ece.

Llana, Sara Miller. "Sleepless in Spain: The Siesta Recedes." *Christian Science Monitor*, 19 January 2006. http://www.csmonitor.com/2006/0119/p20s01-woeu .html.

Lubman, David. "Acoustics of the Great Ball Court at Chichen Itza." *Journal of the Acoustical Society of America* 120, no. 5 (November 2006): 3279.

Lucadamo, Kathleen. "Since New Noise Code Began in July, 135,589 Noise Complaints Made to 311." *New York Daily News*, 4 December 2007.

MacInnis, Laura. "U.S. Most Armed Country with 90 Guns per 100 People." *Reuters*, 28 August 2007. http://www.Reuters.com/article/wtMostRead/edUSL2834893820070828.

Maguire, Kevin. "BBC Cheers Up Lonely Staff with the Chit-Chat Machine." *Guardian*, 14 October 1999. http://www.guardian.co.uk/media/1999/oct/14/bbc.uknews.

Mahler, Richard. "The Human Cost of Silence Lost: How a Noisy Environment Hurts Our Health." In Wuerthner, *Thrillcraft*.

Malcolm X. *Autobiography*. 1964; New York: Grove, 1966.

Mamarbachi, Raya. "Cultural Differences and Taboos in Syrian Business Situations." *Going Global*. http://old.goinglobal.com/hot_topics/syria_business_mamarb.asp.

Manguel, Alberto. *History of Reading*. New York: HarperCollins, 1996.

Manning, Robert. *Parks and Carrying Capacity: Commons Without Tragedy*. Washington, DC: Island Press, 2007.

Marcus, Greil. *Like a Rolling Stone: Bob Dylan at the Crossroads*. New York: PublicAffairs, 2005.

Marinetti, Fillipo Tommaso. *The Founding and Manifesto of Futurism 1909*. In Apollonio, *Futurist Manifestos*.

Maris, Eveline. "The Social Side of Noise Annoyance." Doctoral dissertation, University of Leiden, 17 December 2008. https://www.openaccess.leidenuniv.nl/handle/1887/13361?mode=more.

Marqusee, Mike. *Wicked Messenger: Bob Dylan and the 1960s*. New York: Seven Stories Press, 2005.

Martin, Frankie. "Muslim Man Imprisoned During Rescue of Katrina Victims." *Journey Into America*, 16 March 2009. http://journeyintoamerica.wordpress.com/2009/03/16/muslim-imprisoned-during-rescue-of-katrina-victims/.

Martin, William Hal, Judy Sobel, Susan E. Griest, Linda Howarth, and Yong-bing Shi. "Noise Induced Hearing Loss in Children: Preventing the Silent Epidemic." *Journal of Otology* 1 (2006): 11–21.

Marx, Karl. *Capital*. Ed. Friedrich Engels. *The Great Books of the Western World* 50. Chicago: Britannica, 1952.

Marx, Karl, and Friedrich Engels. *Manifesto of the Communist Party*. In *The Great Books of the Western World*, Vol. 50.

———. *The Marx-Engels Reader*, 2nd ed. Ed. Robert C. Tucker. New York: Norton, 1978.

Marx, Leo. *The Machine in the Garden: Technology and the Pastoral Ideal in America*. New York: Oxford University Press, 1964.

Mauer, Marc, and Ryan Scott King. "Schools and Prisons: Fifty Years After *Brown v. Board of Education*." The Sentencing Project. http://www.sentencingproject.org/doc/publications/rd_brownvboard.pdf.

McEwan, Ian. *Atonement*. 2001; New York: Anchor, 2003.

McKee, Maggie. "Big Bang Waves Explain Galaxy Clustering." *New Scientist*, 12 January 2005. http://www.newscientist.com/article/dn6871.

McKibben, Bill. *Deep Economy: The Wealth of Communities and the Durable Future*. New York: Holt, 2007.

Mcllwain, D. Scott, Kathy Gates, Donald Ciliax. "Heritage of Army Audiology and the Road Ahead: The Army Hearing Program. *American Journal of Public Health* 98, no. 12 (December 2008): 2167–2172.

Melone, Deborah, and Eric W. Wood. *Sound Ideas: Acoustical Consulting at BNN and Acentech*. Cambridge, MA: Acentech, 2005.

Meyers, Jeffrey. *Samuel Johnson: The Struggle*. New York: Basic Books, 2008.

Miller, Nicholas P. "The Effects of Aircraft Overflights on Visitors to U.S. National Parks. *Institute of Noise Control Engineering* 47, no. 3 (May-June 1999): 112–117.

———. "Transportation Noise and Recreational Lands." March 2003. http://www.noisenewsinternational.net.

Miller, Stephen M. "A Conversation with Nicholas Miller, November 2005." *Soundscape: The Journal of Acoustic Ecology* 7, no. 1 (Fall-Winter 2007): 19–22.

Milmo, Cahal. "Airline Emissions 'Far Higher Than Previous Estimates.'" *The Independent*, 6 May 2008. http://www.independent.co.uk/environment/climate-change/airline-emissions-far-higher-than-previous-estimates-821598.html.

Milton, John. *Paradise Lost*. In *Complete Poems and Major Prose*. Ed. Merritt Y. Hughes. Indianapolis: Bobbs-Merrill, 1957.

Miyamoto, Mary. "Speaking Up on Noise." *Kansai Scene Magazine*, 3 February 2009. http://www.kansaiscene.com/2009_01/html/update.shtml.

Monbiot, George. *Heat: How to Stop the Planet from Burning*. 2007; Cambridge, MA: South End Press, 2009.

———. "We Are All Killers." *Monbiot.com*. 28 February 2006.

Moon, Krystyn. *Yellowface: Creating the Chinese in American Popular Music and Performance, 1850s–1920s*. New Brunswick, NJ: Rutgers University Press, 2005.

Moreno, Joseph J. "Orpheus in Hell: Music in the Holocaust." In *Music and Manipulation: On the Social Uses and Social Control of Music*. Ed. Stephen Brown and Ulrik Volgsten. New York: Berghahn Books, 2006.

Mosher, Howard Frank. "Thunder from a Cloudless Sky." *Washington Post Magazine*, 12 July 1998.

Murphy, Roger P. *Antisocial Housing*. London: UK Noise Association, 2004.

———. *Socially Unsound.* London: UK Noise Association, 2003.

Nagahata, Koji. "What Do Citizens Imagine Is a Level of 80 DB?: A Basic Study of Environmental Communication on Soundscape Issues." Paper presented at Inter-Noise, Honolulu, 3–6 December 2006.

"National Noise Awareness Day Launched." *ModernGhana.com,* 16 April 2008. http://www.modernghana.com/news/162814/1/national-noise-awareness-day-launched.html.

National Park Foundation. *Mirror of America: Literary Encounters with the National Parks.* Boulder, CO: Roberts Rinehart, 1989.

National Resources Defense Council. *Flying Off Course: Environmental Impacts of America's Airports,* 1996. http://www.eltoroairport.org./issues/nrdc-flying.html.

National Roads Authority of Ireland. *Guidelines for the Treatment of Noise and Vibration in National Road Schemes.* Revision 1, 25 October 2004.

Nayanatusita, Bhikkhu. "Buddhism and Sound Pollution." http://www.bps.lk/other_library/buddhism_and_sound_pollution.pdf.

Netherlands Environmental Assessment Agency. "Area and Dwellings in the Netherlands Exposed to Noise Levels in Excess of 50 db(A), 2002." Environmental Data Compendium. http://www.mnp.nl/mnc/i-en-0295.html.

Nettle, Daniel. *Happiness: The Science Behind Your Smile.* Oxford: Oxford University Press, 2005.

Neusner, Jacob. *Judaism's Theological Voice: The Melody of the Talmud.* Chicago: University of Chicago Press, 1995.

Neville, Susan. "Mystic or Maniac." *Oprah.com,* May 2007. http://www.oprah.com/article/omagazine/200705_omag_mysticvsmaniac/3.

Nibley, Lydia. "All Apologies." *Westword.com,* 16 December 2004. http://westword.com/2004-12-16/new-year-s-eve/all-apologies/.

Niemann, Hildegard, and Christian Maschke. *WHO LARES* [World Health Organization Large Analysis and Review of European Housing and Health Status]: *Final Report.* Berlin: Berlin Center of Public Health, 2004. EUR/04/5047477.

Nietzsche, Friedrich. *Twilight of the Idols.* In *The Portable Nietzsche.* Ed. Walter Kaufman. New York: Viking, 1968.

"9-11 Research Eye Witness Accounts and Oral Histories." http://911research.wtc7.net/wtc/evidence/eyewitnesses.html and http://911research.wtc7.net/wtc/evidence/oralhistories/explosions.html.

Nisenson, Eric. *Ascension: John Coltrane and His Quest.* New York: St. Martin's, 1993.

"No Horn Day a Success: Mumbai Cops." *Times of India,* 8 April 2008. http://timesofindia.indiatimes.com/articleshow/2934032.cms.

"Noise." *Oxford English Dictionary*, 2nd ed. New York: Oxford University Press, 1989.

"Noise Control Act." Noise Pollution Clearinghouse Law Library. http://www.nonoise.org/epa/act.htm.

"Oakland Police Department: Roaring into Illegality." Noise Free America Press Release, 1 July 2008. http://www.noisefree.org/newsroom/pressreleases.php.

Obama, Barack. *Dreams of My Father: A Story of Race and Inheritance.* New York: Random House, 1995.

O'Connell, Robert L. *Of Arms and Men: A History of War, Weapons, and Aggression.* New York: Oxford University Press, 1989.

Office of the Scientific Assistant, Office of Noise Abatement and Control. *Noise Effects Handbook: A Desk Reference to Health and Welfare Effects of Noise* (EPA 550-9-82-106). Fort Walton Beach, FL: National Association of Noise Control Officials, 1981.

Ogale, Sudhakar B. "South-East Asian Region." In Berglund, Lindvall, and Schwela, *Guidelines for Community Noise*, 129–134.

"Older Adults' Speech-Processing Difficulties May Stem from 'Fast, Noisy Talk,' Not Deafness." *ScienceDaily.com*. http://www.sciencedaily.com/releases/1998/10/981023073930.htm.

Olmos, Daniel. "A Ban on a Noisy Existence: The Los Angeles Leaf Blower Ban, Spatialized Whiteness and the Gardeners' Struggle for Dignity." Paper presented at the annual meeting of The American Studies Association, Philadelphia, 11 October 2007.

Onah, Godfrey Igwebuike. "The Meaning of Peace in African Traditional Religion and Culture." *AfrikaWorld.net*. http://www.afrikaworld.net/afrel/goddionah.htm.

Orwell, George. "The Lion and the Unicorn: Socialism and the English Genius" (1941). In *My Country Right or Left 1940–1943*, Vol. 2, *The Collected Essays, Journalism and Letters of George Orwell*. Ed. Sonia Orwell and Ian Angus. New York: Harcourt Brace Jovanovich, 1968.

Owomoyela, Oyekan. *Yoruba Trickster Tales.* Lincoln: University of Nebraska Press, 1997.

Pedersen, Eja. *Human Response to Wind Turbine Noise: Perception, Annoyance and Moderating Factors.* Göteborg: Sahlgrenska Academy, 2007. http://gupea.ub.gu.se/dspace/bitstream/2077/4431/1/gupea_2077_4431_1.pdf.

Pedersen, Eja, Frits van den Berg, Roel Bakker, and Jelte Bourma. "Response to Noise from Modern Wind Farms in the Netherlands." *Journal of the Acoustical Society of America* 126, no. 2 (August 2009): 634–643.

Pedersen, Eja, and Kerstin Persson Waye. "Perception and Annoyance Due to Wind Turbine Noise—a Dose-Response Relationship." *Journal of the Acoustical Society of America* 116, no. 6 (December 2004): 3460–3470.

———. "Wind Turbine Noise, Annoyance and Self-Reported Health and Well-Being in Different Living Environments." *Occupational and Environmental Medicine* 64, (2007): 480–486.

Penn, William. "For of Light Came Sight." In *The Quaker Reader*. Ed. Jessamyn West. 1962; Wallingford, PA: Pendle Hill, 1992.

Perrin, Noel. *Giving Up the Gun: Japan's Reversion to the Sword, 1543–1879*. 1979; Jaffrey, NH: David R. Godine, 1989.

Picker, John. *Victorian Soundscapes*. New York: Oxford University Press, 2003.

Pierpont, Nina. *Wind Turbine Syndrome: A Report on a Natural Experiment*. Santa Fe, NM: K-Selected Books, 2009.

Pinter, Harold. *The Homecoming*. New York: Grove Press, 1965.

Plunkett Research, Ltd. "Automobiles and Trucks Overview." http://www.plunkettresearch.com/Industries/AutomobilesTrucks/AutomobileTrends/tabid/89/Default.aspx.

Pollak, Michael. "Silence of the Cranks." *New York Times*, 13 February 2005. http://www.nytimes.com/2005/02/13/nyregion/thecity/13fyi.html?_r=1.

Porter, Lewis. *John Coltrane: His Life and Music*. Ann Arbor: University of Michigan Press, 1998.

"Prison Statistics." United States Department of Justice, 30 June 2009. http://www.ojp.usdoj.gov/bjs/prisons.htm.

"Pull the Plug on Torture Music: Binyam Mohamed." http://www.reprieve.org.uk.

Qamar, Saadia. "Noise Pollution Creating Problems for Karachiites." *The Nation*, 26 April 2009. http://www.nation.com.pk/pakistan-news-newspaper-daily-english-online/Regional/Karachi/26-Apr-2009/Noise-pollution-creating-problems-for-Karachiites.

Rabin, Lawrence, Richard G. Coss, and Donald H. Owings. "The Effects of Wind Turbines on Antipredator Behavior in California Ground Squirrels (*Spermophilus beecheyi*). *Biological Conservation* 131, no. 3 (August 2006): 410–420.

Radle, Autumn Lyn. "The Effect of Noise on Wildlife: A Literature Review." March 2007. http://interact.uoregon.edu/MediaLit/wfae/library/articles/radle_effect_noise_wildlife.pdf.

Ramakrishna, Sri. *The Gospel of Sri Ramakrishna*. Trans. Swami Nikhilananda. New York: Ramakrishna-Vivekananda Center, 1977.

Ramsbotham, David. "Why I Will Vote Against the 42-Day Law." *The Guardian*, 13 October 2008. http://www.guardian.co.uk/commentisfree/2008/oct/13/terrorism-lords.

Ravindran, Rajan, Rathinasamy Sheela Devi, James Samson, and Manohar Senthilvelan. "Noise-Stress-Induced Brain Neurotransmitter Changes and the

Effect of *Ocimum sanctum* (Linn) Treatment on Albino Rats." *Journal of Pharmacological Sciences* 98, no. 4 (2005): 354–360.

Rawls, John. *A Theory of Justice*. Cambridge, MA: Harvard University Press, 1971.

"Reducing the Level of Noise." *China Daily*, 23 November 2007. http://www.china.org.cn/english/China/232916.htm.

"Repentance." In *The Oxford Companion to the Bible*. Ed. Bruce M. Metzger and Michael D. Coogan. New York: Oxford University Press, 1993.

"Rhodes Culture: Social Conventions." *Monarch Holidays*. http://holidays.monarch.co.uk/destinations/Greece/Rhodes/culture.

Richie, Donald. "Hisako Shiraishi." In *The Donald Richie Reader: 50 Years of Writing on Japan*. Ed. Arturo Silva. Berkeley, CA: Stone Bridge Press, 2001.

Rilke, Rainer Maria. "Archaic Torso of Apollo." In *The Essential Rilke*. Trans. Galway Kinnell and Hannah Liebmann. New York: Ecco Press, 1999.

Roach, Jim. "Military Sonar May Give Whales the Bends, Study Says." *National Geographic News*, 8 October 2003. http://news.nationalgeographic.com/news/2003/10/1008_031008_whalebends.html.

Robbins, Jim. "For Sacred Indian Site, Neighbors Are Far from Welcome." *New York Times*, 4 August 2006. http://www.nytimes.com/2006/08/04/us/04sacred.html.

Rodrigue, Jean-Paul. "The Geography of Transport Systems." Hofstra University. http://www.people.hofstra.edu/geotrans/eng/ch6en/conc6en/ch6c3en.html.

Rojas, Rick. "Hybrid Cars May Pose Silent Threat to the Blind: Advocates Want Vehicles to Make Some Sound." *Courier-Journal* (Louisville, KY), 8 July 2008.

Rollin, Bernard E., Carolyn M. Gray, Kerri Mommer, and Cynthia Pineo, eds. *Harley-Davidson and Philosophy: Full Throttle Aristotle*. Chicago: Open Court, 2006.

Rorty, Richard. *Contigency, Irony, and Solidarity*. Cambridge, UK: Cambridge University Press, 1989.

Rose, Tricia. *Black Noise: Rap Music and Black Culture in Contemporary America*. Hanover: University Press of New England, 1994.

Rosenberg, John D. *The Darkening Glass: A Portrait of Ruskin's Genius*. New York: Columbia University Press, 1961.

Ross, Alex. *The Rest Is Noise: Listening to the Twentieth Century*. New York: Farrar, Straus and Giroux, 2007.

Ross, Jeffrey Ian, and Stephen C. Richard. *Behind Bars: Surviving Prison*. New York: Alpha, 2002.

Roth, Philip. *Everyman*. New York: Random House, 2006.

Rothschild, Emma. "Can We Transform the Auto-Industrial Society?" *New York Review of Books*, 26 February 2009.

Runte, Alfred. *National Parks: The American Experience*, 3rd ed. Lincoln: University of Nebraska Press, 1997.

Russolo, Luigi. *The Art of Noises* (extracts) 1913. In Apollonia, *Futurist Manifesto.*

Ryals, Brenda. "Hair Cell Regeneration: How It Works and What It Means for Human Beings." *ASHA Leader*, 5 May 2009.

Sacks, Oliver. *Musicophilia: Tales of Music and the Brain.* New York: Alfred A. Knopf, 2007.

Safe, Mike. "Bad Vibrations." *The Australian*, 31 January 2009. http://www.theaustralian.com.au/news/features/bad-vibrations/story-e6frg8h6-1111118667268.

St. Clair, Jeffrey. "Glory Boy and the Snail Darter." *CounterPunch*, 3–4 March 2007. http://www.counterpunch.org/stclair03032007.html.

Salerno, Roger A. *Beyond the Enlightenment: Lives and Thoughts of Social Theorists.* Westport, CT: Greenwood, 2004.

Sales, Rosemary. *Women Divided: Gender, Religion and Politics in Northern Ireland.* London: Routledge, 1997.

Salmon, Felix. "How Driving a Car into Manhattan Costs $160." *Reuters Blogs*, 3 July 2009. http://blogs.Reuters.com/felix-salmon/2009/07/03/how-driving-a-car-into-manhattan-costs-160/.

Saunders, John Joseph. "Did the Mongols Use Guns? Appendix 2 in *The History of the Mongol Conquests.* Philadelphia: University of Pennsylvania Press, 2001.

Sbihi, Hind, Hugh Davies, and Paul Demers. "Hypertension in Noise-Exposed Sawmill Workers: A Cohort Study." *Occupational and Environmental Medicine* 65 (2008): 643–646.

Scarry, Elaine. *The Body in Pain: The Making and Unmaking of the World.* New York: Oxford University Press, 1985.

Schafer, R. Murray. *The Book of Noise.* 1970.

———. *The Soundscape: Our Sonic Environment and the Tuning of the World.* 1977; Rochester, VT: Destiny Books, 1994.

Scholem, Gershom G., ed. *Zohar: The Book of Splendor.* 1949; New York: Schocken Books, 1963.

Schomer, Paul. "Criteria for Assessment of Noise Annoyance." *Noise Control Engineering Journal* 53, no. 4 (July-August 2005): 132–144.

Schonsheck, Jonathan. "Rudeness, Rasp, and Repudiation." In *Civility and Its Discontents: Civic Virtue, Toleration, and Cultural Fragmentation.* Ed Christine T. Sistare. Lawrence: University Press of Kansas, 2004.

Schopenhauer, Arthur. "On Noise." In *Great Essays.* Ed. Houston Peterson. New York: Washington Square Press, 1960.

Schreibman, Laura. *The Science and Fiction of Autism.* Cambridge, MA: Harvard University Press, 2005.

Schwartz, Hillel. "Noise and Silence: The Soundscape of Spirituality." Talk delivered for the Inter-Religious Federation for World Peace, Seoul, South

Korea, August, 1995. http://www.noisepollution.org/library.noisesil/noisesil.htm.

Selander, J., M. E. Nilsson, G. Bluhm, M. Rosenlund, M. Lindqvist, G. Nise, and G. Pershagen. "Long-Term Exposure to Road Traffic Noise and Myocardial Infarction." *Epidemiology* 20, no. 2 (March 2009): 272–279.

Seneca. "On Noise." In *The Art of the Personal Essay.* Ed. Phillip Lopate. New York: Doubleday, 1994.

Shapiro, Gary. *Alcyone: Nietzsche on Gifts, Noise, and Women.* Albany: SUNY Press, 1991.

Shapiro, Sidney A. *The Dormant Noise Control Act and Options to Abate Noise Pollution.* Administrative Conference of the United States, November 1991.

Shelton, Jill T., Emily M. Elliott, Sharon D. Eaves, and Amanda L. Exner. "The Distracting Effects of a Ringing Cell Phone: An Investigation of the Laboratory and the Classroom Setting." *Journal of Environmental Psychology* 29, no. 4 (December 2009): 513–521.

Shi, David E. *The Simple Life: Plain Living and High Thinking in American Culture.* 1985; Athens: University of Georgia Press, 2007.

Shklar, Judith. *Ordinary Vices.* Cambridge, MA: Harvard University Press, 1984.

Siegler, Richard, and Eva Talel. "Impact of New York City's Amended Noise Control Code." *New York Law Journal*, 2 July 2008.

Silver, Scott. "From Recreation to Wreckreation: Efforts to Commercialize, Privatize, and Motorize the Great Outdoors." In Wuerthner, *Thrillcraft.*

Singer, Isaac Bashevis. *The Collected Stories.* New York: Farrar, Straus and Giroux, 1982.

Singleton, Herb. "Community Perceptions and Their Role in Noise Control: A Tale of One City." Paper presented at Inter-Noise, Honolulu, 3–6 December 2006.

Siniscalchi, Col. Joseph. *Non-Lethal Technologies: Implications for Military Strategy.* Occasional Paper No. 3. Maxwell Air Force Base, AL: Center for Strategy and Technology, March 1998.

"Sleep Deprivation Leads to Impaired Risky Decision Making." National University of Singapore, 20 January 2008. http://www.nus.edu.sg/research/rg124.php.

Smilor, Raymond W. "American Noise, 1900–1930." In Mark M. Smith, *Hearing History.*

Smith, Adam. *Theory of Moral Sentiments.* 1st American Edition. 1759; Philadelphia: Anthony Finley, 1817.

Smith, Bruce R. *The Acoustic World of Early Modern England: Attending to the O-Factor.* Chicago: University of Chicago Press, 1999.

Smith, Mark M., ed. *Hearing History: A Reader.* Athens: University of Georgia Press, 2004.

———. "Listening to the Heard Worlds of Antebellum America." In Mark M. Smith, *Hearing History.*

———. *Listening to Nineteenth Century America.* Chapel Hill: University of North Carolina Press, 2001.

Smith, Sean, and Carl Schneebeck, "Troubled Waters: Protecting Our Communities and Ourselves from Jet Ski Noise." In Wuerthner, *Thrillcraft.*

Smollett, Tobias. *The Expedition of Humphry Clinker.* 1771; New York: Rinehart, 1950.

Sobotta, Robin R., Heather E. Campbell, and Beverly Owens. "Aviation Noise and Environmental Justice: The Barrio Barrier." *Journal of Regional Science* 47, no. 1 (2007) 125–154.

Song, Sora. "Nighttime Noise and Blood Pressure." *Time,* 13 February 2008. http://www.time.com/time/health/article/0,8599,1713178,00.html.

Sontag, Susan. "The Aesthetics of Silence." In *Styles of Radical Will.* New York: Farrar, Straus and Giroux, 1969.

Staudenmaier, John M. "Denying the Holy Dark: The Enlightenment Ideal and the European Mystical Tradition." In *Progress: Fact or Illusion?* Ed. Leo Marx and Bruce Mazlish. Ann Arbor: University of Michigan Press, 1996.

Stephens, S.D.G. "The Treatment of Tinnitus—a Historical Perspective." *Journal of Laryngology and Otology* 98 (October 1984): 963–972.

Stevens, S. Smith. "The Science of Noise." *Atlantic Monthly,* July 1946.

Still, Henry. *In Quest of Quiet.* Harrisburg, PA: Stackpole Books, 1970.

Strout, Elizabeth. *Amy and Isabelle.* New York: Random House, 1998.

Stuijt, Adriana. "Noise Pollution Kills 600 Dutch a Year." *Digital Journal,* 23 February 2009. http://www.digitaljournal.com/article/267835.

Suter, Alice H. *Noise and Its Effects.* Cincinnati, OH: Administrative Conference of the United States, November 1991.

Swift, Jonathan. *Jonathan Swift: Poems.* Ed. Derek Mahon. London: Faber and Faber, 2006.

Taylor, Rodney S. *The Confucian Way of Contemplation: Okaha Takehiko and the Tradition of Quiet Sitting.* Columbia, SC: University of South Carolina Press, 1988.

Thalheimer, Erich, and Charles Shamoon. "New York City's New and Improved Construction Noise Regulation." Paper presented at Noise-Con, Reno, 22–24 October 2007.

Thangham, Chris V. "U.S. Supreme Court Rules in Favor of Navy over Whales." *Digital Journal,* www.digitaljournal.com/article/262302.

The Epic of Gilgamesh. Trans. N. K. Sandars. Harmondsworth, UK: Penguin, 1972.

"The World's Most Densely Populated Countries." *Telegraph.co.uk.* 23 October 2009. http://www.telegraph.co.uk/news/worldnews/6413308/The-worlds-most-densely-populated-countries.html.

Thibodeau, Patrick. "Data Center Noise Levels Rising." *Computer World*, 3 September 2007. http://computerworld.com/s/article/299947/Can_You_Hear_Me_Now.

Thompson, Emily. *The Soundscape of Modernity: Architectural Acoustics and the Culture of Listening in America 1900–1933*. Cambridge, MA: MIT Press, 2002.

Thoreux, Paul. "America the Overfull." *New York Times*, 31 December 2006.

Tocqueville, Alexis de. *Democracy in America*. 1835; New York: Vintage, 1945.

Torigoe, Keiko. "Insights Taken from Three Visited Soundscapes in Japan." *Soundscape: The Journal of Acoustic Ecology* 6, no. 2 (Fall/Winter 2005): 9–12.

Truax, Barry. *Acoustic Communication*. Norwood, NJ: Ablex, 1984.

Tyldesley, Joyce. *Tales from Ancient Egypt*. Bolton, UK: Rutherford Press, 2004.

UK Noise Association. *Location, Location, Location: An Investigation into Wind Farms and Noise by the Noise Association*. London: UK Noise Association, July 2006.

Underdown, David. *Revel, Riot, and Rebellion: Popular Politics and Culture in England 1603–1660*. Oxford: Clarendon Press, 1985.

United Nations. *The World at Six Billion*, 1999. www.un.org/esa/population/publications/sixbillion/sixbilpart1.pdf.

United Nations Declaration of Human Rights. Cited in Simon Blackburn, *Being Good: An Introduction to Ethics*. Oxford: Oxford University Press, 2001.

United States Census Bureau, World POPclock Projection. http://www.census.gov/ipc/www/popclockworld.html.

United States Department of Labor. "Occupational Noise Exposure—1910.95" http://www.osha.gov/pls/oshaweb/owadisp.show_document?p_table=standards&p_id=9735.

United States Department of Transportation. *Exterior Sound Level Measurements of Over-Snow Vehicles at Yellowstone National Park*. Cambridge, MA: Research and Innovative Technology Administration, John A. Volpe National Transportation Systems Center, 2008.

Upton, Dell. *Another City: Urban Life and Urban Space in the New American Republic*. New Haven: Yale University Press, 2008.

van den Berg, Godefridus Petrus. "The Sound of High Winds: The Effect of Atmospheric Stability on Wind Turbine Sound and Microphone Noise." Doctoral dissertation, University of Groningen, Netherlands, 12 May 2006. http://irs.ub.rug.nl/ppn/294294104.

van Kamp, Irene, and Hugh Davies. "Environmental Noise and Mental Health: Five Year Review and Future Directions." Paper presented at 9th International Congress on Noise as a Public Health Problem, ICBEN, Foxwoods, CT, 2008.

Vidal, Gore. "The Meaning of Timothy McVeigh." In *The Best American Essays 2002*. Ed. Stephen Jay Gould and Robert Atwan. Boston: Houghton Mifflin, 2002.

Vieru, Tudor. "Searching for the Loudest Band in the World." *Softpedia*, 21 February 2009. http://news.softpedia.com/news/Searching-for-the-Loudest-Band-in-the-World-105105.shtml.

Vintiñi, Leonard. "Tracing the 'Hum.'" *The Epoch Times*, 19 July 2008. http://en.epochtimes.com/n2/content/view/7074/.

Vipperman, Jeffrey S., Eric R. Bauer, and Daniel R. Babich. "Survey of Noise in Coal Preparation Plants." *Journal of the Acoustical Society of America* 121, no. 1 (January 2006): 206–212.

Vlek, C. "Could We Have a Little More Quiet, Please? A Behavioral-Science Commentary on *Research for a Quieter Europe in 2020*." *Noise & Health* 7, no. 26 (2005): 59–70.

Vronksy, Peter. *Serial Killers: The Method and Madness of Monsters*. New York: Berkley Books, 2004.

Waggoner, Andrew. "The Colonization of Silence." 2007, http://www.newmusicbox.org.

Wain, John. *Samuel Johnson*. New York: Viking, 1974.

Walder, Bethanie. "Miles from Everywhere: Roads, Off-Road Vehicles, and Watershed Restoration on Public Lands." In Wuerthner, *Thrillcraft*.

Ward, W. Dixon. "Noise-Induced Hearing Loss." In Jones and Chapman, *Noise and Society*.

Was, Liz. "knoise pearl #1." In *Sounding Off! Music as Subversion/Resistance/Revolution*. Ed. Ron Sakolsky and Fred Wei-han Ho. Brooklyn: Autonomedia, 1995.

West, James E., Ilene J. Busch-Vishniac, Mark MacLeod, Jonathan Kracht, Douglas Orellana, and Jeffrey Dunn. "Characterizing Noise in Hospitals." Paper presented at Inter-Noise, Honolulu, 3–6 December 2006.

Wetherell, W. D. "Quiet." In *North of Now: A Celebration of Country and the Soon to Be Gone*. Guilford, CT: Lyons Press, 1998.

Whitman, Walt. "Song of Myself." In *Leaves of Grass*. 1855; New York: Mentor, 1954.

"Who Invented Radio?" PBS. http://www.pbs.org/tesla/ll/ll_whoradio.html.

Wilder, Laurel Ingalls. *Little House in the Big Woods*. 1932; New York: Harper & Row, 1971.

Williams, Charles. *The Descent of the Dove*. Grand Rapids, MI: William B. Eerdmans, 1939.

Williams, William Carlos. "To Elsie." In *Selected Poems*. 1949; New York: New Directions, 1969.

Willingham, Elizabeth. "Tinnitus." Baylor College of Medicine website, http://ww.bcm.edu/oto/grand/07_22_04.htm.

Wilson, Stan. "Behind the Scenes: Life after San Quentin." *CNN.com*, 26 February 2009. http://www.cnn.com/2009/LIVING/wayoflife/02/26/btsc.life.after.prison/index.html.

Winerman, Lea. "Brain, Heal Thyself: Researchers Are Finding That Sleep May Provide a Crucial Time for the Brain to Perform Biochemical Housekeeping." *Monitor on Psychology* 37, no. 1 (January 2006). http://www.apa.org/monitor/jan06/brain.html.

Wisman, Phil. "EPA History (1970–1985)." http://www.epa.gov/history/topics/epa/15b.htm.

Woolf, Virginia. "Carlyle's House." In *Carlyle's House and Other Sketches*. Ed. David Bradshaw. London: Hesperus Press, 2003.

———. "Great Men's Houses." In *The London Scene*. New York: Random House, 1982.

Woolgar, C. M. *The Senses in Late Medieval England*. New Haven: Yale University Press, 2006.

Wuerthner, George, ed. *Thrillcraft: The Environmental Consequences of Motorized Recreation*. White River Junction, VT: Chelsea Green; Foundation for Deep Ecology, 2007.

"You're Getting Very Sleepy." *American Psychological Association Online*, 12 May 2004. http://psychologymatters.org/sleep.html.

Yu, Lei, and Jian Kang. "Effects of Social, Demographical and Behavioral Factors on the Sound Level Evaluation in Urban Open Spaces." *Journal of the Acoustical Society of America* 123, no. 2 (February 2008): 772–783.

Zaidi, Shabih H. "Eastern Mediterranean Region." In Berglund, Lindvall, and Schwela, *Guidelines for Community Noise*, 123–128.

Zinko, Carolyn. "Gardeners in California City to Protest Leaf Blower Ban, Claiming Ban Is Racist." *San Francisco Chronicle*, 3 March 1998.

ACKNOWLEDGMENTS

This book owes a great deal to the faith of several individuals and two foundations. A Guggenheim Fellowship, generously supplemented by the Leon Levy Foundation, gave me support without which I doubt I would or could have undertaken a project of this scope. My agent Peter Matson never lost heart in his representation of my proposal, with the result that I would find an editor who believed in the book and would stand by it through several major changes, both in my original conception and in her own career. I am in your debt, Morgen Van Vorst.

The idea for the book grew out of my first essay for *Harper's Magazine*, the result of another act of faith, this time on the part of Lewis Lapham and Ellen Rosenbush, both of whom have my abiding thanks, as does the magazine for continuing to nurture my work.

In addition to the good advice of my editor, I had the benefit of early, close, and helpful readings by Kathy Keizer and Howard Frank Mosher. For technical questions I turned most often to Les Blomberg of the Noise Pollution Clearinghouse and to Nick Miller of Harris Miller Miller & Hanson. For topical information and global perspective I owe much to Charlie Shamoon. Others to whom I always could and sometimes did turn for expert counsel, and who were ever ready to encourage my progress, include Karin Bijsterveld, Arline Bronzaft, Jamie Kassler, Charles Komanoff, David Lubman, Herbert Singleton, and Frits van den Berg. Any errors in this book are mine.

For a brief but critical period I was assisted in my researches by Simone Rowen, whose careful and unstinting help was of great value.

A portion of this book pertaining to U.S. national parks appeared in altered form on *Smithsonian.org*; I am grateful to Laura Helmuth for editing that effort and to Jim Rutman for introducing us. For help in the preparation of that piece and this book as well, I am grateful to Karen Trevino and Kurt Fristrup of the Natural Sounds Program.

Along with those mentioned above, many others added considerably both to my understanding of noise and to the pleasure of my work by consenting to interviews. I regret that space does not permit me to pay more specific acknowledgment to Antonio

Acuna, Jay Allen, "Anderson," James Bailey, Mother Barbara, Douglas E. Barrett, Chris Berg, Ella Boyd, Matthew Brelis, Tamra Brennan, Steve Burchell, Timothy Chamberlin, Charlotte Clark, Stephanie Clement, Ken Czarnowski, Betty Desrosiers, Chris Devereaux, Mary Ellen Eagan, Gregg Fauth, David S. Ferriero, Teresa LaRue Forbes, Charlotte Formichella, Lawrence Gamble, Marcella Gilbert, Colin Grimwood, Phillip Gullikson, Carl Hanson, Wendy Hartnell, Peter Henry, Richard Hingston, Helen Hondius, Richard D. Horonjeff, Kevin Jackins, Fred Jansen, Terry Jemison, Howard Kelly, John T. Kelley, Mein Kinder, Owen King, Bernie Krause, Irene Little, Tony Lyons, Antonia M., Vicki McCusker, Robert E. Manning, Gwyn Mapp, Helen Matthews, Dorothy and Peter Miles, Robert Miller, Lucy Moore, John Moyers, Mei Musser, Peter Newman, Vincent O. Malley, Chris Peeler, Ericka Pilcher, Frits Platte, Susan Poole, Jay Red Hawk, Annette Reinboud, Ted Rueter, Wim and Inger Ronchetti, Knut Rostad, R. Murray Schafer, Bernard Schuijt, Karl Searson, Bob Selwyn, Christine Silverstein, Linda Skippings, Stephen A. Stansfeld, John Stewart, Wendy Todd, Armando Tovar, David A. Towers, Michael Tranel, Frank Turina, Ina Vonk, Val Weedon, Jock Whitworth, Eric W. Wood, Fred Woudenberg, Michael Wright, Charlotte Wyatt, and Lynn Young. For accurate transcriptions of some of their words I was ably assisted by Karen Myrick.

Among the many generous and knowledgeable men and women who answered queries, offered suggestions, and in many instances were ready to offer more help than I was able to receive, I wish to thank Margaret Adesina, Marie Alexander, Sietske Altinke, Haapakangas Annu, Penny Bergman, Jane Biro, Barry Blesser, Paul Bloom, Christian Broer, Barbara Bristol, Bennett M. Brooks, Phil Budahn, Stephen Carter, Andy Coghlan, Darren Copeland, Nadene Theriault-Copeland, John G. Cramer, Hugh Davies, Deborah F. Eiring, Andreas Fickers, Brigette Schulte-Fortkamp, Roger Gentry, Vic S. Gladstone, Paul Goldschagg, Eric Greenspoon, David L. Griffin, Sam and Paula Guarnaccia, Catherine Guastavino, David Healy, Bryna Hellmann, Phillis Hirshorn, Danny Houthuijis, Peter Ibbotson, Soames F. Jobs, Dylan Jones, Yori Kanekama, Pat Kenschaft, Victor Kestenbaum, Rokho Kim, Yvonne de Kluizenaar, Sharon Kujawa, William W. Lang, Bruce L. Lambert, Tom Laware, Dr. Brendan Lucey, Thomas Lynch, Maria Leon Maimone, Alison Martin, Calvin Luther Martin, William Martin, Nan Merrill, Henry Morgan, Janet Moss, Eja Pedersen, Rob Pforzheimer, Kenneth Plotkin, Robert Priddle, Cristina Pronello, Joel Reynolds, Richard Rodriguez, Peter Schwela, Bob Selwyn, Julia Shipley, Arlene Silverman, James Simmons, Dave Southgate, Pieter Jan Stallen, Mary Stevens, Mike Stinson, Anne L. Strader, Katsuaki Terasawa, Miki Terasawa, Nancy Timmerman, Irene van Kamp, the audiologists and speech pathologists of the Veterans Administration, James E. West, Jennica Wallenborg, Kerstin Persson-Waye, Cheri Whiton, Mark Whittle, and Joe Wolfe.

I am especially thankful to friends and neighbors who, once I'd cocked my ear toward noise, tilted their correspondence in the same direction, among them Paul and Carol Brouha, Stephen Dunn, Elizabeth Galle, Gary Greenberg, Amy Leal, and David Yaffe.

Regrettably, I am not able to name the hundreds of people who sent stories to my website. Just as regrettably, I was able to use only a few, but nearly all of them were helpful in influencing my approach to the subject and in convincing me that the subject was worth pursuing. Thanks and a quieter life to everyone who contributed. Along with several individuals already mentioned, I am grateful for substantive exchanges with Tammy Bailis, Nathan Blum, Dan Bonis, Havi Brooks, Jon Dokter, Gary Figallo, Kristi Grant, Ann Hamilton, Susan Hardenbergh, Pia Levensteins, Jeff LaRive, Patrick Lincoln, Dean Lindberg, Deborah Lonergan, Malcolm Maclean, Ted McCaslin, Grant Millin, Tracy Powell, Aldo and Luisa Rescigno, Luke Reynolds, Wendy Schaal, Leslie Schroeder, Elaine Walker, Lisa Wesseley, and Micah White, and I'd have no trouble tripling the list. I am also grateful to Ethan Marcotte for setting up the site and for advising me on all manner of computer-related questions.

At PublicAffairs, I had a reliable liaison and advocate in Lindsay Jones, attentive oversight of the book's production from Melissa Raymond and Meredith Smith, expert copyediting from Christine Arden, and a propitious introduction to the reading public from Whitney Peeling and Tessa Shanks. My thanks to all.

It will be obvious to any reader that some of my most important acknowledgments are in my bibliography. Less obvious are those who helped maintain my lifeline to books—Kim Crady-Smith at Green Mountain Books and Prints and the long-suffering librarians of the Samuel Read Hall Library at Lyndon State College, the Cobleigh Public Library, and the Bailey-Howe Library at the University of Vermont, with special thanks to Donna Edwards, Garet Nelson, and Pat Webster at LSC.

What raucous fun it would be to have all those I've named at one enormous table—with a few extra place settings and glasses of wine for those I have inadvertently forgotten. Forgive me.

Finally, thank you, Dr. John Ajamie, Chris Carsola, Ellen Doyle, Addison Hall, and Howard Mosher, for your various ministrations throughout this arduous task. Thank you, Henry, Wendy, John, and Joan Keizer for hospitality and support I can always count on. Thank you, Sarah, for always waking me to the moment. Most of all, thank you, Kathy, for your clear voice amid the noise.

INDEX

Kathy Keizer

Garret Keizer is a freelance writer, a contributing editor to *Harper's Magazine*, and a recent Guggenheim Fellow. He is the author of six books, including the critically acclaimed *Help* and *The Enigma of Anger.* His essays and poems have appeared in the *Los Angeles Times*, *Mother Jones*, the *New Yorker*, *The Best American Essays*, and *The Best American Poetry*. He lives with his wife in northeastern Vermont.

PublicAffairs is a publishing house founded in 1997. It is a tribute to the standards, values, and flair of three persons who have served as mentors to countless reporters, writers, editors, and book people of all kinds, including me.

I. F. Stone, proprietor of *I. F. Stone's Weekly*, combined a commitment to the First Amendment with entrepreneurial zeal and reporting skill and became one of the great independent journalists in American history. At the age of eighty, Izzy published *The Trial of Socrates*, which was a national bestseller. He wrote the book after he taught himself ancient Greek.

Benjamin C. Bradlee was for nearly thirty years the charismatic editorial leader of *The Washington Post*. It was Ben who gave the *Post* the range and courage to pursue such historic issues as Watergate. He supported his reporters with a tenacity that made them fearless and it is no accident that so many became authors of influential, best-selling books.

Robert L. Bernstein, the chief executive of Random House for more than a quarter century, guided one of the nation's premier publishing houses. Bob was personally responsible for many books of political dissent and argument that challenged tyranny around the globe. He is also the founder and longtime chair of Human Rights Watch, one of the most respected human rights organizations in the world.

• • •

For fifty years, the banner of Public Affairs Press was carried by its owner Morris B. Schnapper, who published Gandhi, Nasser, Toynbee, Truman, and about 1,500 other authors. In 1983, Schnapper was described by *The Washington Post* as "a redoubtable gadfly." His legacy will endure in the books to come.

Peter Osnos, *Founder and Editor-at-Large*